Lecture Notes in Computer Science   12003

More information about this series at http://www.springer.com/series/7409

Zhiming Zhao · Margareta Hellström (Eds.)

Towards Interoperable Research Infrastructures for Environmental and Earth Sciences

A Reference Model Guided Approach for Common Challenges

 Springer

Editors
Zhiming Zhao
University of Amsterdam
Amsterdam, The Netherlands

Margareta Hellström
Lund University
Lund, Sweden

ISSN 0302-9743   ISSN 1611-3349   (electronic)
Lecture Notes in Computer Science
ISBN 978-3-030-52828-7   ISBN 978-3-030-52829-4   (eBook)
https://doi.org/10.1007/978-3-030-52829-4

LNCS Sublibrary: SL3 – Information Systems and Applications, incl. Internet/Web, and HCI

The cover illustration was taken from https://envri.eu/

This Springer imprint is published by the registered company Springer Nature Switzerland AG
The registered company address is: Gewerbestrasse 11, 6330 Cham, Switzerland

Preface

This book summarises the latest developments on data management in the EU H2020 ENVRIplus project, which brought together more than 20 environmental and Earth science research infrastructures into a single community. It provides readers with a systematic overview of the common challenges faced by research infrastructures and how a 'reference model guided' engineering approach can be used to achieve greater interoperability among such infrastructures in the environmental and Earth sciences.

The research problems behind environmental and societal challenges such as climate change, food security, and natural disasters are intrinsically interdisciplinary. Modelling these processes individually is difficult enough, but modelling their interactions is another order of complexity entirely. Scientists are challenged to collaborate across conventional disciplinary boundaries, but must first discover and extract data dispersed across many different sources and in many different formats. Effective research support environments are needed for various user-centralised research activities, from formulating research problems to designing experiments, discovering data and services, executing workflows, and analysing then publishing the final results. Such support environments also have to manage research data during their entire lifecycle, throughout the phases of data acquisition, curation, publication, processing, and use. Moreover, support environments must support the management of underlying infrastructure resources for computing, storage, and networking. In this ecosystem, research infrastructure (RI) is an important form of supportive environment that bridges the gap between the curation of research data and user-centred scientific activity, and also between research data and the underlying physical infrastructure. It brings together facilities, resources, and services used by the scientific community to conduct research, establish best practices for science, and foster innovation.

This book presents the design, development, deployment, operation, and use of research infrastructures as 20 chapters via five parts. Part one provides an overview of the state of the art of research infrastructure and relevant e-Infrastructure technologies, part two discusses the reference model guided engineering approach, the third part presents the software and tools developed for common data management challenges, the fourth part demonstrates the software via several use cases, and the last part discusses the sustainability and future directions.

The main readers of the book will be developers, managers, operators, and potential users of research infrastructures in environmental and earth sciences. This book will provide RI data managers in environmental and earth sciences with a common ontological framework and facilities for modeling data management requirements and practical data management guidelines during entire research life-cycle. It will provide RI stakeholders with very practical case studies on RI architecture design, service interoperability, and system-level environmental research. The book can also be a textbook for training young researchers and data managers in data management skills,

RI service development and operation practices, and using RIs for data-centric research.

In addition to researchers and developers involved in the *data for science* theme, the development of the book has also been greatly supported by the project coordinator and RI partners, in particular those specialists willing to serve in the editorial board. We thank all the authors for contributing to the individual chapters, and reviewers for providing valuable feedback on the content. Without their support, this book would not have been possible.

May 2020   Zhiming Zhao
                                                        Margareta Hellström

Organisation

Editorial Board

Massimo Cocco	INGV, Italy
Nicola Fiore	LifeWatch ERIC, Italy
Paul Martin	University of Amsterdam, The Netherlands
Paola Grosso	University of Amsterdam, The Netherlands
Paolo Laj	Centre National de Recherche Scientifique, France
Peter van Tienderen	University of Amsterdam, The Netherlands
Øystein Godøy	SIOS, Norway
Robert Huber	University of Bremen, Germany
Sanna Sorvari	Finnish Meteorological Institute, Finland
Spiros Koulouzis	University of Amsterdam, The Netherlands
Sylvie Pouliquen	Ifremer, France
Thierry Carval	Ifremer and EuroArgo, France
Werner L. Kutsch	ICOS-RI, Finland
Wouter Los	University of Amsterdam, The Netherlands
Xiaofeng Liao	University of Amsterdam, The Netherlands
Yannick Legre	EGI Foundation, The Netherlands
Yin Chen	EGI Foundation, The Netherlands

Contents

Data Management in Environmental and Earth Sciences

Supporting Cross-Domain System-Level Environmental and Earth Science

Alex Vermeulen[1](✉) (iD), Helen Glaves[2] (iD), Sylvie Pouliquen[3] (iD), and Alexandra Kokkinaki[4] (iD)

[1] ICOS ERIC - Carbon Portal, 22362 Lund, Sweden
alex.vermeulen@icos-ri.eu
[2] British Geological Survey, Keyworth, Nottingham NG12 5GG, UK
hmg@bgs.ac.uk
[3] IFREMER, 29280 Plouzané, France
sylvie.pouliquen@ifremer.fr
[4] BODC, National Oceanography Centre, Liverpool L3 5DA, UK
alexk@bodc.ac.uk

Abstract. Answering the key challenges for society due to environmental issues like climate change, pollution and loss of biodiversity, and making the right decisions to tackle these in a cost-efficient and sustainable way requires scientific understanding of the Earth System. This scientific knowledge can then be used to inform the general public and policymakers. Scientific understanding starts with having available the right data, often in the form of observations. Research Infrastructures (RIs) exist to perform these observations in the required quality and to make the data available to first of all the researchers. In the current Big Data era, the increasing challenge is to provide the data in an interoperable and machine-readable and understandable form. The European RIs on environment formed a project cluster called ENVRI that tackles these issues. In this chapter, we introduce the societal relevance of the environmental data produced by the RIs and discuss the issues at hand in providing the relevant data according to the so-called FAIR principles.

Keywords: Research Infrastructure · FAIR · Data management · Environmental and earth science · Societal challenges

1 Data-Centric Science in Environmental and Earth Sciences

1.1 Relevance to the Big Questions of Science and Society

Our society is becoming increasingly complex, and human interaction with the natural systems is intensifying due to population growth and increased usage of energy and resources in nearly all parts of the world. These interactions increase the pressure on natural systems and have serious consequences for the environment, which in turn affect the quality of life for both humans and the whole biosphere.

Z. Zhao and M. Hellström (Eds.): Towards Interoperable Research Infrastructures for Environmental and Earth Sciences, LNCS 12003, pp. 3–16, 2020.
https://doi.org/10.1007/978-3-030-52829-4_1

In August 2016, the Anthropocene Working Group of the Sub commission on Quaternary Stratigraphy[1] of the International Commission on Stratigraphy[2] officially voted to define our time as the Anthropocene in the Geological Time Scale. The ratification of this Anthropocene era by the International Commission on Stratigraphy of the International Union of Geological Sciences[3] is pending due to a discussion on where this period should begin (between the beginning of Agricultural Revolution about 12000 years ago or only since the so-called Great Acceleration (1945 A.D.), but nevertheless we can safely say that we are now in a period where mankind is the main determinant in the fate of Earth [1].

Human impacts on climate and biodiversity are the most striking illustrations of the Anthropocene, as demonstrated by the UN IPCC programme in its most recent Fifth Assessment Report on climate [2], and by the very recent 2019 IPBES Global Assessment Report on Biodiversity and Ecosystem Services[4]. Global rates of extinction are shown to have been on the rise since at least 1500 and are now accelerating at an unparalleled pace. A recent estimate is that since the rise of human civilisation 83% of wild mammals and 50% of plants have already been lost [8]. The use of fossil fuels since the industrial revolution has now increased the CO_2 global atmospheric average atmosphere from the normal 180–280 ppm in the past million years to more than 405 ppm in 2017[5].

The human influence on natural resources is increasing due to population and economic growth but in return the natural processes in solid Earth, climate, ecosphere, terrestrial and marine domains have an increasing effect on mankind and society due to the increasing complexity and capital intensity of our society and economies. Understanding and quantifying these pressures and resulting changes is a requirement for the sustainable development of our societies using fact-based decision making. Assessments of changes in environmental conditions and their relationship with the driving forces must be based on trustworthy and well-documented observations. This is not an easy task as there are many interactions between the changes in the atmosphere, land and hydrosphere, and the resulting impacts on ecosystems all need special and focused high-quality long-term observations. This requires us to have better observations and data on these important pre-conditions in order to better inform decision makers to take the measures needed to maintain a thriving society. Research infrastructures are an important element in providing the information required to support science and fact-based policy development.

1.2 Supporting Sustainable Development with Data

The United Nations Sustainable Development Goals are a call for action by all countries – poor, rich and middle-income – to promote prosperity while protecting the planet.

They recognise that ending poverty must go hand-in-hand with strategies that build economic growth and address a range of social needs including education, health, social

---

[1] http://quaternary.stratigraphy.org/working-groups/anthropocene/.

[2] http://stratigraphy.org/.

[3] https://www.iugs.org/.

[4] https://ipbes.net/ipbes-global-assessment-report-biodiversity-ecosystem-services.

[5] https://public.wmo.int/en/media/press-release/greenhouse-gas-levels-atmosphere-reach-new-record (checked Feb 2020).

protection, and job opportunities while tackling climate change and environmental protection. The UN defined a set of 17 Sustainable Development Goals (SDG) where data
is required in order to develop policies and evaluate and track the progress of the developments, as shown in Fig. 1. For the environmental research infrastructures (ENVRI)
to be discussed in this book, most SDGs are very relevant but particularly relevant are
Climate Action (Goal 13), Life Below (in) Water (Goal 14) and Life On Land (Biodiversity, Forests and land degradation) (Goal 15). Of course, all these SDGs are also closely
related to SDGs like Energy (Goal 7), Sustainable production and consumption (Goal
12), Cities (Goal 11) and Water and sanitation (Goal 6). One of the global partnerships in
the framework of the UN SDGs is the Global Partnership for Sustainable Development
Data with motto: BETTER DATA. BETTER DECISIONS. BETTER LIVES[6].

Fig. 1. The 17 sustainable development goals from the United Nations depicted as icons (https://
www.un.org/sustainabledevelopment/wp-content/uploads/2019/01/SDG_Guidelines_AUG_
2019_Final.pdf).

1.3 The Role of Research Infrastructures

Research Infrastructures (RI) of the Environment Domain as defined by ESFRI[7] cover
the main four subdomains of the complex Earth system (Atmosphere, Marine, Solid
Earth, and Biodiversity/Terrestrial Ecosystems), thus forming the cluster of European
Environmental and Earth System Research Infrastructures (ENVRI)[8]. Environmental
Research Infrastructures are crucial pillars for environmental scientists in their quest for
understanding and interpreting the complex Earth System. They are the larger producers
and providers of Environmental Research data in Europe collected from in-situ and
space-based observing systems. ENVRIs all contribute to global observing systems and
they generate relevant information for Europe and worldwide.

---

[6] http://www.data4sdgs.org/.

[7] European Strategy Forum on Research Infrastructures https://www.esfri.eu/.

[8] ENVRI: ENVironmental Research Infrastructures.

The RI facilities were developed to respond to the needs of specific research communities, following individual requirements and methods of specific disciplines. However, the necessity of interdisciplinary cooperation has been evident for decades. Therefore, the ENVRI community has increasingly cooperated within the cluster projects ENVRI (2011–2014, FP7) [9], which paved the way for the ENVRIplus[9] project (2015–2019, H2020) [9, 10] and the ENVRI-FAIR[10] project (2019–2022, H2020) [11]. ENVRIplus gathered all subdomains of the Earth system science to work together, capitalise the progress made in the various disciplines, and strengthen interoperability amongst RIs and subdomains.

In Sect. 3, three example cases will be shown where Research Infrastructures from ENVRI provide data to inform policy and society for better decision making with regards to reaching the Sustainable Development Goals.

2 The ENVRIplus Objectives

The objective of ENVRIplus was to provide common solutions to shared challenges for European Environmental and Earth System Research Infrastructures (RIs) in their efforts to deliver new services for science and society.

To reach this overall goal, ENVRIplus brought together the environmental RIs included in the ESFRI Roadmap, leading preparatory projects, key developing RI networks and specific technical specialist partners to build common synergistic solutions for pressing issues in RI construction and implementation. ENVRIplus was organised around six key objectives, identified as "Themes" as shown in Fig. 2:

1. Improve the ability of RIs to observe the Earth System, in particular through development and testing of new sensor technologies, harmonizing observation methodologies and developing techniques to overcome common problems associated with distributed remote observation networks;
2. Generate common solutions for shared information technology and data related challenges of the environmental RIs, especially in data and service discovery and use, workflow documentation, mechanisms for data citations, service virtualization, and user characterization and interaction;
3. Develop harmonised policies for access (physical and virtual) for the environmental RIs, including access services for multidisciplinary users;
4. Investigate the interactions between RIs and society that includes: finding common approaches and methodologies for assessing the ability of an RI to address economic and societal challenge; developing ethics guidelines for RIs, and investigating the possibility of enhancing the use of Citizen Science in RI products and services;
5. Ensure the cross-fertilisation and knowledge exchange between RIs on new technologies, best practices, approaches and policies by generating training material for RI personnel to provide instruction on using the new observational, technological and computational tools, as well as facilitating inter-RI knowledge transfer via a staff exchange program;

---

[9] http://www.envriplus.eu.
[10] http://www.envri-fair.eu/.

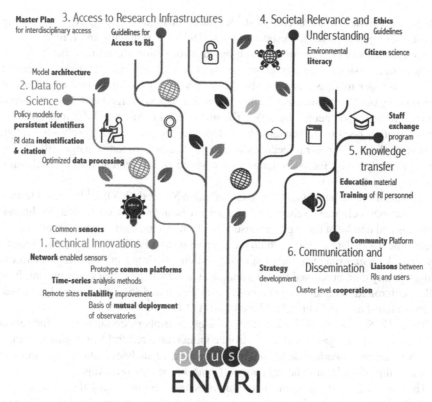

Fig. 2. The six themes in the ENVRIplus project.

6. Create a communication and cooperation framework to coordinate the activities of the environmental RIs for the purposes of common strategic development, improved user interaction and interdisciplinary cross-RI products and services.

3   Example Science Cases Related to Environmental Research Infrastructures

3.1   Climate Change and Atmospheric Composition Research (ICOS, ACTRIS and IAGOS)

Climate Change has been recognised by the United Nations and the European Union as the major environmental challenge for mankind. Research is needed on future scenarios on climate change that will have a dramatic effect on natural environments, plants and animals, leading to an acceleration in biodiversity loss in some areas. The impacts will have knock-on effects for many communities and sectors that depend on natural resources, including agriculture, fisheries, energy, tourism and water. The Stern Review [3] stated as early as 2007 that climate change is the greatest and widest-ranging market failure ever seen, presenting a unique challenge for economics. According to the Stern

Review, without action, the overall costs of climate change will be equivalent to losing at least 5% of global gross domestic product (GDP) each year, now and forever.

Another important area for research-based information for climate policy is the validation of emission reductions required as part of the COP21 Paris Climate Agreement of 2015. In order to keep climate change as a consequence of increased emissions of greenhouse gases due to human activities under 2.0 °C and preferably 1.5 °C the world will need to be carbon neutral by 2050. The mitigation measures and the speed of their implementation need to be validated by independent methods and closely monitored, while the influence of natural feedback due to the ongoing climate change will require attention, as this may force a change in the speed of implementation of mitigation measures and adaptation.

The data from the Integrated Carbon Observation Network (ICOS)[11] Research Infrastructure supports climate science to inform scientists and society on natural and human emissions and uptake of these greenhouse gases from ocean, land ecosystems and atmosphere. The ICOS data portal[12], which has been setup as a FAIR[13] [4] compliant repository, provides data from over 130 monitoring stations, as shown in Fig. 3. It gives access to high-quality data processed by the Thematic Centers as raw, near real-time and final quality-controlled data, and supplemented with elaborated (model) data and analyses, which is almost always licensed under a CC4BY[14] license.

The IAGOS[15] research infrastructure provides atmospheric composition information including greenhouse gas observations from commercial aircraft. IAGOS data are being used by researchers worldwide for process studies, trend analysis, validation of climate and air quality models, and the validation of spaceborne data retrievals.

The ACTRIS[16] research infrastructure observes aerosols and their precursors. Aerosols also have a large influence on the earth's radiation balance and thus climate, and their concentrations are tightly connected to human activities and emissions.

All of these infrastructures are part of a global endeavour to advance science-based high-quality observations that ultimately allow for better decisions. Therefore, the methods and data are based on global, often community-based standards. Interoperability on the global scale with, for example, the World Meteorological Organisation (WMO)[17].

3.2 Mitigating the Societal and Economic Impacts of Future Volcanic Eruptions and the Role of the European Plate Observing System (EPOS)

The eruption of the Icelandic Eyjafjallajökull volcano in 2010 yielded an estimated 250 million cubic metres (0.25 km^3) of ejected tephra, with the resulting ash plume rising

---

[11] https://www.icos-ri.eu.

[12] https://www.icos-cp.eu and https://data.icos-cp.eu/portal.

[13] FAIR principles: Findable, Accessible, Interoperable and Reusable: https://www.go-fair.org/fair-principles/, further explained in Sect. 5

[14] https://creativecommons.org/licenses/by/4.0/.

[15] In-service Aircraft for a Global Observing System https://www.iagos.org/.

[16] European Research Infrastructure for the observation of Aerosol, Clouds and Trace Gases https://www.actris.eu.

[17] World Meteorological Organisation, part of the United Nations. https://public.wmo.int/en.

Fig. 3. Overview of the ICOS monitoring station network.

to a height of around 9 km into the atmosphere. Due to the potential damage to aircraft engines from the ash, the ongoing eruption of Eyjafjallajökull (see Fig. 4) from April to June 2010 led to the largest suspension of commercial air traffic since World War II. This closure of European airspace led to the cancellation of large numbers of flights that left millions of passengers stranded and cost airlines an estimated \$200 million per day in lost revenue. The total global losses in GDP due to the prolonged inability to move people or goods have been estimated at approximately \$4.7 billion. This figure incorporates both net airline industry and destination losses, along with general productivity losses [5]. The long-term effects of the eruption also continue to impact local inhabitants and the environment due to the potential toxicity to humans, animals and plant life either by direct inhaling the particulates or due to the acid rain that can result from the sulphur in the ash.

Eruptions of Icelandic volcanoes are relatively frequent with events similar to that of the Eyjafjallajökull volcano occurring, on average, every 20–40 years. In this case, the

Fig. 4. The eruption of the Eyjafjallajökull volcano in May 2010 that disturbed air traffic in Europe for a sustained period, leading to large economic losses (photo credits: M. Rietze) (http://www.tboeckel.de/EFSF/efsf_wv/island_10/Eyjafoell/may_10/may_10_e.htm).

combination of a volcanic event with the prevailing weather conditions caused significant disruption both within Europe and beyond, with major economic and societal impacts. However, the potential for this type of event had been previously been recognised but precautionary measures to limit the impact of such an event had been limited [6].

To mitigate for future volcanic eruptions and reduce the potential impact of these events, enhanced monitoring of Icelandic volcanoes combined with the increased availability of the data for integrated use by multiple agencies, and to provide timely information to local inhabitants has become a priority. Enhanced monitoring of volcanoes also allows better disaster response planning at the local, national and international level in an effort to minimise the impact of future events on both local inhabitants and the wider population.

The European Plate Observing System (EPOS)[18] Research Infrastructure has integrated various solid Earth research facilities, the so-called thematic core services (TCS), into a single framework that facilitates sharing of various data for the solid Earth domain. These facilities range from monitoring networks such as those delivering real-time seismic data from Icelandic volcanoes to Global Navigation Satellite System (GNSS) data used for global positioning and navigation.

Data services made available by the EPOS research infrastructure, such as those delivered by the Icelandic FUTUREVOLC[19] supersite initiative, can be used by various agencies in Iceland to provide real-time monitoring information for the approximately 130 Icelandic volcanoes currently known to be either currently or potentially active. This information can be used to provide early warning of an eruption for local inhabitants

---

[18] https://www.epos-ip.org.
[19] http://futurevolc.hi.is/.

and can also be used in combination with other types of data such as meteorological information to predict the likely impact of an eruption. For example, the Icelandic Met Office provides information on volcanic activity using colour coding that conforms with the International Civil Aviation Organisation (ICAO)[20] to inform the aviation industry of potential risks to aircraft due to ash plumes associated with an eruption event[21]. This allows better modelling of the potential disruption that may be caused by an eruption depending on different combinations of prevailing winds, type and volume of ejecta, and the duration of any eruption.

The ENVRI community brings together environmental research infrastructures from different domains. Integration of EPOS with those RIs focused on atmospheric data and data products provide the necessary framework for modelling the potential impacts and informing the mitigation strategies for the various agencies that require timely information to inform disaster response and remediation strategies following a major volcanic event.

3.3 The Importance of Data Management to Solve Societal and Scientific Questions for the Oceans (SeaDataNet)

The ocean plays a central role in regulating the Earth's climate [12]. As the International Oceanographic Data and Information Exchange (IODE)[22] has announced: "The timely, free and unrestricted international exchange of oceanographic data is essential for the efficient acquisition, integration and use of ocean observations gathered by the countries of the world for a wide variety of purposes including the prediction of weather and climate, the operational forecasting of the marine environment, the preservation of life, the mitigation of human-induced changes in the marine and coastal environment, as well as for the advancement of scientific understanding that makes this possible"[23].

Marine data are important and relevant for many uses such as:

- Scientific research to gain knowledge and insight
- Monitoring and assessment (water quality, climate status, stock)
- Coastal Zone management
- Modelling (including hindcast, now-cast, forecast)
- Dimensioning and supporting operations and activities at sea (shipping, offshore industry, and dredging industry)
- Implementation and execution of marine conventions for the protection of the seas, including aligning with international legislation such as the European Marine Strategy Framework Directive (MSFD).

Acquisition of marine data is expensive: annual cost in Europe estimated at 1.4 Billion € (1 for in-situ data, 0.4 for satellite data). In order to achieve IODE's goals for unrestricted exchange of oceanographic data, professional data management is essential

---

[20] https://www.icao.int.

[21] https://en.vedur.is/earthquakes-and-volcanism/volcanic-eruptions/.

[22] http://www.iode.org/.

[23] https://www.iode.org/index.php?option=com_content&view=article&id=51&Itemid=95.

with agreements on standardisation, quality control procedures, long term archiving, catalogue and access. The main objective of data management was to ensure safe and long-term storage of data and metadata so that present and future users are able to use all of the data that have been collected over time.

SeaDataNet[24] is a pan-European infrastructure set up and operated for managing marine and ocean data in cooperation with the National Oceanographic Data Centre (NODCs) and data focal points of 34 countries bordering the European seas, as shown in Fig. 5. SeaDataNet's significant contribution to the ocean data landscape is through the establishment of collaboration across the partners and the agreements on the consistent use of standards and controlled vocabularies for data annotation, formatting and discovery. SeaDataCloud, the EU project currently driving the further development of the SeaDataNet infrastructure will deliver a collaborative and high-performing cloud and virtual research environment (VRE), configured with tools and services for processing essential marine data. Using Open Geospatial Consortium (OGC), ISO, and World Wide Web Consortium (W3C) standards and incorporating scientific expertise, dynamic workflows are configured for analysing, processing, and combining subsets of data. The VRE and workflows will allow data product teams to work more efficiently for processing large amounts of input datasets and generating data products collaboratively, while also adopting innovations like machine learning for QA/QC of large data collections. This way, the production cycle for data products can be reduced in duration and higher-quality products can be achieved. One of the challenges is to make the SeaDataNet data, metadata and related services more FAIR [4]. This focuses on improving and optimising Findability, Accessibility, Interoperability, and Re-usability, both for machines and for people, with emphasis on machines. As part of improving FAIRness of SeaDataNet services, several activities are planned and some have already been undertaken.

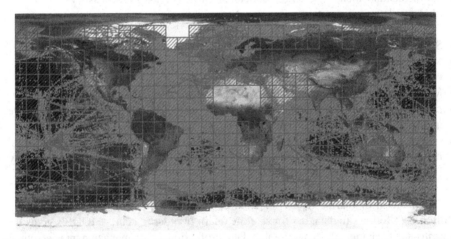

Fig. 5. Overview of SeaDataNet CDI entries per July 2017: >1.97 million data sets from 600+ originators and 100+ connected data centres.

---

[24] https://www.seadatanet.org/.

4   The ENVRIplus Data to Science Theme

Environmental Research infrastructures are important pillars not only for supporting their own communities, but also (a) for interdisciplinary research, (b) for the European Earth Observation Program COPERNICUS[25], and (c) as a contribution to the Global Earth Observation System of Systems (GEOSS[26]). As such, it is very important that the data-related activities of the environmental RIs are well integrated. This requires common policies, models and e-infrastructure to optimise technological implementation, define workflows; and ensure coordination, harmonization, integration and interoperability of data, applications and other services between ESFRI and other research infrastructure initiatives.

The key is common metadata systems that utilise a rich metadata model with formal syntax and declared semantics, which acts as the 'switchboard' for interoperation. Metadata is used to characterise data, services, users and ICT resources (including sensors and detectors). This approach provides an e-infrastructure that is virtualised for end-users but within which expert domain users and ICT experts can work to provide improved services as requirements evolve.

The objectives of this ENVRIplus Data to Science theme were to:

- optimise data processing and to develop common models, rules and guidelines for research data workflow documentation;
- facilitate data discovery and use, and to provide integrated end-user information technology to access heterogeneous data sources;
- make data citable by developing existing approaches with practical examples, exchange of expertise, and agreements with publishers;
- facilitate the discovery of software services and their composition;
- characterise users and build a community evolving from current RI communities;
- characterise ICT resources (including sensors and detectors) to allow virtualisation of the environment (for instance onto Grid- or Cloud-based platforms) such that data and information management and analysis is optimised in use of resources and energy usage;
- facilitate the connection of users, composed software services, appropriate data and necessary resources in order to meet end-user requirements.

To maximise re-use of existing technologies and solutions, this theme conducted an in-depth review of the results from the ESFRIs (such as ICOS, Euro-Argo, EPOS and SIOS) [7], and interacted closely with computational e-Infrastructures (such as EGI and CLOUD Nebula, platforms (such as DIRAC), data infrastructures (such as EUDAT CDI and D4Science), and other initiatives working on related issues, such as the European Open Science Cloud (EOSC) that was initiated during the ENVRIplus project.

---

[25] https://www.copernicus.eu/en.
[26] https://www.earthobservations.org/geoss.php.

5   The FAIR Principles as Guidelines for Data Management

The term FAIR, a set of guiding principles to make data Findable, Accessible, Interoperable, and Reusable was developed in 2014 and published two years later [4].

Based on these 15 principles, a set of 14 metrics have been defined to quantify levels of FAIRness. The latest developments on FAIR are available at GO-FAIR[27]. The FAIR principles are characterised as:

Findable

- F1. (meta)data are assigned a globally unique and eternally persistent identifier.
- F2. data are described with rich metadata.
- F3. (meta)data are registered or indexed in a searchable resource.
- F4. metadata specify the data identifier.

Accessible

- A1 (meta)data are retrievable by their identifier using a standardised communications protocol.

  - A1.1 the protocol is open, free, and universally implementable.
  - A1.2 the protocol allows for an authentication and authorization procedure, where necessary.
- A2 metadata are accessible, even when the data are no longer available.

Interoperable

- I1. (meta)data use a formal, accessible, shared, and broadly applicable language for knowledge representation.
- I2. (meta)data use vocabularies that follow FAIR principles.
- I3. (meta)data include qualified references to other (meta)data.

Re-usable

- R1. meta(data) have a plurality of accurate and relevant attributes.

  - R1.1. (meta)data are released with a clear and accessible data usage license.
  - R1.2. (meta)data are associated with their provenance.
  - R1.3. (meta)data meet domain-relevant community standards.

Although good data management is not a goal in itself, it is a necessary condition that enables innovation, knowledge creation, data and knowledge integration, and reuse of data by other users. There are currently many factors missing or inadequately implemented, and also many institutional barriers that limit the deployment of research data. This situation can be improved using a systematic approach in applying these principles in order to maximise the FAIRness of data management.

---

[27] https://www.go-fair.org/.

6   Challenges

There are many challenges for ENVRIs on the way to becoming fully FAIR compliant. To begin with, the concept of FAIRness is still evolving and has different interpretations depending on the community of practice that continues to be discussed in different fora such as the Research Data Alliance (RDA[28]) and the GoFAIR[29] initiative.

One of the biggest challenges for RIs is that most of them are already (partly) operational and rely for a large part on legacy database and metadata systems that were built years or, in some cases, decades ago, and that are based on highly specialised and sometimes informal and dynamically generated community standards. They cannot simply redesign existing systems, and cannot afford system downtime, as this would interrupt their services to users and might even lead to unacceptable data losses.

In addition, the underlying databases are often rigid relational database systems that have been optimised for performance to serve the designated user community of the RI, and in some cases utilise proprietary software that requires authentication and authorisation through custom systems. This complicates the accessibility of the systems and hampers the linking to external catalogues necessary for enhanced findability of the data. These challenges will be discussed further in Chapter 3 of this book.

Interoperability has many facets and one of these involves the translation of community standards to more generally usable metadata standards. This translation from one metadata standard into another (machine operable) metadata standard will potentially lead to risks of loss of information or even errors, which will hamper the acceptance by the involved scientific communities. An important first step on this route to interoperability is the development of controlled vocabularies and data type registries, that document and stabilise the community standards.

Acknowledgements. This work was supported by the European Union's Horizon 2020 research and innovation programme via the ENVRIplus project under grant agreement No 654182.

References

1. Steffen, W., Grinevald, J., Crutzen, P., McNeill, J.: The Anthropocene: conceptual and historical perspectives. Philos. Trans. R. Soc. A Math. Phys. Eng. Sci. **369**(1938), 842–867 (2011). https://doi.org/10.1098/rsta.2010.0327
2. Stocker, T.F., et al.: Climate Change 2013: The Physical Science Basis. Contribution of Working Group I to the Fifth Assessment Report of the Intergovernmental Panel on Climate Change. Cambridge University Press, Cambridge (2013)
3. Stern, N.: The Economics of Climate Change (2007). https://doi.org/10.1017/cbo9780511817434
4. Wilkinson, M., Dumontier, M., Aalbersberg, I., et al.: The FAIR guiding principles for scientific data management and stewardship. Sci. Data **3**, 160018 (2016). https://doi.org/10.1038/sdata.2016.18

---

[28] RDA: Research Data Alliance, https://www.rd-alliance.org/.

[29] https://www.go-fair.org/.

5. Mazzocchi, M., Hansstein, F., Ragona, M.: The 2010 volcanic ash cloud and its financial impact on the European airline industry. CESifo Forum **11**(2), 92–100 (2010)
6. Sammonds, P., McGuire, W., Edwards, S. (eds.): Volcanic Hazard from Iceland: Analysis and Implications of the Eyjafjallajökull Eruption. UCL Institute for Risk and Disaster Reduction, London (2010)
7. Atkinson, M.H., et al.: D5.1 – A consistent characterisation of existing and planned RIs. Retrieved from ENVRIplus website (2016). http://www.envriplus.eu/wp-content/uploads/2016/06/A-consistent-characterisation-of-RIs.pdf
8. Bar-On, Y.M., Phillips, R., Milo, R.: The biomass distribution on Earth. Proc. Natl. Acad. Sci. U.S.A. **115**, 6506–6511 (2018). https://doi.org/10.1073/pnas.1711842115
9. Chen, Y., et al.: A common reference model for environmental science research infrastructures. In: Proceedings of EnviroInfo 2013 (2013). http://enviroinfo.eu/sites/default/files/pdfs/vol7995/0665.pdf
10. Zhao, Z., et al.: Reference model guided system design and implementation for interoperable environmental research infrastructures. In: 2015 IEEE 11th International Conference on e-Science, Munich, Germany, pp. 551–556. IEEE (2015). https://doi.org/10.1109/eScience.2015.41
11. Petzold, A., et al.: ENVRI-FAIR - interoperable environmental FAIR data and services for society, innovation and research. In: 2019 15th International Conference on eScience (eScience), San Diego, CA, USA, pp. 277–280. IEEE (2019). https://doi.org/10.1109/escience.2019.00038. https://zenodo.org/record/3462816
12. Tanhua, T., et al.: Ocean FAIR data services. Front. Mar. Sci. **6**, 440 (2019). https://doi.org/10.3389/fmars.2019.00440

ICT Infrastructures for Environmental and Earth Sciences

Keith Jeffery[1(✉)] , Antti Pursula[2] , and Zhiming Zhao[3]

[1] Keith G Jeffery Consultants, Faringdon, UK
keith.jeffery@keithgjefferyconsultants.co.uk
[2] CSC - IT Center for Science, Espoo, Finland
antti.pursula@csc.fi
[3] Multiscale Networked Systems, University of Amsterdam,
1098XH Amsterdam, The Netherlands
z.zhao@uva.nl

Abstract. E-Infrastructures play an increasingly important part in the provision of digital services to environmental researchers and other users. The availability of reliable networks, storage facilities, high performance and high throughput computers and associated middleware and services to ease their utilisation all contribute to enabling research and its exploitation. Their relevance, possible use and utilisation to date are described.

Keywords: Infrastructure · Open Science · Networking · Computers · Cloud computing

1 Introduction

To tackle the scientific challenges discussed in the previous chapter, researchers need access to sophisticated *research support environments* that enable efficient discovery, access, interoperation and re-use of the data, tools, etc. available for advanced data science and provide a platform for the integration of all resources into cohesive observational, experimental and simulation investigations with replicable workflows. Examining current initiatives in Europe and beyond, we have identified three main types of research support environment [1]:

e-Infrastructures. Unified computing, storage and network infrastructures provided via initiatives such as EGI[1], GEANT[2], and EUDAT[3]. The e-Infrastructure providers manage the *service lifecycle* of computing, storage and network resources, and enable research communities to provision dedicated infrastructure and to manage persistent services and their underlying storage, data processing and networking requirements.

---

[1] http://www.egi.eu/.
[2] http://www.geant.org/.
[3] http://www.eudat.eu/.

© The Author(s) 2020
Z. Zhao and M. Hellström (Eds.): Towards Interoperable Research
Infrastructures for Environmental and Earth Sciences, LNCS 12003, pp. 17–29, 2020.
https://doi.org/10.1007/978-3-030-52829-4_2

Public e-Infrastructures typically offer their services based on service-level agreements (SLAs) established at the institutional level or negotiated with specific groups [5]. Such services are now predominantly Cloud-based, using virtual machines or containers that can be easily migrated and scaled across clusters of generic hardware.

Research Infrastructures (RIs). Dedicated data infrastructures constructed by specific scientific communities for combining scientific data collections with integrated services for accessing, searching and processing research data within specific scientific domains; examples include the Integrated Carbon Observation System (ICOS)[4] for carbon monitoring in atmosphere, ecosystems and marine environments, the European Plate Observing System (EPOS)[5] for solid Earth science and Euro-Argo[6] for collecting environmental observations from large-scale deployments of robotic floats in the world's oceans. RIs play a key role in the *research data lifecycle*, providing standard policies, protocols and best practices for the acquisition, curation, publication, processing and further usage of research data and other assets such as tools and simulation/modelling platforms. They typically work closely with (or effectively subsume) individual data centres dedicated to research data, sensor networks, laboratories and experimental sites.

Virtual Research Environments (VREs). Platforms providing user-centric support for discovering and selecting data and software services from different sources, and composing and executing application workflows [3], also referred to as Virtual Laboratories [2] or Science Gateways [3]. Examples include VRE4EIC[7], D4Science[8] and EVER-EST[9]. VREs play a direct role in the *activity lifecycle* of research activities performed by scientists, for example, the planning of experiments, search and discovery of resources from different sources (notably including RIs), integration of services into cohesive workflows and collaboration with other scientists [4]. Graphical environments, workflow management systems, and data analytics tools are typical components of such environments.

While the roles and functions of these different kinds of environment may substantially overlap, none individually fulfil all the requirements of data-centric research; in practice, all these types of research support environment must be tightly integrated (and their overlapping functions reconciled and duly delegated). In particular, e-infrastructures focus on generic ICT (Information and Communication Technologies) resources (e.g. computing or networking), RIs manage data and services focused on specific scientific domains, and VREs support the lifecycle of specific research activities. Although, as already noted, the boundaries between these environments are not always entirely clear (often sharing services for infrastructure and data management), collectively they represent an important trend in many international research and development projects. Figure 1 shows the abstract logical relationship between e-infrastructures, RIs and VRE.

---

[4] https://www.icos-ri.eu/.

[5] https://www.epos-ip.org/.

[6] http://www.euro-argo.eu/.

[7] http://www.vre4eic.eu/.

[8] https://www.d4science.org/.

[9] https://ever-est.eu/.

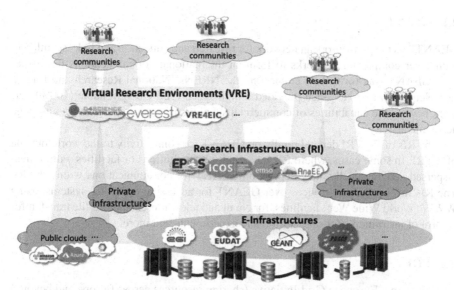

Fig. 1. A layered view of the different kinds of research support environment used by research communities.

Like other domains of research, environmental science has progressively adopted ICT. Perhaps more than other domains, environmental science has complexity because it encompasses observational, experimental and modelling/simulation methods across complex natural systems which have a past, a present and a predicted future. The RIs in environmental and Earth sciences commonly have their own ICT infrastructures but increasingly utilise e-Infrastructures external to the RI and shared commonly among multiple domains of research. This chapter characterises those e-Infrastructures and places them within the ENVRI framework.

In this chapter, we will introduce some typical examples of e-infrastructures. Based on those low-level ICT technologies and infrastructures, we will discuss the research infrastructures and Virtual Research Environments in the later chapters.

2   The e-Infrastructures

This section outlines the e-Infrastructures of relevance to ENVRI, their characteristics and offerings and how they have been used by RIs in ENVRI.

2.1 GEANT

GEANT[10] is the pan-European network for research and education and links seamlessly with other continental networks to form an international communications infrastructure. GEANT was formed by connecting the NRENs (National Research and Education Networks) and has since provided a high speed (100 Gb/s), reliable (100%) network beyond the capabilities of commercial suppliers in order to support leading-edge academic activity.

The RIs of ENVRI depend totally on GEANT for connectivity to the world outside of the RI. In some cases, where RIs have multiple institutions or facilities within them dispersed geographically, they depend on GEANT for communications within the RI. The RIs in ENVRI use services over GEANT for accessing computer systems, using WWW (World Wide Web) facilities, for email and teleconferencing, for file transfer, for control of instruments for observation and experiments and more.

2.2 EGI

Arising from a European Grid Initiative (sharing resources across Europe and beyond) EGI[11] is a federation and not-for-profit organisation providing virtualised access to multiple e-Infrastructures providing computing resources (through HTC and Cloud computing) and storage (online and archival), and services for data processing (i.e., Jupyter Notebook), data management (i.e., Datahub), and AAI (i.e., Check-in).

Various RIs in ENVRI have used EGI facilities to provide computing and storage resources beyond the capability of the RI itself. EGI staff involved in ENVRIplus have supported joint pilot projects with RIs to demonstrate the capabilities of the EGI facilities.

2.3 EUDAT

EUDAT[12] offers an e-Infrastructure for storage and associated services. The EUDAT CDI (Collaborative Data Infrastructure) is essentially a European e-infrastructure of integrated data services and resources to support research. This infrastructure and its services have been developed in close collaboration with over 50 research communities spanning across many different scientific disciplines and involved at all stages of the design process. The establishment of the EUDAT CDI is timely with the imminent realisation of the European Open Science Cloud (EOSC)[13], which aims to offer open and seamless services for storage, management, analysis and re-use of research data, across borders and scientific disciplines.

EUDAT services include B2FIND for searching a catalogue of available datasets described by CKAN[14] with its metadata schema[15] (although commonly enlarged);

[10] https://www.geant.org/Networks.
[11] http://www.egi.eu/.
[12] https://eudat.eu/.
[13] https://ec.europa.eu/research/openscience/index.cfm?pg=open-science-cloud.
[14] https://ckan.org/.
[15] https://ckan.org/portfolio/metadata/.

B2SHARE and B2DROP for data deposit and B2ACCESS for access control. B2STAGE transfers a dataset to local storage for processing while B2SAFE provides storage and curation facilities.

The EUDAT services for data management are utilised to a various extent by a number of ENV RIs, including eLTER, ICOS and Euro-Argo. The capabilities of B2FIND were demonstrated in ENVRI with a catalogue utilising the CKAN metadata schema (extended) providing access to datasets. Some of the pilot projects performed jointly with EGI staff within ENVRIplus utilised EUDAT, for instance the B2SAFE data storage used by Euro-Argo was extended with a EUDAT Data subscription functionality in an ENVRIPlus use case (ref. Chapter 16).

2.4 PRACE

PRACE[16] (Partnership for Advanced Computing in Europe) is an e-Infrastructure consisting of supercomputer facilities in Europe. The computer systems and their operations accessible through PRACE are provided by 5 PRACE members (BSC representing Spain, CINECA representing Italy, ETH Zurich/CSCS representing Switzerland, GCS representing Germany and GENCI representing France). Four hosting members (France, Germany, Italy, and Spain) secured funding for the initial period from 2010 to 2015. In 2016 a fifth Hosting Member, ETH Zurich/CSCS (Switzerland) opened its system via the PRACE Peer Review Process to researchers from academia and industry. In pace with the needs of the scientific communities and technical developments, systems deployed by PRACE are continuously updated and upgraded to be at the apex of HPC technology. Applications to use PRACE are peer-reviewed to provide project access for typically 3 years. Preparatory projects (to prepare for project access) are supported.

Individual researchers from various RIs in ENVRI have used PRACE facilities for particular research activities but there is no wholesale use of PRACE by ENVRI RIs at present.

2.5 OpenAIRE

OpenAIRE[17] has grown through a series of project phases funded by the European Commission: from the DRIVER projects to link Europe's scholarly publication repository infrastructure, to the first OpenAIRE project aimed to assist the EC in implementing its initial pilot for Open Access (OA) to publications, and, through several further phases which have extended and consolidated the OpenAIRE mission to implement Open Science policies. OpenAIRE has been providing the standards and services (e.g. harvesting, retrieval) to allow a catalogue of research assets to be built and used based on CERIF[18] under an agreement with euroCRIS[19]. CERIF provides the fully connected graph model with base entities and linking (relationship) entities with the role and temporal duration required for describing accurately the word of research.

---

[16] http://www.prace-ri.eu/.

[17] https://www.openaire.eu/.

[18] https://www.eurocris.org/cerif/main-features-cerif.

[19] https://www.eurocris.org/.

Many researchers in RIs within ENVRI use OpenAIRE directly for searching for relevant publications or other research assets (e.g. datasets) or - indirectly via their institutional repository - through harvesting of metadata on scholarly publications or other research assets to the catalogue. OpenAIRE has another lesson for ENVRI: because of the heterogeneity of metadata formats in the various repositories of research assets, the project discovered that simple metadata schemes were inadequate and chose to use the rich metadata model of CERIF to allow ingestion of the various heterogeneous metadata models describing the distributed institutional assets.

2.6 EOSC

EOSC[20] (European Open Science Cloud) is an initiative funded by the EC to provide a 'commons' for networking, computing resources, storage, services and assets useful to research, industry and citizens. Feasibility has been demonstrated through the EOSC Pilot[21]. EOSC is still under construction and is centred around the EOSC Hub[22] but there are also other more recent projects for constructing the EOSC such as EOSC Secretariat supporting the EOSC governance as well as facilitating a number of European working groups. The facilities are provided by EGI, EUDAT, Indigo Data Cloud[23] and OpenAIRE utilising GEANT.

RIs in ENVRI have participated, first in some joint work with EGI and then in the EOSC Pilot where work was concentrated on metadata and interoperability of data and services. Currently, ENVRI RIs interact with building the EOSC through the ENVRI-FAIR project [8]. A key point about EOSC is that it is built around the concept of services and provides a catalogue of services. Most ENVRI RIs provide catalogues of datasets and so there is a mismatch. Uniquely, EPOS within ENVRI designed and built its catalogue of assets to encompass services, datasets, data products, workflows, software modules, equipment and other research assets, concentrating first on services to align with the evolving EOSC. Furthermore, EPOS uses CERIF and so has a rich metadata format allowing interconversion with less rich metadata formats and also ensuring compatibility with OpenAIRE.

2.7 Sensor Networks

Sensor networks are essential for observation in environmental science. Modern networks are digital with local processing power - sometimes referred to as Fog or Edge Cloud Computing. Many modern sensors can be configured remotely to detect one or more physical attributes (e.g. temperature, pressure, salinity and pH) and to adjust precision and accuracy. By their nature, many sensor networks are specific to a particular RI within ENVRI but some sensor networks are shared among several RIs.

A specialised kind of sensor is earth observation satellites. In this case, the RIs in ENVRI receive data products particularly images in various wavebands (after sensing,

---

[20] https://www.eosc-portal.eu/.

[21] https://eoscpilot.eu/.

[22] https://www.eosc-hub.eu/.

[23] https://www.indigo-datacloud.eu/.

calibration and any necessary corrections and further processing) from agencies such as ESA (European Space Agency). Many RIs in ENVRI use such services. Similarly, geodesy services utilising satellites including GPS (Global Positioning System) provide information on surface elevation changes. This is used by several RIs in ENVRI from generating 3-D topographic models to detecting earth movements e.g. earthquakes.

2.8 Laboratory Equipment

RIs in ENVRI use laboratory equipment for a variety of purposes from chemical analysis and work on DNA to flumes for hydrological studies and pressure cells for rock mechanics. By their nature, they tend to be specialised to a particular RI although it is possible to utilise external commercial services for some equipment use where the equipment cost is not justified by the amount of likely use. The equipment is usually commercially produced with proprietary formats for data and metadata recording the experiment. Increasingly the equipment has digital capabilities for output and also increasingly for input to control the equipment during the experiment. This opens the possibility of a researcher sending a sample to a particular laboratory and both monitoring/controlling the experiment and collecting the experimental data remotely.

RIs in ENVRI have a large variety of equipment utilised within each RI.

2.9 Computing

RIs in ENVRI have computing equipment within their institutions, and in addition, they may be utilising local or national computing centres for this. These are used for data collection and processing. There is little sharing of such facilities among RIs, nor much sharing of software or even best practice in the use of such equipment across RIs. It is to be hoped that progressively the RIs in ENVRI will appreciate the benefits of shared best practice and software (decreasing costs, increasing professionalism, permitting interoperability) and even sharing of computing resources so that idle computing capacity may be utilised. However, it may be that the cost of data transfer and potential security/privacy risks outweigh the cost savings.

3 Access to the e-Infrastructures

The e-Infrastructures are to be used for research, education and wealth creation and - in the case of ENVRI - there is an opportunity to take advantage of the facilities. However, access to e-Infrastructures requires passing some controls.

3.1 AAAI

AAAI (Authentication, Authorisation, Accounting Infrastructure) refers to the process whereby an end-user gains access to computing and other digital facilities. Typically, from a non-commercial background, a researcher applies to the local institution which authenticates her manually (usually with an email address and password) which in turn provides access with online authentication via EduGAIN[24] to GEANT and thence - subject to authorisations - to other e-Infrastructures (federated identity management). The authorisation is more complex and is e-Infrastructure-specific (or, for that matter, RI-specific). The RI defines policy and this is then enacted. If the policy is for total open access no authorisation is required although accounting will be required to record accesses as needed by GDPR (General Data Protection Regulation) [6]. Usually, the RI catalogue provides the relationship (authorisation) between an authenticated user and research assets; the relationship being the actions authorised within a role (e.g. execute, read, update, write, and delete) and referred to as RBAC (Role-based access control) [7]. The access may be temporally limited e.g. to ensure no overuse of computing resource or to embargo access to a research asset while the lead researcher(s) publish based on that asset. This is temporally bound RBAC.

3.2 TNA

TNA (Trans-National Access) is a scheme designed to allow researchers from one RI or community to utilise equipment at another. The TNA process is essentially matching a researcher requirement to perform an observation or experiment with a RI that has the appropriate equipment available. It may be compared to hotel reservation systems, although the specifications tend to be more complex and the governance and funding arrangements need to be agreed - ideally generally and in advance. It is expected that the use of the equipment is acknowledged and - in some cases - that publications based on the results are joint between the researcher and staff at the RI owning the equipment, especially if the equipment requires complex and expensive set-up.

Within ENVRI there appears to be little use of TNA. In EPOS a TNA system - accessed from the EPOS portal - is being implemented (currently being tested) to try to optimise the use of expensive laboratory equipment.

4 Aspects of Future Infrastructure

The technologies are evolving constantly. Here some significant developments are outlined and their importance to ENVRI estimated.

4.1 Smart Networks

Smart Networks, commonly known as SCN (Software Controlled Networks) are becoming a reality increasingly. They have the ability to manage the available bandwidth on a network segment to obtain maximum throughput together with recording monitoring information to enable dynamic improvements. This is important for RIs in ENVRI, especially for data collection from observations (sensor networks) or experiments (equipment) where there may be very high data rates.

---

[24] https://edugain.org/.

4.2 Cloud Dynamic Resource Allocation

Cloud computing virtualises computing resources so that the end-user neither knows nor cares where their computing is being done. Building upon the concept of GRIDs developed through an EC Expert Group 2000–2006[25] Cloud Computing was considered by another EC Expert Group[26]. The major obstacles to Cloud Computing were identified as (a) security, privacy and trust; (b) availability; (c) lock-in to one supplier. Despite some sensational difficulties over availability (when a large computer centre was out of action due to a security attack or power outage) in general (a) and (b) have been overcome. (c) was overcome by techniques to describe an application workflow such that it (or semi-independent components of it) could be deployed by a controlling middleware across one or more Cloud suppliers using VMs (Virtual Machines). Further work led to the optimisation of deployments depending on cost, elapsed time, Cloud supplier computer characteristics (e.g. kind of processor). All of this depended on containerisation – using containers (typically Docker[27]) and a container management environment (with scaling and deployment) such as Kubernetes[28]. Various ENVRI RIs have been experimenting with using such computing environments and it is expected that such architectures will become prevalent in the future.

5 Looking Backward and Forward

The RIs within ENVRI - like all RIs within the ESFRI family - have been on a journey over the last few years, increasing their capabilities and knowledge and adapting to the opportunities provided by the new, emerging technologies and the ever-increasingly ambitious requirements of the researchers and other users. Here we assess the journey during the ENVRIplus project and suggest some future projections.

5.1 Shared Experience

There has been much sharing of experience during ENVRIplus. Now each RI in ENVRI has an appreciation of the way each other RI has developed its ICT. There has been an increasing realisation that there are opportunities for sharing of more tangible assets such as software and leading to the end-goal of interoperability so that a researcher in one domain can utilise the assets of other domains to form a more comprehensive understanding of the environment.

5.2 Shared Best Practice

Different ENVRI RIs started at different stages of development of their ICT. There was an expectation that associated best practice could be shared to improve the offerings and utilisation of each RI by cross-adoption among them of appropriate better techniques.

---

[25] https://www.ercim.eu/publication/Ercim_News/enw66/jeffery.html.

[26] https://ercim-news.ercim.eu/en80/es/the-future-of-cloud-computing.

[27] https://www.docker.com/.

[28] https://kubernetes.io/.

In some areas this has been demonstrated: there is much more awareness of the need for curation of assets, for example, using the DCC (Digital Curation Centre) model and DMP (Data Management Plan) template. Similarly, awareness has been raised in the areas of use of PIDs (Permanent Identifiers), Citation and rich metadata including for provenance [9]. It is to be expected that in future further convergence of best practice – but specialised for each RI domain – will occur leading to greater opportunities for sharing and an overall raising of standards of research support in all RIs.

5.3 Shared Sensor Networks

Some ENVRI RIs share already sensor networks especially when the equipment is expensive or located remotely – examples are some oceanic instruments whether associated with a particular research vessel cruise or (semi-)permanently positioned. Several RIs use data products derived from satellite sensors. It is to be anticipated that such shared use will increase in the future as the costs of deploying the sensors increases, the sophistication of the network control (through Fog/Edge Cloud computing) - allowing autonomic operation - increases and the research requirements demand more shared use of multiple sensor networks to produce a multidomain environmental analysis.

5.4 Shared Equipment

Some RIs are or have institutions which own and use particular experimental equipment. While some equipment is inexpensive and there is no real advantage in sharing, in other cases not only is the equipment expensive but the experienced technicians and researchers needed to operate the equipment are expensive. Therefore, there is merit in sharing. While in some cases a service may be offered such that a sample may be sent, analysed at the equipment and the results returned digitally in other cases it is recommended that the researcher attend the equipment themselves to fully understand the capabilities and limitations (including accuracy, precision and calibration) of the equipment. It is to be expected that there is more equipment sharing in the future; the 'remote service' kind being more common than the 'attend and operate the equipment' kind. In EPOS, for example, a prototype system for TNA (Trans-National Access) is being tested where the researcher request for access to equipment is matched with suitable equipment availability and the agreement facilitated.

5.5 Shared RI Computing

The ENVRI RIs have their own computing equipment, usually used as servers to perform computing tasks and to provide data storage. Some have their computing resources operating as a cluster and a few have utilised Cloud middleware to provide an in-house private Cloud service. It is true generally that these separate, distributed and distinct computing resources are not utilised fully. If they could be coupled together, appropriate middleware installed and canonical systems for security, privacy, trust and resource management introduced then there would be two benefits: (1) each RI would have available more compute and storage capacity; (2) the assets of any RI would become more easily interoperable since the asses would be virtualised in the Cloud environment. However, RIs would also perceive disbenefits: (1) there is clearly a security risk in opening up computing resources previously private to a wider networking environment; (2) local management of the resources of a single RI would no longer necessarily have precedence so an urgent task may not be allowed to run immediately. This could be particularly important if real-time data is being streamed to the RI computing centre. (1) could be partially overcome by 'sandboxing' any executable software deployed to computing resources other than the RI of origin. (2) could be overcome by system management overrides in the resource allocation system. This may be a possible route forward for the future but would require a degree of cooperation in governance as yet not foreseen.

5.6 Shared External Computing

As indicated earlier, several ENVRI RIs have experimented with using external services such as those provided by EGI and PRACE in order to gain more computing power than available at the local RI computer centre. Similarly, some have experimented with using EUDAT for data storage. Many have used OpenAIRE for its curation and provenance capabilities (a central canonical catalogue pointing to open repositories) for documents and increasingly for datasets. During ENVRIplus no consistent policy shared across the ENVRI RIs emerged. However, the emergence of the EOSC concept provides an architectural basis for the utilisation of external resources since the above external e-Is are collected under that umbrella. Various ENVRI RIs were involved in the EOSC Pilot project particularly considering interoperation sanctioned by conversion of heterogeneous metadata schemas to a canonical rich metadata format. It is expected that during the ENVRI-FAIR project that a coherent policy for external access covering governance, sustainability and FAIR principles[29] as well as technical architecture based on a common, logically-centralised rich metadata catalogue will be adopted by ENVRI RIs.

5.7 Shared Datasets

ENVRIplus concentrated on datasets as primary assets of the RIs. Some scientific use cases had requirements for datasets from several RIs and thus some datasets were shared. However, the datasets usually required some management and manipulation in order to make them reusable: this usually involved unit conversions or adjustment of spatial coordinates. The overall aim of ENVRI is to make datasets (and other assets) shareable so that more comprehensive environmental analyses may be achieved. In the future, it is to be anticipated that datasets described by rich metadata (as recommended by FAIR and a being parameterised by the FAIR Data Maturity Working Group of RDA[30]) will be shareable.

5.8 Shared Workflows

Many workflows are – by their nature – specific not only to a domain but to a particular part of a subdomain. However, for purposes of reproducibility of research results, it is

---

[29] https://www.force11.org/group/fairgroup/fairprinciples.

[30] https://www.rd-alliance.org/groups/fair-data-maturity-model-wg.

important that workflows be stored, characterised by rich metadata and made available. Furthermore, taking a pre-existing and available workflow and modifying it for a new purpose may save much research effort. This depends – of course – on the assets utilised by the workflow being FAIR. Once a workflow is shared comes the challenge of workflow deployment. This is where the work on interoperable multi-Clouds described above becomes especially valuable.

5.9 Shared Software

One of the aims of ENVRIplus was to share software – either pre-existing at one or more RIs or developed within the project. The software envisaged was of two types: (1) software that every RI needed to manage its assets including software for curation, cataloguing, asset access, provenance; (2) software required to interoperate data across RIs. During the project both types were specified formally with the RM (Reference Model) but (a) no existing software met exactly the specification; (b) the project did not have the resources to develop the required software.

5.10 Shared Services

EPOS made the decision to first catalogue services as assets rather than datasets (although datasets, equipment and other assets are being catalogued now). This was for several reasons: (1) it was clear that the proposed EOSC was going to be based on services and EPOS wished to have EOSC interoperability; (2) by offering a service, a provider implicitly also offers (a) access to the dataset(s) utilised; (b) to the computing resources required to execute the service; (c) data management services to reduce the dataset(s) only to those records that meet the parameters input by the user. Progressively some ENVRI RIs offer services through their portals as well as access to datasets for download. Longer-term the concept of data download (analogous to using a library catalogue card to find a book then taking it home to read) may become unviable since datasets are growing larger and network speeds are not increasing at the equivalent rate. Hence the concept of user-controlled (or control by software acting on behalf of the user) data management at a remote RI becomes necessary – and this implies access through a service.

5.11 Interoperation - Shared Metadata (FAIR)

The vision of ENVRI is that a researcher, policymaker, commercial user or citizen at any location can 'see' through a catalogue a homogeneous view over the heterogeneous assets available at the ENVRI RI sites. The optimal way to achieve this is through homogenised rich metadata (as discussed in Chapter 8) derived from the heterogeneous local metadata standards utilised at each RI. Mechanisms to achieve this matching and mapping of metadata schemas are discussed in Chapter 8, as is the need for and benefits of rich metadata. In order for the assets to be FAIR, they need to meet certain standards or achieve appropriate scores against parameters currently being defined by the RDA FAIR Data Maturity Working Group. The current ENVRI RIs are all - to some extent- FAIR but few reach the more advanced aspects of FAIR, The ENVRI-FAIR project should assist in improving the FAIRness of assets in all ENVRI RIs.

Acknowledgements. This work was supported by the European Union's Horizon 2020 research and innovation programme via the ENVRIplus project under grant agreement No 654182.

References

1. Koulouzis, S., et al.: Time critical data management in clouds: challenges and a Dynamic Infrastructure Planner (DRIP) solution. Concurr. Comput. Pract. Exp. e5269 (2019). https://doi.org/10.1002/cpe.5269
2. Martin, P., Remy, L., Theodoridou, M., Jeffery, K., Zhao, Z.: Mapping heterogeneous research infrastructure metadata into a unified catalogue for use in a generic virtual research environment. Int. J. Future Gen. Comput. Syst. **101**, 1–13 (2019). https://doi.org/10.1016/j.future.2019.05.076
3. Miller, M.A., Pfeiffer, W., Schwartz, T.: The CIPRES science gateway: enabling high-impact science for phylogenetics researchers with limited resources. In: Proceedings of the 1st Conference of the Extreme Science and Engineering Discovery Environment: Bridging from the Extreme to the Campus and Beyond, Chicago, IL (2012)
4. Remy, L., et al.: Building an integrated enhanced virtual research environment metadata catalogue. J. Electron. Libr. (2019). https://zenodo.org/record/3497056
5. Skene, J., Emmerich, W., Raimondi, F.: Service-level agreements for electronic services. IEEE Trans. Softw. Eng. **36**(02), 288–304 (2010)
6. Tene, O., Evans, K., Gencarelli, B., Maldoff, G., Zanfir-Fortuna, G.: GDPR at year one: enter the designers and engineers. IEEE Secur. Priv. **17**(06), 7–9 (2019)
7. Ghafoor, A., Joshi, J., Latif, U., Bertino, E.: A generalized temporal role-based access control model. IEEE Trans. Knowl. Data Eng. **17**, 4–23 (2005)
8. Petzold, A., et al.: ENVRI-FAIR - interoperable environmental fair data and services for society, innovation and research. In: 2019 15th International Conference on eScience (eScience), San Diego, CA, pp. 277–280. IEEE (2019). https://doi.org/10.1109/escience.2019.00038. https://zenodo.org/record/3462816
9. Wofford, M., Boscoe, B., Borgman, C., Pasquetto, I., Golshan, M.: Jupyter notebooks as discovery mechanisms for open science: citation practices in the astronomy community. Comput. Sci. Eng. **22**(01), 5–15 (2020)

Common Challenges and Requirements

Barbara Magagna[1]([⊠]) [iD], Paul Martin[2] [iD], Abraham Nieva de la Hidalga[3] [iD],
Malcolm Atkinson[4] [iD], and Zhiming Zhao[2] [iD]

[1] Environment Agency Austria, Vienna, Austria
barbara.magagna@umweltbundesamt.at
[2] Multiscale Networked Systems, University of Amsterdam,
1098XH Amsterdam, The Netherlands
pwmartin.research@gmail.com, z.zhao@uva.nl
[3] Cardiff University, Cardiff, UK
nievadelahidalgaa@cardiff.ac.uk
[4] Edinburgh University, Edinburgh, UK
mpa@staffmail.ed.ac.uk

Abstract. Research infrastructures available for researchers in environmental and Earth science are diverse and highly distributed; dedicated research infrastructures exist for atmospheric science, marine science, solid Earth science, biodiversity research, and more. These infrastructures aggregate and curate key research datasets and provide consolidated data services for a target research community, but they also often overlap in scope and ambition, sharing data sources, sometimes even sites, using similar standards, and ultimately all contributing data that will be essential to addressing the societal challenges that face environmental research today. Thus, while their diversity poses a problem for open science and multidisciplinary research, their commonalities mean that they often face similar technical problems and consequently have common requirements when addressing the implementation of best practices in curation, cataloguing, identification and citation, and other related core topics for data science.

In this chapter, we review the requirements gathering performed in the context of the cluster of European environmental and Earth science research infrastructures participating in the ENVRI community, and survey the common challenges identified from that requirements gathering process.

Keywords: Requirements · Environmental science · Research infrastructure

1 Introduction

Today's societal challenges such as hazard mitigation and sustainable resource provision need new interdisciplinary approaches pooling, resources, insights, data methods and models. The transformation of the way to analyse natural phenomena through the advent of digital instruments and the intensive use of computers also known as the "Fourth Paradigm" [1] came along with a wealth of data that poses immense challenges in how to handle and exploit it. This transition is common to all environmental and Earth sciences

Z. Zhao and M. Hellström (Eds.): Towards Interoperable Research
Infrastructures for Environmental and Earth Sciences, LNCS 12003, pp. 30–57, 2020.
https://doi.org/10.1007/978-3-030-52829-4_3

and many research communities are engaged in generating and exploiting that data. To optimise the working practices and the platforms that support the data pipelines from distributed data sensors to storage, use, presentation and analysis, it is crucial to study the condition of research infrastructure and systems currently available.

In the context of the ENVRI community, this becomes even more important as all the RI initiatives share common challenges, both in their construction and operation. Although each RI has its specific ICT strategy, there are potential benefits from defining, designing, and developing shared solutions for common problems. The use of common solutions also fosters strategies for sharing data or reducing the barriers to data interoperability. The ENVRI community thus encourages and facilitates joint work to develop synergies, learn from each other, harmonise the RI service landscape, and share best practices. Since 2011, the ENVRI community has been supported by multiple projects aiming to improve the collaboration within the European Research Area and beyond. Thus, requirements elicitation has a long tradition within the ENVRI community and with each ENVRI supporting project this effort has been repeated with the aim to record the last status of the involved RIs in the light of emerging technologies and newly pressing challenges.

The first ENVRI project (FP7 funded, 2011–2014) analysed the design of six participating ESFRI environmental research infrastructures (ICOS, Euro-Argo, EISCAT_3D, LifeWatch, EPOS, and EMSO) to identify common computational characteristics and requirements. The ISO/IEC standard Open Distributed Processing (ODP), a multi-viewpoint conceptual framework for building distributed systems, was used as a common platform for interpretation and discussion to ensure a unified understanding. The outcome of this endeavour proved to be helpful for the understanding of strengths and weaknesses in the outline and planned developments of the RIs.

Within the time frame of the ENVRIplus project (2015–2019) [2], the status of involved RIs (which now numbered over 20) along the dimensions of data, users, software services and resources were re-analysed. The aim was to identify commonalities and differences between RIs and to point to the state-of-the-art of RI technologies. The characterisation of RIs under a standard documentation method with vocabulary defined in the ENVRI Reference Model (see chapter 4) allowed comparison and discussion leading to best practice and consistent development plans for RI improvements.

For the requirements study, a standard method for describing all relevant ICT aspects needed to provide the facilities and capabilities required by RI researchers was defined. This process led to the identification of seven specific topics identified in the project [2]:

- **Identification and citation.** Mechanisms to provide and retrieve durable references to data objects and collections of data objects.
- **Curation.** Processes to assure the availability and quality of data over the long term.
- **Cataloguing.** The construction of discoverable and searchable indexes of datasets, processes, software and other resources made available by RIs.
- **Processing.** Computational transformations of data, including but not restricted to processing and selection of raw data close to instruments, signal processing, analysis of data for quality assurance (QA) purposes, simulation runs with a subsequent comparison with observations, and statistical analyses.

- **Provenance.** Processes to record information about how data, code and working practices were created and were transformed into their current form.
- **Optimisation.** Methods for improving the quality of service offered to researchers as data and processing requirements increase, focusing mainly on the underlying movement and processing of data.
- **Community support.** Addressing all aspects of the use of resources by researchers and their relationships with resource providers.

The requirements of the ENVRI RIs were distilled and assessed in the context of these seven topics; from this assessment, a number of general requirements can be seen that remain broadly applicable even as the research landscape continues to evolve, incorporating the concepts of FAIR [3] data and European Open Science Clouds [4]. Part of this chapter is drawn from the project requirement analysis deliverable [5].

2 Requirements Collection in ENVRI

The ENVRI community brings together environmental and Earth science RIs, projects, networks and technical specialists with the joint ambition to create a holistic, coherent, interdisciplinary and interoperable cluster of environmental research infrastructures across Europe (and beyond). To do this, the ENVRI community brings together roughly four different environmental domains: *atmospheric, marine, biosphere/ecosystem* and *solid earth*. By working together, the idea is to capitalise on the progress made in various disciplines and strengthen interoperability amongst RIs and domains to better support more interdisciplinary research of the sort needed to address modern environmental grand challenges.

Table 1 gives an overview of the RIs that participated in the requirements gathering activity during the ENVRIplus project (2015–2019). The table indicates their organisation type, domain and involved data life cycles, as defined by the ENVRI RM. These RIs are typically composed of distributed entities (e.g. data generators, data processors and data sharers) and thus federations of often diverse autonomous organisations. These organisations have established roles, cultures, working practices and resources, and they often play roles in different federations of international infrastructures. The organisations' roles thus must remain unperturbed.

The original analysis of requirements gathered by the ENVRIplus project is provided by Atkinson et al. [5], but in this chapter we review some of the details that remain most pertinent to continued RI development in the environmental sciences and beyond. Most RIs focus on one domain, but several address multiple domains, typically with a common factor in mind (e.g. carbon observation or observations based on a specific region of interest, such as the Arctic). We look at each domain in turn.

2.1 Atmospheric Domain

Atmospheric science RIs typically seek to provide scientists and other user groups with free and open access to high-quality data about atmospheric aerosols, clouds, and trace

Table 1. Overview of RIs contributing to requirements gathering.

RI	Type of RI	Domain	Data Lifecycle
ACTRIS	Distributed	Atmospheric	Use to publishing
AnaEE	Distributed	Ecosystem	Curation to processing
EISCAT-3D	Single RI, multi-site	Atmospheric	Use to publishing
ELIXIR	Distributed	Ecosystem	Acquisition to publishing
EMBRC	Distributed	Marine, Ecosystem	Use to publishing
EMSO	Single RI, multi-site	Marine	Acquisition to publishing
EPOS	Distributed	Solid Earth	Acquisition to publishing
Euro-Argo	Distributed	Marine	Production to publishing
EuroGOOS	Distributed	Marine	Production to publishing
FixO3	Distributed	Marine	Acquisition to publishing
IAGOS	Distributed	Atmospheric	Acquisition to processing
ICOS	Distributed	Atmospheric, Marine, Ecosystem	Acquisition to publishing
INTERACT	Distributed	Ecosystem	Acquisition to publishing
IS-ENES2	Virtual	Multi-domain	Acquisition to publishing
LTER	Distributed	Ecosystem	Use to publishing
SeaDataNet	Virtual	Marine	Acquisition to publishing
SIOS	Distributed	Multi-domain	Publishing

gases from coordinated long-term observations, complemented with access to innovative and mature data products, together with tools for quality assurance, data analysis and research. Data are typically gathered by a mix of both ground and in-air sensors (e.g. aircraft-mounted), with different focuses on particular types of observation (e.g. greenhouse gases) or specific regions (e.g. the arctic, or the upper atmosphere).

RIs in ENVRI focusing on the atmospheric domain include: ACTRIS (Aerosols, Clouds, and Trace gases Research Infrastructure), which integrates European ground-based stations for long-term observations of aerosols, clouds and short-lived gases; EISCAT_3D, which operates next-generation incoherent scatter radar systems for observation of the middle and upper atmosphere, ionosphere and near-Earth objects such as meteoroids; ICOS, which provide long-term observations to understand the behaviour of the global carbon cycle and greenhouse gas emissions and concentrations and IAGOS (In-service Aircraft for a Global Observing System), which implements and operates a global observation system for atmospheric composition by deploying autonomous instruments aboard commercial passenger aircraft.

ACTRIS has requirements for 1) improving interoperability so as to make their data as accessible and understandable as possible to others; 2) understanding best practices when researchers need to discover data, particularly given the experiences of other RIs; 3) planning and managing the activity of sensors; 4) developing understanding of how instruments work in extreme conditions; and 5) improving the capabilities of small sensors. ACTRIS is therefore concerned with such issues as data visualisation, data provisioning and interoperability between data centre nodes.

EISCAT_3D will contribute to our understanding of the near-Earth space environment for decades to come; whereas the domain of research is common with existing environmental RIs, EISCAT_3D will differ from most of those in its modes of operation and volumes of data. It is planned that at least 2 petabytes (PB) of data per year will be selected for long-term curation and archival. EISCAT_3D will thus present both high throughput computing (HTC) and big data analysis and curation challenges similar to those encountered in the particle physics and astrophysics communities. A common and increasingly important issue is citability and reusability of the data, as embodied in the FAIR concept. To keep track of data use is also a requirement from many funding bodies, including the member science councils and institutes of EISCAT. Therefore EISCAT is preparing to adopt a common scheme of persistent IDs for Digital Objects, such as fractional Digital Object Identifiers (DOI). EISCAT thus has requirements for workflow specification, data access with search and visualisation, interoperability with other RIs and instruments via virtual observatories.

ICOS aims to provide effective access to a single and coherent data set to enable research into the multi-scale analysis of greenhouse gas emissions, sinks and the processes that determine them. ICOS is concerned with 1) metadata curation, 2) data object identification and citation and 3) data collection and management of provenance information.

IAGOS provides freely accessible data for users in science and policy, including air quality forecasting, verification of CO_2 emissions and Kyoto monitoring, numerical weather prediction, and validation of satellite products. IAGOS expected through its participation in ENVRIplus to improve data discovery, metadata standardisation, interoperability and citation and DOI management. It also expected ENVRIplus to provide services for, citation, cataloguing and provenance.

2.2 Marine Domain

Marine domain RIs are concerned with observations and measurements of marine environments both near the coast and out to sea, near the oceans' surface and on the seafloor, and everywhere in-between [6]. As for atmospheric domain RIs, different RIs may focus on specific types of measurement or on specific sub-environments. Marine RIs have been observed to have the greatest maturity overall in terms of established data standards and curation practices for certain classes of dataset when compared to the other environmental science domains.

RI projects in ENVRI focused on the marine domain include: EMBRC (European Marine Biological Resource Centre), a distributed European RI which is set up to become the major RI for marine biological research, covering everything from basic biology, marine model organisms, biomedical applications, biotechnological applications, environmental data and ecology; EMSO (the European multidisciplinary seafloor & water column observatory), a large-scale RI for integrating data gathered from a range of ocean observatories; Euro-Argo, the European contribution to the Argo initiative, which provides data service based on a word-wide deployment of robotic floats in the ocean; EuroGOOS (European Global Ocean Observing System), an international Not-for-Profit organisation promoting operational oceanography, i.e., the real-time use of oceanographic information; FixO3 (Fixed Open Ocean Observatory network), a research project that integrates oceanographic data gathered from a number of ocean observatories and provides open access to that data to academic researchers; and SeaDataNet, a Pan-European infrastructure for ocean & marine data management, which provides on-line integrated databases of standardised quality.

In what concerns data, the role of EMBRC is to generate and make it available. It does not usually do any analysis of those data, unless it is contracted to do so. Data is usually generated through sensors in the site in the sea or samples that are collected and then measured in the lab. EMBRC aimed to achieve several objectives through participation in ENVRIplus: establishing collaborations with the environmental community, which would benefit from their environmental and ecological data; developing and learning about new standards and best practices in terms of standards; developing new standards within INSPIRE, which can be used for other datasets; exploring new data workflows, which make use of marine biological and ecological data; and networking with other RIs. EMBRC requires common standards and workflows, harmonisation of data between labs, backup systems, maintenance of software and their integration into a single platform.

A goal of EMSO is to harmonise data curation and access, while averting the tendency for individual institutions to revert to idiosyncratic working practices after any particular harmonisation project has finished. There is a notable overlap between EMSO and FixO3 data (i.e., some FixO3 data is provided within the EMSO infrastructure). EMSO would like to obtain with the help of ENVRIplus better mechanisms for ensuring harmonisation of datasets across their distributed networks. Heterogeneous data formats increase the effort that researchers must invest to cross discipline boundaries and to compose data from multiple sources. Improved search is also desirable; currently expert knowledge is required, for example to be able to easily discover data stored in the MyOcean environment. Furthermore, EMSO is investigating collaborations with data processing infrastructures such as EGI for providing resources for infrastructure-side data processing. EMSO requires data interoperability across distributed networks and data search.

Like EMSO, FixO3 requires better mechanisms for ensuring harmonisation of datasets across their distributed networks. Heterogeneous data formats make life difficult for researchers. Improved search is also desirable; currently expert knowledge is required, for example to be able to easily discover data stored in the MyOcean environment.

FixO3 also requires harmonisation of data formats and protocols across their distributed networks, as well as harmonisation of data curation and access.

The Euro-Argo research infrastructure comprises a central facility and distributed national facilities. Euro-Argo aims at developing the capacity to procure and deploy and monitor 250 floats per year and ensure that all the data can be processed and delivered to users (both in real-time and delayed-mode). Euro-Argo sought within ENVRIplus to design and pioneer access to and use of a cloud infrastructure with services close to European research data to deliver data subscription services. Users would provide their criteria: time, spatial, parameter, data mode, update period for delivery (daily, monthly, yearly, near real-time).

EuroGOOS strives to improve the coordination between their different member research institutes. Another important role of EuroGOOS is that of facilitating access to data for their community. Through participation in ENVRIplus, EuroGOOS valued: learning about other European RIs and getting inspiration from them for deciding on the general objectives and services that they could provide at European level; from a technological perspective, getting recommendations about the design of their common data system, including formats or data platforms and data treatments; getting inspiration from other RIs about ways to distribute the data to end users using applications which are more focused in this respect; and improved data assimilation.

Regarding SeaDataNet, the on-line access to in-situ data, metadata and products is provided through a unique portal interconnecting the interoperable node platforms constituted by the SeaDataNet data centres. SeaDataNet wanted to enhance the cross-community expertise on observation networks, requirements support and data management expertise by participating in ENVRIplus. More specifically, SeaDataNet needs technical support for cross-community (ocean, solid earth and atmosphere) visibility of information provided by SeaDataNet (platforms, metadata, datasets and vocabulary services), as well as expertise on interoperability services and standards.

SeaDataNet requires data policy to involve data providers in the publication of their own datasets.

2.3 Ecosystem Domain

Ecosystem or biodiversity RIs focus on the study and monitoring of biological ecosystems both large and small with the objectives of both providing accurate and in-depth information about the condition and spread of various species throughout the Earth and their interactions, and developing models that can predict how ecosystems will evolve under various conditions or may have evolved in the past.

RI projects in ENVRI focused on the ecosystem/biodiversity domain include: AnaEE (Analysis and Experimentation on Ecosystems), which focuses on providing innovative and integrated experimentation services for ecosystem research; ELIXIR, a European infrastructure for biological information that unites Europe's leading life-science organisations in managing and safeguarding the massive amounts of data being generated every day by publicly funded research; INTERACT (International Network for Terrestrial Research and Monitoring in the Arctic), a circum-arctic network of 76 terrestrial field stations in various northern nations for identifying, understanding, predicting and responding to diverse environmental changes; and LTER (Long-Term Ecosystem Research), a long-standing alliance of researchers and research sites invested in better understanding the structure, functions, and long-term response of ecosystems to environmental, societal and economic drivers.

AnaEE aims to provide excellent platforms with clear accessibility conditions and service descriptions, and a clear offering to researchers. The gathering of information in a common portal should help with this. Experiences gathered from the construction and operation of other platforms would be helpful to shape this development. Within the context of ENVRIplus, AnaEE was particularly interested in participating in the work on identification and citation and on cataloguing, as these were of fairly immediate concern to their infrastructure, and consequently, it was useful to synchronise their approach with other RIs. Processing is of some interest as well, in particular the interoperability between models and data, and the quality control of data produced by platforms.

ELIXIR will provide the facilities necessary for life-science researchers—from bench biologists to chemo-informaticians—to make the most of our rapidly growing store of information about living systems, which is the foundation on which our understanding of life is built. By participating in ENVRIplus, ELIXIR aimed to establish a closer collaboration with other environmental RIs and improve their access to life science data. An enhanced interaction, a better insight into data structures and relevant data standards widely adopted across environmental RIs can facilitate an effective evaluation of areas of collaboration for development of new tools, services and training. Ultimately, this can lead to better interoperability and discoverability of environmental and life science data by users across atmospheric, marine, solid earth and biosphere domains.

INTERACT is keen on working on homogenisation with other infrastructures. The most important bilateral benefits of NordGIS (the INTERACT geographical metadata information system[1]) versus ENVRIplus are the broad European standards exposed to NordGIS, as well as the grass-root requirements exposed to ENVRIplus. INTERACT is open for new interactive solutions, and recognises that standards on how to turn primary data into data products suitable for OPEN dissemination need to be adopted. INTERACT needs to move into the realm of handling actual data concerning active field stations.

Due to the fragmented character of LTER Europe, harmonised data documentation, real-time availability of data as well as harmonisation of data and data flows are the overarching goals. As of the most recent review, LTER Europe is developing a Data Integration Portal (DIP, e.g. including a time series viewer) and is working on the integration of common data repositories into their workflow system (including metadata documentation with LTER Europe DEIMS[2]). Therefore, based on the common reference model, the outputs of ENVRIplus can provide development advice on those matters, which would be appreciated by LTER.

2.4   Solid Earth Domain

RIs in the solid earth domain are concerned with the study of seismology, volcanology, geodesy and other research disciplines focused on the Earth beneath our feet. The RI project focused on the solid earth domain in ENVRI is EPOS (European Plate Observing System), a long-term plan for the integration of Research Infrastructures for solid Earth science in Europe. Its main aim is to integrate communities to make scientific discovery in the domain of solid Earth science, integrating existing (and future) advanced European facilities into a single, distributed, sustainable infrastructure (the EPOS Core Services).

EPOS will enable the Earth Science community to make a significant step forward by developing new concepts and tools for accurate, durable, and sustainable answers to societal questions concerning geo-hazards and those geodynamic phenomena (including

---

[1] http://www.nordgis.org/.
[2] https://deims.org/.

geo-resources) relevant to the environment and human welfare. EPOS wanted advice from ENVRIplus to improve the Interoperable AAAI (Authentication, Authorisation, Accounting and Identification) system (federated & distributed), taking already existing software and make it available and scalable across communities.

2.5 Cross-Domain Concerns

Not all RIs neatly fit their activities into a particular domain as defined above. Many RIs address cross-cutting environmental concerns such as greenhouse gas emissions, and so integrate facilities and data sources with very different characteristics that might simultaneously contribute to more dedicated RIs as well.

RI projects participating in the ENVRI community that have a notable cross-domain focus include: ICOS (Integrated Carbon Observation System), an RI providing the long-term observations required to understand the present state and predict future behaviour of the global carbon cycle and greenhouse gas emissions and concentrations; IS-ENES2, contributing to the European Network for Earth System Modelling; and SIOS (Svalbard Integrated Earth Observing System), an integral Earth Observing System built to better understand the on-going and future climate changes in the Arctic.

The objectives of ICOS are to provide effective access to a single and coherent dataset to facilitate research into the multi-scale analysis of greenhouse gas emissions, sinks and the processes that determine them, and to provide information, which is profound for research and for the understanding of regional budgets of greenhouse gas sources and sinks, their human and natural drivers, and the controlling mechanisms. This requires insight into the interaction between atmospheric, marine and ecosystem datasets and services in particular.

IS-ENES encompasses climate models and their environment tools, model data and the interface of the climate modelling community with high-performance computing, in particular the European RI PRACE. Its requirements were mainly collected from the climate-modelling community, two data-dissemination systems (ESGF for project run time; LTA as long-term archiving), CMIP5 as climate modelling data project (2010–2015) and CMIP6 (2016–2021). By participating in ENVRIplus IS-ENES2 expected to obtain a better understanding of interdisciplinary use cases and end-user requirements, as well as advice for data catalogues to compare their model data with other data (e.g. observations). IS-ENES2 requires sharing of best practices as new nodes are integrated into the RI federation; keeping data near to processing; handling of volume and distribution of data; replication and versioning; and providing related information for data products (e.g. provenance, annotations and metadata).

Currently, SIOS is building a distributed data management system called SIOS Knowledge Centre, to develop methods for how observational networks are to be designed and implemented. The centre will lay the foundation for better-coordinated services for the international research community with respect to access to infrastructure, data and knowledge management, sharing of data, logistics, training and education.

2.6   Overall Requirements

It can be seen that there is substantial variability by RI and by topic. For every RI, a significant effort was made to develop communication and obtain information about requirements for all relevant topics. In some cases the RI was mature, in the sense that the RI or those involved in the work being done within the RI had been active in the particular domain for a significant number of years; the marine RIs that are already sharing data, such as Euro-Argo and SeaDataNet are good examples. Such maturity leads to an appreciation of the complexities and significance of various requirements. In other cases, the RI concerned was in a consortium of interacting, often global, related communities that share data and hence appreciate many of the issues; EPOS is one such example. For such RIs, it was possible to gather good input on virtually every topic. For all of the RIs, contact was made and information was gathered for at least the general requirements. In some cases, an RI deemed their interests were already covered by another RI known to be similar with which they worked closely.

The variation between topics is also a manifestation of variation in RI maturity. Topics such as identification and citation, cataloguing and processing are all encountered at the early stages of an RI's development and at early stages of the data lifecycle. Conversely, the value of topics such as curation and provenance become much more apparent after running a data gathering and sharing campaign for long periods or from being involved in the later stages of the data lifecycle. Optimisation is an extreme example of this effect; only when production and a large community of users demand more resources than an RI can afford does optimisation become a priority; before that the focus is on delivering the breadth of functionality users require and promoting adoption by a substantive community.

The overall conclusion would be that there are many opportunities for benefit from sharing ideas, methods and technologies between RIs, that there is much potential for using their data in combination and that there is a general need for awareness raising and training. However, these high-level consistencies have to be treated with great care; there are many lower level details where differences are significant. Continued work is needed to tease out those differences that are fundamentally important and which are coincidental results from the different paths that RI projects have already taken to date. Fundamental differences need recognition and support with well-developed methods for linking across them founded on scientific insights. The unforeseen differences may in time be overcome by incremental alignment; however, great care must be taken to avoid unnecessary disruption to working practices and functioning systems.

3   Requirement Analysis

We can now consider the requirements of RIs on a per-topic basis.

3.1   Identification and Citation

Identification of data (and associated metadata) throughout all stages of processing is central in any RI and can be ensured by allocating unique and persistent digital identifiers

(PIDs) to data objects throughout the data lifecycle. PIDs allow unambiguous references to be made to data during curation and cataloguing. They are also a necessary requirement for correct citation (and hence attribution) of data by end-users. Environmental RIs are often built on a large number of distributed observational or experimental sites, run by hundreds of scientists and technicians, financially supported and administered by a large number of institutions. If this data is shared under an open access policy, it becomes therefore very important to acknowledge the data sources and their providers.

The survey of ENVRI RIs found great diversity between RIs regarding their practices. Most apply file-based storage for their data, rather than database technologies, which suggests that it should be relatively straightforward to assign PIDs to a majority of the RI data objects. A profound gap in knowledge about what persistent and unique identifiers are, what they can be used for, and best practices regarding their use, emerged, however. Most identifier systems used are based on handles (DOIs from DataCite[3] most common, followed by ePIC PIDs[4]), but some RIs rely on formalised file names. While a majority see a strong need for assigning PIDs to their "finalised" data (individual files and/or databases), few apply this to raw data, and even fewer to intermediate data—indicating PIDs are not used in workflow administration. Also, metadata objects are seldom assigned PIDs.

Currently, researchers refer to datasets in publications using DOIs if available, and otherwise provide information about producer, year, report and number either in the article text or in the bibliography. A majority of RIs feel it is absolutely necessary to allow unambiguous references to be made to specified subsets of datasets, preferably in the citation, while few find the ability to create and later cite collections of individual datasets as important. Ensuring that credit for producing (and to a lesser extent curating) scientific datasets is "properly assigned" is a common theme for all RIs—not least because funding agencies and other stakeholders require such performance indicators, but also because individual PIs want and need recognition of their work. Connected to this, most RIs have strategies for collecting usage statistics for their data products, i.e., through bibliometric searches (quasi-automated or manual) from scientific literature, but thus often rely on publishers indexing also data object DOIs.

The use of persistent and unique identifiers for both data and metadata objects throughout the entire data lifecycle needs to be encouraged, e.g. by providing training and best-use cases. There is strong support for promoting "credit" to data collectors, through standards of data citation supporting adding specific sub-setting information to a basic (e.g. DOI-based) reference. Demonstrating that this can be done easily and effectively, and that data providers can trust that such citations will be made, will be a priority, as it will lead to adoption and improvement of citation practices.

A key issue is adoption of appropriate steps in working practices. Where these are exploratory or innovative the citation of underpinning data may be crucial to others verifying the validity of the approach and to later packaging for repeated application. Once a working practice is established, it should be formalised, e.g. as a workflow, and packaged, e.g. through good user interfaces, so that as much of the underpinning record keeping e.g. citation, cataloguing and provenance is automated. This has two positive

---

[3] https://datacite.org/.

[4] https://www.pidconsortium.eu.

effects: it enables practitioners to focus on domain-specific issues without distracting record keeping chores, and it promotes a consistent solution to be incrementally refined. For these things to happen there have to be good technologies, services and tools supporting each part of these processes, e.g. data citations being automatically and correctly generated as suggested by Buneman et al. [7]. Similarly, constructing immediate payoffs for practitioners using citation, as suggested by Myers et al. [8], will increase the chances of researchers engaging with identification at an earlier stage.

Many researchers today access and therefore consider citing individual files. This poses problems if the identified files may be changed, the issue of fixity. Many research results and outputs depend on very large numbers of files and simply enumerating them does not yield a comprehensible citation. Many derivatives depend on (computationally) selected parts of the input file(s). Many accesses to data are via time varying collections, e.g. catalogues or services, that may yield different results or contents on different occasions—generically referred to as databases. Some results will deal with continuous streaming data. Often citations should couple together the data sources, the queries that selected the data, the times at which those queries were applied, the workflows that processed these inputs and parameters or steering actions provided by the users (often during the application of the scientific method) that potentially influenced the result. All of these pose more sophisticated demands on ata identification and citation systems. At present they should at least be considered during the awareness raising proposed above. In due course, those advanced aspects that would prove useful to one or more of the RI communities should be further analysed and supported.

3.2 Curation

When the RIs in the ENVRI community were surveyed during the ENVRIplus project (in 2016), many did not have fully detailed plans for how they would curate their data and did not yet have complete Data Management Plans; in the years since, many of these RIs have made significant progress as they move into the implementation phases of their respective developments.

Only one RI mentioned OAIS[5] (the ISO/IEC 14721 standard for curation); this may be because it is not much used, and when it is the implementations are very varied since it is really an overview architecture rather than a metadata standard. With regard to the metadata standards used or required by the RIs, several used ISO19115/INSPIRE, one used CERIF, and one used Dublin Core; of these standards, only CERIF explicitly provides curation information. None mentioned metadata covering software and its curation except EPOS (using CERIF). A few use Git[6] to manage software. Most have no curation of software nor plans for this.

Possibly due to the early stage of some RIs, the requirements for curation were not made explicit, for example, none of the RIs (who responded) had appropriate metadata and processes for curation. It is known that EPOS has plans in place and there are indications of such planning for some of the others. Since curation often underpins

---

[5] http://www.dcc.ac.uk/resources/curation-reference-manual/chapters-production/using-oais-ref
erence-model-curation.

[6] https://git-scm.com/.

validation of the quality of scientific decisions and since environmental sciences observe phenomena that do not repeat in exactly the same form, the profile of curation needs raising.

Curation requirements validate the need to develop curation solutions but do not converge on particular technical requirements. Some further issues arise. These are enumerated below:

- The need for intellectual as well as ICT interworking between these closely related topics: Identification and Citation, Curation, Cataloguing and Provenance is already recognised. Their integration will need to be well supported by tools, services and processing workflows, used to accomplish the scientific methods and the Curation procedures. The need for this combination for reproducibility is identified by Belhaj-jame et al. with implementations automatically capturing the context and synthesising virtual environments [9].
- As above, it is vital to support the day-to-day working practices and innovation steps that occur in the context of Curation with appropriate automation and tools. This is critical both to make good use of the time and effort of those performing Curation, and to support innovators introducing new scientific methods with consequential Curation needs.
- Curation needs to address preservation and sustainability; carefully preserving key information to underwrite the quality and reproducibility of science requires that the information remains accessible for a sufficient time. This is not just the technical challenge of ensuring that the bits remain stored, interpretable and accessible. It is also the socio-political challenge of ensuring longevity of the information as communities' and funders' priorities vary. This is a significant step beyond archiving, which is addressed in EUDAT for example with the B2SAFE service.

One aspect of the approach to sustainable archiving is to form federations with others undertaking data curation, as suggested by OAIS. Federation arrangements are also usually necessary in order that the many curated sources of data environmental scientists need to use are made conveniently accessible. Such data-intensive federations (DIF) underpin many forms of multi-disciplinary collaboration and supporting them well is a key step in achieving success. As each independently run data source may have its own priorities and usage policies, often imposed and modified by its funders, it is essential to set up and sustain an appropriate DIF for each community of users. Many of the RIs deliver such federations, today without a common framework to help them, and many of the ENVRIplus partners are members of multiple federations.

3.3 Cataloguing

Regarding the possible items to be managed in catalogues, the RIs surveyed showed interest in:

- Observation systems and lab equipment: most RIs manage equipment which requires management (e.g. scheduling, maintenance and monitoring) and some of them are managing or would like to manage this with an information system. Some are already using a standardised approach (OGC/SWE and SSN).

- Data processing procedures and systems, software: a very few or none mentioned an interest in supporting this in a catalogue. We observe, however, that this may be necessary as part of the provision for provenance and as an aid for those developing or formalising new methods.
- Observation events: not explicitly mentioned as a requirement most of the time. Again, this need may emerge when provenance is considered.
- Physical samples: mentioned by a few especially in the biodiversity field.
- Processing activities: not explicitly mentioned.
- Data products or results: widely mentioned as being done by existing systems (EBAS, EARLINET, CLOUDNET, CKAN, MAdrigal and DEIMS) and widely standardised (ISO/IEC 191XX). Compliance is sometimes required with the INSPIRE directive; support for this in the shared common subsystems would prove beneficial.
- Publications: widely mentioned. However, very few manage the publications on their own. Links for provenance between publications and datasets are quite commonly required.
- Persons and organisations: not explicitly mentioned. However, this is reference information, which is required for the other described items (e.g. datasets and observation systems) and for provenance (contact points).
- Research objects or features of interest: mentioned once as feature of interest (airports for IAGOS).

As a consequence, the following three categories of catalogues are cited in the requirements collection:

- Reference catalogues, which are not developed by ENVRIplus or within RIs but are pre-existing infrastructures containing reference information to be used. They can also be considered as gazetteer, thesaurus or directories. Among them we consider catalogues for people and organisations, publications, research objects, and features of interest.
- Federated catalogues, which are pre-existing and partly harmonised in an RI but could be federated by ENVRIplus. Among them we consider data products, results, observation systems and lab equipment, physical samples, data processing procedures and systems, and software components metadata.
- Finally, activity records, observation events, processing activities, and usage logs can be considered.

There are a wide variety of items that could be catalogued, from instruments and deployments at the data acquisition stage, right through every step of data processing and handling, including the people and systems responsible, up to the final data products and publications made available for others to use [10]. Most responding RIs pick a small subset of interest, but it is possible that a whole network of artefacts needs cataloguing to facilitate Provenance, and many of these would greatly help external and new users find and understand the research material they need. There is a similar variation in the kinds of information, metadata, provided about catalogue entries. Only EPOS has a systematic approach in its use of CERIF, though many have commonalities developing out of the INSPIRE directive (EU Directive 2007/2/EC)[7]. So again we will consider a few implications:

- A critical factor that emerged in general requirements discussions was the need to easily access data. This clearly depends on good query systems that search the relevant catalogues and couple well with data handling and provenance recording. The query system is closely coupled with catalogue design and provision, but it also needs integration with other parts of the system. Euro-Argo identified a particular version of data access—being able to specify a requirement for a repeating data feed.
- Catalogues are a key element in providing convenient use of federations of resources. It is probably necessary to have a high-level catalogue that identifies members of the federation and the forms of interaction, preferably machine-to-machine, they support. Initially users may navigate this maze and handle each federation partner differently, but providing a coherent view and a single point of contact has huge productivity gains. It is a moot point whether this requires an integrated catalogue or query systems that delegate sub-queries appropriately. This is another example where effective automation can greatly improve the productivity of all the RI's practitioners; those that support the systems internally and maintain quality services, and those who use the products for research and decision making. It is anticipated that federations will grow incrementally and that the automation will advance to meet their growing complexity and to deliver a holistic and coherent research environment where the users enjoy enhanced productivity. This will depend on catalogues holding the information needed for that automation as well as the information needed for RI management and end-user research.
- Once again there may be some merit in making the advantages of catalogues evident in the short-term, e.g. by coupling catalogue use with operations that user want to perform, such as: having selected data via a catalogue, moving it or applying a method to each referenced item. Similarly, allowing the users some free-form additions and annotations to catalogue entries that help them pursue their own goals may be helpful.

3.4 Processing

Data processing (or analytics) is an extensive domain, including any activity or process that performs a series of actions on dataset(s) to distil information [11]. It may be applicable at any stage in the data lifecycle from quality assurance and event recognition close to data acquisition to transformations and visualisations to suit decision makers as results are presented. Data analytics methods draw on multiple disciplines, including statistics, quantitative analysis, data mining, and machine learning. Very often these methods require compute-intensive infrastructures to produce their results in a suitable time, because of the data to be processed (e.g. huge in volume or heterogeneity) and/or because of the complexity of the algorithm/model to be elaborated/projected. Moreover, these methods being devised to analyse datasets and produce other "data"/information (than can be considered a dataset) are strongly characterised by the "typologies" of their

---

[7] https://eur-lex.europa.eu/legal-content/EN/ALL/?uri=CELEX:32007L0002.

inputs and outputs. In some data-intensive cases, the data handling (access, transport, IO and preparation) can be a critical factor in achieving results within acceptable costs.

As largely expected, RIs' needs with respect to datasets to be processed are quite diverse because of the diversity in the datasets that they deal with. Datasets and related practices are diverse both across RIs and within the same RI. For instance, in EPOS there are many communities each having its specific typologies of data and methodologies (e.g. FTP) and formats (e.g. NetCDF) for making them available. Time series and tabular data are two very commonly reported types of dataset to be processed yet they are quite abstract. In what concerns "volume", datasets vary from a few kilobytes to terabytes and beyond. In the large majority of cases datasets are made available as files while few infrastructures have plans to make or are making their data available through OGC services, e.g. ACTRIS. The need to homogenise and promote state-of-the-art practices for data description, discovery and access is of paramount importance to providing RIs with a data processing environment that makes it possible to easily analyse dataset(s) across the boundaries of RI domains.

Considering actual processing itself, it emerged that RIs are at diverse levels of development and that there is a large heterogeneity. For instance, the programming languages currently in use by the RIs range from Python, Matlab and R to C, C++, Java, and Fortran. The processing platforms available to RIs range from a few Linux servers to the HPC approaches exploited in EPOS. Software in use or produced tends to be open source and freely available. In the majority of cases there is almost no shared or organised approach to make available the data processing tools systematically both within the RI and outside the RI. One possibility suggested by some RIs is to rely on OGC/WPS for publishing data processing facilities.

Any common processing platform should be open and flexible enough to allow: (a) scientists to easily plug-in and experiment with their algorithms and methods without bothering with the computing platform; (b) service managers to configure the platform to exploit diverse computing infrastructures; (c) third-party service providers to programmatically invoke the analytics methods; and (d) engineers to support scientists executing existing analytic tasks eventually customising/tuning some parameters without requiring them to install any technology or software.

Regarding the output of processing tasks, we can observe that the same variety characterising inputs as being there for outputs also. In this case, however, it is less well understood that there is a need to make these data available in a systematic way, including information on the entire process leading to the resulting data. In the case of EMBRC it was reported that the results of a processing task are to be made available via a paper while for EPOS it was reported that the dataset(s) are to be published via a shared catalogue describing them by relying on the CERIF metadata format.

In many cases, but by no means all, output resulting from a data processing task should be 'published' to be compliant with Open Science practices. A data processing platform capable of satisfying the needs of scientists involved in RIs should offer an easy to use approach for having access to the datasets that result from a data processing task together. As far as possible it should automatically supply the entire set of metadata characterising the task, e.g. through the provenance framework. This would enable scientists to properly interpret the results and reduce the effort needed to prepare for curation. In cases where aspects of the information are sensitive, could jeopardise privacy, or have applications that require a period of confidentiality, the appropriate protection should be provided.

Only a minority of the RIs within ENVRIplus contributed information about statistical processing during requirements gathering. Unsurprisingly given the diversity of the component RIs, there were a variety of different attitudes to the statistical aspects of data collection and analysis. One RI (IS-ENES-2) felt that data analysis (as opposed to collection) was not their primary mission, whereas for others (e.g. within EMBRC) reaching conclusions from data is very much their primary purpose.

As environmental data collection is the primary aim of many of the RIs it appears that day-to-day consideration of potential hypotheses underlying data collection is not undertaken. Hypothesis generation and testing is for scientific users of the data and could take many forms. However, some RIs (e.g. LTER and ICOS) stressed that general hypotheses were considered when the data collection programmes and instruments were being designed especially if the data fed into specific projects. Hypotheses could be generated after the fact by users after data collection and indeed this would be the norm if data collection is a primary service to the wider scientific community.

RIs can be collecting multiple streams of data often as time series, thus there is the potential to undertake multivariate analysis of the data. Again unsurprisingly given the diversity of science missions, there was no consistency in approaches. Data could be continuous and discrete, be bounded by its very nature or have bounds enforced after collection. Datasets are potentially very voluminous; total datasets with billions of sample points might be generated. Most analysers will be engaging in formal testing of hypotheses rather than data mining, although the latter was not necessarily ruled out. Many RIs had or are going to implement outlier or anomaly detection on their data.

The wide scope of potential contexts in which processing could be applied: from quality assurance close to data acquisition to transformations for result presentation (and every research, data-management or curation step in between) makes this a complex factor to consider. User engagement with this topic also varies validly between two extremes: those who use a pre-packaged algorithm in a service almost unknowingly as part of a well-formalised, encapsulated, established method they use, to those who are engaged in creating and evaluating new algorithms for innovative ways of combining and interpreting data. Clearly, both continua are valid and any point in each continuum needs the best achievable support for the context and viewpoint. With such diversity it is clear that a one-size-fits-all approach is infeasible. This conclusion is further reinforced by the need to exploit the appropriate computational platforms (hardware architectures, middleware frameworks and provision business models) to match the properties of the computation, and the priorities of the users given their available resources. If such matching is not considered it is unlikely that all of the developing research practices will be sustainable in an affordable way. For example, too much energy may be used or the call on expert help to map to new platforms may prove unaffordable. Such issues hardly rise to the fore in the early stages of an RI or a project. So again, we note forces that will cause the understanding and nature of requirements to evolve with time. This leads to the following follow-up observations:

- The packaging of computations and the progressive refinement of scientific methods are key to productivity and to the quality of scientific conclusions. Consequently, as far as possible processing should be defined and accessed by high-level mechanisms. This allows a focus on the scientific domain issues and it leaves freedom for optimised mappings to multiple computational platforms. This protects scientific intellectual investment, as it then remains applicable as the computational platforms change. This will happen as their nature is driven by the much larger entertainment, media, leisure and business sectors. The higher-level models and notations for describing and organising processing also facilitates optimisation and automation of chores that otherwise will distract researchers and their supporters.
- Providing support for innovation in this context is critical. Without innovation the science will not advance and will not successfully address today's societal challenges. It requires support for software development, testing, refinement, validation and deployment conducted by multi-site teams engaging a wide variety of viewpoints, skills and knowledge. For the complex data-intensive federations the environmental and Earth sciences are dealing with, this involves new intellectual and technological territory. Alliances involving multiple RIs and external cognate groups such as EUDAT, PRACE and EGI, may be the best way of gathering sufficient resources and building the required momentum.

3.5 Provenance

For modern data-driven science there is a pressing need to capture and exploit good provenance data. Provenance, the records about how data was collected, derived or generated, is crucial for validating and improving scientific methods and is a key aspect of making data FAIR. It enables convenient and accurate replay or re-investigation and provides the necessary underpinning when results are presented for judging the extent to which they should influence decisions whether for hazard mitigation or paper publication. It provides a foundation for many activities of import to RIs, such as attributing credit to individuals and organisations, providing input to diagnostic investigations, providing records to assist with management and optimisation and preparing for curation. All RIs will need to perform these functions and consequently the e-infrastructures they depend on will need to support provenance collection and use as well.

Most RIs already consider provenance data as essential and are interested in using a provenance recording system. Among all of the nine RIs who gave feedback about provenance only two already had a data provenance recording system embedded in their data processing workflows. EPOS uses the dispel4py workflow engine in VERCE[8], which is based on and is able to export to PROV-O whereas in future it is planned to use the CERIF data model and ontology instead [12]. IS-ENES2 instead does not specify which software solution is applied but mentions: the use of community tools to manage what has been collected from where, and what is the overall transfer status to generate provenance log files in workflows [13]. Some, such as SeaDataNet and Euro-Argo, interpret provenance as information gathered via metadata about the lineage

---

[8] http://www.verce.eu/AboutVerce/RelatedProjects.php.

data with tools like Geonetwork based on metadata standards like ISO19139[9], but the information gathered is not sufficient to reproduce the data as the steps of processing are not documented in enough detail. Other RIs, such as ICOS and LTER, are already providing some provenance information about observation and measurement methods used within the metadata files but are aware that a real tracking tool still needs to be implemented. IAGOS is using the versioning system GIT for code but not the data itself. A versioning system can only be seen as a part of the provenance information sought.

On which information is considered to be important, the answers range from versioning of data to the generation of data and modification of the data as well as on who, how and why data is used. So there seems to be two interpretations about what provenance should comprise: should it enable the community to follow the data 'back in time' and see all the steps that happened from raw data collection, via quality control and aggregation to a useful product, or should it enable the data provider as a means of tracking the usage of the data, including information about users in order to understand the relevance of the data and how to improve their services? These two roles for metadata may be served by the same provenance collecting system. The provenance data is then interpreted via different tools or services.

Regarding the controlled vocabulary used for the descriptions of the steps for data provenance, some RIs already use research specific reference tables and thesauri like EnvThes and SeaDataNet common vocabularies. There is great interest among the RIs to get clear recommendations from ENVRIplus about the information range provenance should provide. This includes drawing an explicit line between metadata describing the 'dataset' and provenance information. Also it should be defined clearly whether usage tracking should be part of provenance. It is considered as being very important to get support on automated tracking solutions and or provenance management APIs to be applied in the specific e-science environments. Although there are some thesauri already in use there is a demand for getting a good overview of the existing vocabularies and ontologies that are ready to use or that need to be slightly adapted for specific purposes.

At present, the need for and benefits of provenance provision are only recognised by some RIs. In abstract, we are sure that most scientists appreciate the value of provenance, but they tend to think of it as a painful chore they have to complete when they submit their final, selectively chosen data to curation. They often only do this when their funders or publishers demand it. That culture is inappropriate. For many RIs they are in the business of collecting and curating primary data and commonly required derivatives. Clearly, they want to accurately record the provenance of those data, as a foundation for subsequent use and to achieve accountable credit. For environmental and Earth scientists use of provenance throughout a research programme can have significant benefits. During method development it provides ready access to key diagnostic and performance data, and greatly reduces the effort required to organise exactly repeated re-runs; a frequent chore during development. As they move to method validation they have the key evidence to hand for others to review. When they declare a success and move to production, the provenance data informs the systems engineers about what is required and can be exploited by the optimisation system. Once results are

---

[9] https://www.iso.org/standard/32557.html.

produced using the new method these development-time provenance records underpin the provenance information collected during the production campaign.

The RIs survey reported very different stages of adoption, and when there was adoption it did not use the same solutions or standards—this was almost always related to data acquisition rather than the use of data for research. The change in culture among researchers may be brought about by ENVRIplus through a programme of awareness raising and a well-integrated compendium of tools. The latter may be more feasible if the development of the active provenance framework is amortised over a consortium of RIs. This leads to similar observations to those given above:

3.6 Optimisation

Environmental and Earth sciences now rely on the acquisition of great quantities of data from a range of sources for building the complex models. The data might be consolidated into a few very large datasets, or dispersed across many smaller datasets; it may be ingested in batch or accumulated over a prolonged period. Although efforts are underway to store data in common data stores, to use this wealth of data fast and effectively, it is important that the data is both optimally distributed across a research infrastructure's data stores, and carefully characterised to permit easy retrieval based on a range of parameters. It is also important that experiments conducted on the data can be easily compartmentalised so that individual processing tasks can be parallelised and executed close to the data itself, so as to optimise the use of resources and provide swift results for investigators.

Perhaps more so than the other topics, optimisation requirements are driven by the specific requirements of those other topics, e.g. time-critical requirements for data processing and management [14]. For each part of an infrastructure in need for improvement, we must consider what it means for the part to be optimal and how to measure that optimality.

In the context of common services for RIs, it is necessary to focus on certain practical and broadly universal technical concerns, generally those being to do with the movement and processing of data. This requires a general understanding of what bottlenecks exist in the functionality of (for example) storage and data management subsystems, understanding of peak volumes for data access, storage and delivery in different parts of the infrastructure, understanding of computational complexity of different data-processing workflows, and understanding of the quality of service requirements researchers have for data handling in general [15]. Many optimisation problems can be reduced down to ones of data placement, in particular of data staging, whereby data is placed and prepared for processing on some computational service (whether that is provided on a researcher's desktop, within an HPC cluster or on a web server), which in turn concerns the further question of whether data should be brought to where they can be best computed, or instead computing tasks be brought to where the data currently reside. Given the large size of many RI's primary datasets, bringing computation to data is appealing, but the complexity of various analyses also often requires supercomputing-level resources, which require the data be staged at a computing facility such as are brokered in Europe by consortia such as PRACE. Data placement is reliant however on data accessibility, which is not simply based on the existence of data in an accessible location, but is also

based on the metadata associated with the core data that allows it to be correctly interpreted; it is based on the availability of services that understand that metadata and can so
interact (and transport) the data with a minimum of manual configuration or direction.

Experience shows that as data-handling organisations transition from pioneering to
operations, many different reasons for worrying about optimisation emerge. These are
addressed by a wide variety of techniques, so that investment in optimisation is usually
best left until the RI or RI community can establish what they want to be optimised
and the trade-offs that they would deem acceptable. Very often there are significantly
different answers from different members of a community. The RI's management may
need to decide on compromises and priorities.

Optimisation needs to look beyond individuals and single organisations. When looking at overall costs or energy consumption in a group of RIs or the e Infrastructures they
use, tactics may consider the behaviour of a data-intensive federation. For example,
when data is used from remote sites, or is prepared for a particular class of uses, the use
of caching may save transport and re-preparation costs, and accelerate the delivery of
results. However, the original provider organisation needs to have accountable evidence
that their data is being used indirectly, and the caching organisation needs its compute
and storage costs amortised over the wider community.

3.7 Community Support

We define community support as part of the RI concerned with managing, controlling
and tracking users' activities within an RI and with supporting all users to conduct their
roles in their communities. It includes many miscellaneous aspects of RI operations,
including for example authentication, authorisation and accounting, the use of virtual
organisations, training, and help-desk activities.

The questions asked of RIs focused on 3 aspects: a) functional requirements; b)
non-functional requirements (e.g. privacy, licensing and performance); and c) training.

The following is a summary of the main functional requirements expressed by the
RIs (not all apply to all RIs):

- Data Portal: a data portal was frequently requested by RIs. Many RIs already have
  their own data portal, and some of the others are in the process of developing one.
  Data portals provide (a single point of) access to the system and data both for humans
  and machines (via APIs). Commonly requested functionalities include access control (for example, IS-ENES2 currently uses OAuth2, OpenID, SAML and X.509 for
  AAI management) and discovery of services and data facilities (e.g. metadata-based
  discovery mechanisms).
- Accounting: the tracking of user activities, which is useful for analysing the impact of
  the RI, is commonly requested. For example, EMBRC records where users are going,
  what facilities they are using, and the number of requests. The EMBRC head office
  will in the future provide a system to analyse resource DOIs, metrics for the number
  of yearly publications and impact factor, and questionnaires submitted by users about
  their experience with their services. LTER plans to track the provenance of the data,
  as well as its usage (e.g. download or access to data and data services). DEIMS, for
  example is planning that statistics about users will be implemented, mainly to allow

for a better planning of provided services. Features will be implemented by exploiting EUDAT services, e.g. provenance support of B2SHARE to track data usage. Google analytics is currently used to track the usage of the DEIMS interface.

- Issue tracker: ACTRIS has recently introduced an issue tracker to link data users and providers, and to follow up on feedback on datasets in response to individual requests.
- Community software: EPOS is in the process of deciding which private software to use and how to integrate it in the data portal. In LTER, the R statistical software and different models (e.g. VSD+ dynamic soil model, LandscapeDNDC regional scale process model for simulating biosphere-atmosphere-hydrosphere exchanges.) are provided.
- Wiki: a wiki is often used to organise community information, and as a blackboard for collaborative work for community members (e.g. to add names and responsibilities to a list of tasks to be done). Sometimes, it is also used to keep track of the progress on a task, both for strategic and IT purposes. FAQ pages (and other material targeting a more general audience, or outreach materials for educational institutes) are a special type of wiki page describing more technical aspects of data handling and data products, and also a system for collecting user feedback.
- Mailing lists, twitter & Forums are intended to facilitate communication to and from groups of community members. Forums and mailing lists can be interlinked so that any message in the mailing list is redirected to the forum and vice-versa.
- Files and image repositories represent shared spaces where members and stakeholders can upload/download and exchange files. They are also a fundamental tool for storing and categorising images and other outreach materials.
- Shared calendars keep track and disseminate relevant events for community members.
- Tools to organise meetings, events and conferences should handle all the aspects of a conference/meeting: programme, user registration, deadlines, document submission and dissemination of relevant material. Tools like Indico are currently popular.
- Website: The purpose of the website is to disseminate community relevant information to all stakeholders. The website should not contain reserved material but only publicly accessible material (e.g. documents and presentations for external or internal stakeholders, images for press review). The website should also include news and interactions from social networks. The website should be simple enough to allow almost anyone with basic IT skills to add pages, articles, images. A simple CMS (content management system) is the most reasonable solution (e.g. Wordpress, Joomla).
- Teleconferencing tools: Communication with all stakeholders (internal and external) is also carried on through teleconferencing. For this purpose, good quality tools (e.g. screen sharing, multi-user, document exchange, and private chat) are needed. Popular tools include Adobe Connect, Web Ex, GoToMeeting, Google Hangout and Skype.
- Helpdesk & Technical support: For example, the data products that ICOS produces are complex and often require experience of, and detailed knowledge about the underlying methods and science to be used in an optimal way. Technical support must be available to solve any problem. The ICOS Thematic Centres (for Atmosphere, Ecosystems and Ocean) are ready to provide information and guidance for data users. If needed, requests for information may also be forwarded to the individual observation stations. The mission of ICOS also comprises a responsibility to support producers of derived

products (typically research groups performing advanced modelling of greenhouse gas budgets) by providing custom-formatted "data packages".

The non-functional requirements of the RIs that were most frequently referred to were:

- Performance: RIs need robust, fast-reacting systems, which offer security and privacy. Moreover, they need good performance for high data volumes.
- Data policy and licensing constraints: The data produced by some communities has licensing constraints that restrict access to a certain group of users. For example, while ICOS will not require its users to register in order to use the data portal or to access and download data, it plans to offer an enhanced usage experience to registered users. This will include automatic notifications of updates of already downloaded datasets, access to additional tools, and the possibility to save personalised searches and favourites in a workspace associated with a user's profile. Everyone who wishes to download ICOS data products must also acknowledge the ICOS data policy and data licensing agreement (registered users may do so once, while others must repeat this step every time).

Training activities within ENVRIplus communities can be categorised as follows:

- No training plan: The majority of ENVRIplus RI communities do not have a common training plan at the moment.
- No community-wide training activities: For example:
  - Within SIOS, many organisations have their own training activities. Training is provided to students or scientists. For example, The University Centre in Svalbard (UNIS) has its own high-quality-training programme on Arctic field security (i.e., how to operate safely in an extreme cold climate and in accordance with environmental regulations) for students and scientists.
  - Within ACTRIS, each community has its own set of customised training plans. Courses and documentation are made available online, for example for training on how to use the data products. Their preferred methods for delivering training are through the community website or through targeted sessions during community specific workshops. ACTRIS also considers organising webinars.
  - ICOS does not have a common training plan at the moment. The Carbon Portal organises occasional training events, e.g. on Alfresco DMS (Document Management System used by ICOS RI). The different Thematic Centres periodically organise training for their respective staff and in some cases also for data providers (station PIs). ICOS also (co-)organises and/or participates in summer schools and workshops aimed at graduate students and postdocs in the relevant fields of greenhouse gas observational techniques and data evaluation. Representatives of ICOS have participated in training events organised by EUDAT, e.g. on PID usage and data storage technology. The method of delivering training through one- or two-day face-to-face workshops concentrated on a given topic and with a focus on hands-on activities is probably the most effective. This should also be backed up by webinars (including recordings from the workshops) and written materials.

- A community training plan is under development: A number of communities are in the process of developing a community training plan. For example:
  - LTER plans the development of a community-training plan. Within LTER Europe, the Expert Panel on Information Management is used to exchange information on a personal level and to guide developments such as DEIMS to cater for user needs. LTER Europe also provides dissemination and training activities to selected user groups. Training activities will enhance the quality of the data provided, by applying standardised data quality control procedures for defined data sets.
  - For EPOS, training is part of its communication plan.

- An advanced system is in place for training activities:
  - Within IS-ENES2, workshops are organised from time to time. Also, communities communicate about the availability of training courses and workshops organised by HPC centres (PRACE) or EGI.
  - Within EMBRC, a Training web portal is provided, offering support to training organisers to advertise and organise courses.

The above list covers virtually all of the facilities for communication, information sharing, organisation and policy implementation that a distributed community of collaborating researchers and their support teams might expect—and they normally expect those facilities to be well integrated and easily accessed wherever they are from a wide range of devices. However, care should be taken to consider the full spectrum of end users. A few may be at the forefront of technological innovations but the majority may be using very traditional methods, because they work for them. Investment is only worthwhile if it is adopted and benefits the greater majority of such communities, taking into account their actual preferences.

There may be two key elements missing in the context of ENVRIplus, which focuses on achieving the best handling and use of environmental data:

- Workspaces that can be accessed from anywhere and are automatically managed, in which individuals or groups can store and organise the data concerned with their work in progress: e.g. test data sets, sample result sets, intermediate data sets, results pending validation, results pending publication. Since environmental researchers have to work in different places, such as in field sites, in different laboratories and institutions, they need to control these logical spaces, which may be distributed for optimisation or reliability reasons. These are predominantly used to support routine work but can also be used for innovation. This includes intelligent sensors requiring access to a variety of logical spaces for their operations.
- Development environments that can be accessed from most workstations and laptops, and that facilitates collaborative innovation and refinement of the scientific methods and data handling. Sharing among a distributed community, testing, management of versions and releases and deployment aids would be expected.

To a lesser or greater extent every RI will depend on a mix of roles and viewpoints. Community support needs to recognise and engage with these multiple viewpoints as well as help them to work together. This is particularly challenging in the distributed environments and federated organisations underpinning many RIs. At least training and help desk organisation will need to take these factors into account. Productivity will come from each category being well supported. Significant breakthroughs will depend on the pooling of ideas and effort across category boundaries.

3.8   Cross-Cutting Requirements

There are a few additional requirements that appear in the analysis of RI needs that have aspects of improving usability to improve the experience and productivity of users and the teams who support them. In part, they are better packaging of existing or planned facilities and in part they are intended to deliver immediate benefits to keep communities engaged and thereby, improve take up and adoption of RI products.

- Boundary crossing. The participating communities experience boundaries between the different roles identified above (see Sect. 3.7), between disciplines, sub-disciplines and application domains, and between organisations. This can be stimulated by:
    - Organising ad hoc think tanks so that it brings together (virtually) participants from across the boundaries and stimulates them to think and work together on relevant topics, e.g. by bringing in suitable experts and setting up suitable practical challenges to be addressed during the course. This requires elapsed time, and allocation of both training effort and trainee time, so the target understanding that the course will deliver has to be carefully chosen.
    - Establishing suitable agile development processes where people work intensely together on a common issue with a carefully set goal. Then assimilating the results and building on the networks provided.
    - Delivering services and tools well suited to each role and organisational context.
    - Arranging workspaces that facilitate such collaborative behaviour while ideas are being developed and formulated. This requires those involved to have control over the release and sharing of the material they work on. Individuals may be involved in several groups, probably with different roles.

- Integrated communication facilities. The individual elements of communication for distributed participants in an RI need to be conveniently integrated. There are several potential solutions in this area. It may help if at least one well-integrated one were run to be available for RIs, project participants and ENVRIplus. This needs to present views that work well for each category of practitioner. Some of the selected use cases in Theme 2 may serve to achieve this.
- Exemplars and early benefits. The development of exemplars of effective methods and software or services that support them is key to spreading ideas, testing them in new contexts and developing buy in. This will be helpful in the training and outreach programme. It is also vital as part of the process of delivering as early as possible benefits to the active researchers and other practitioners. If we can deliver immediate benefits they will not have to struggle for so long investing unproductive time in tedious workarounds.

- Data access interfaces. Researcher and others managing data-driven processes spend a great deal of time, identifying data they want, arranging to be permitted access, arranging transfers, arranging local storage, arranging onward shipment to computation resources if necessary and returning storage resources when they have finished. If this is packaged as a convenient operation their work is simplified and more productive. The parts of such a process are all being built, but delivering an integrated solution that just works would be a large benefit. It needs the provision of a user's or group's workspace. It needs a means of identifying the required data. Once deployed, it can be grown in small increments, taking the users along an improving path. They might prioritise some of the following:

  – Identification using queries over associated metadata (in the identity registries or in catalogues (see Sect. 3.2)).
  – Extension of the operations that are easily applied to the accessed data (we have found visualisation particularly relevant).
  – Handling batches of data consistently at the same time (the tea tray metaphor).
  – Handling intermediate (transient) results with various aids for handling them in bulk and for clearing up afterwards.
  – Promoting selected results to properly identified and citable.
  – Arranging for their data to be published or curated.

4 Conclusion

A general conclusion that can be drawn from the information acquired from the RIs is that there are more differences than commonalities between the RIs—the RIs are all at different stages in their development and have different organisational status, from well-established and operational to RIs still in their definition phase. Moreover, some are heavily distributed with heterogeneous networks of sensors and network services in different countries making different kinds of measurement or observation, while other RIs have one single observing platform and one central hub for data. Nevertheless, it is still possible to identify a number of key common concerns. These include:

- The need to achieve data harmonisation, i.e., consistency of representation, interpretation and access, both within and between RIs.
- The need for RIs to learn from one another and pool efforts in order to accelerate and harmonise delivery of data services and working practices that efficiently support each stage of the scientific data lifecycle, from data acquisition to delivery of actionable derived information.
- Help with facing the challenge of sustainably delivering data services immediately to meet current RI priorities while considering longer-term issues and technology trends.

The ability to describe different processes from multiple viewpoints in a standard way helps facilitate the collaboration between RIs and alignment of their activities. The ENVRI Reference Model (RM) [16] provides a conceptual model to this, enabling RI communities to discuss and see where improvements in data processing in the RIs are possible and required. The RM is a living model that has been developed on the basis of evaluating RIs within the ENVRI community. Data science solutions that can fulfil the identified requirements can be expressed in terms of the RM and then projected onto RIs in order to help in optimising their data lifecycles.

Atkinson et al. [5] provide an in-depth analysis of the state of the RIs and the technologies they used as of mid 2016, providing a number of recommendations. Atkinson et al. stress that the diversity within and between RIs and the complexity of the RIs, involving many different roles, require effective communication and collaboration to address. Data sharing and governance of the data is essential to RI operation and to the production of valuable science, and needs to be considered with ample allocation of resources and attention as a main priority when setting up and governing RIs. This requires training of staff and education of future scientists. Shared developments in sustainable software and platforms for performing data-driven (environmental) science are also needed to minimise costs and increase the sustainability of the RIs.

Acknowledgements. This work was supported by the European Union's Horizon 2020 research and innovation programme via the ENVRIplus project under grant agreement No 654182.

References

1. Hey, T., Tansley, S., Tolle, K. (eds.): The fourth paradigm: data-intensive scientific discovery. Microsoft Research (2009)
2. Zhao, Z., et al.: Reference model guided system design and implementation for interoperable environmental research infrastructures. In: 2015 IEEE 11th International Conference on e-Science, Munich, Germany, pp. 551–556. IEEE (2015). https://doi.org/10.1109/eScience.2015.41
3. Wilkinson, M., Dumontier, M., Aalbersberg, I., et al.: The FAIR Guiding Principles for scientific data management and stewardship. Sci. Data **3**, 160018 (2016). https://doi.org/10.1038/sdata.2016.18
4. Petzold, A., et al.: ENVRI-FAIR - interoperable environmental FAIR data and services for society, innovation and research. In: 2019 15th International Conference on eScience (eScience), San Diego, CA, USA, pp. 277–280. IEEE (2019). https://doi.org/10.1109/escience.2019.00038. https://zenodo.org/record/3462816
5. Atkinson, M., et al.: D5.1 A consistent characterisation of existing and planned RIs. H2020 ENVRIplus Project (2016). http://www.envriplus.eu/wp-content/uploads/2016/06/A-consistent-characterisation-of-RIs.pdf
6. Tanhua, T., et al.: Ocean FAIR data services. Front. Mar. Sci. **6**, 440 (2019). https://doi.org/10.3389/fmars.2019.00440
7. Buneman, P., Davidson, S., Frew, J.: Why data citation is a computational problem. Commun. ACM **59**(9), 50–57 (2016). https://doi.org/10.1145/2893181
8. Myers, J., et al.: Towards sustainable curation and preservation. In: Proceedings of the IEEE eScience Conference 2015, pp. 526–535 (2016)
9. Belhajjame, K., et al.: A suite of ontologies for preserving workflow-centric research objects. J. Web Semant. **32**, 16–42 (2015)

10. Martin, P., Remy, L., Theodoridou, M., Jeffery, K., Zhao, Z.: Mapping heterogeneous research infrastructure metadata into a unified catalogue for use in a generic virtual research environment. Future Gener. Comput. Syst. **101**, 1–13 (2019). https://doi.org/10.1016/j.future.2019.05.076

11. Bordawekar, R., Blainey, B., Apte, C.: Analysing analytics. SIGMOD Rec. **42**, 4 (2014)

12. Filgueira, R., Krause, A., Atkinson, M., Klampano, I.: dispel4py: a python framework for data-intensive scientific computing. IJHPCA **31**, 316–334 (2016)

13. Ahanach, E., Koulouzis, S., Zhao, Z.: Contextual linking between workflow provenance and system performance logs. In: 15th IEEE International Conference on e-Science, San Diego, US (2019). http://doi.org/10.1109/eScience.2019.00093

14. Hu, Y., et al.: Deadline-aware deployment for time critical applications in clouds. In: Rivera, F.F., Pena, T.F., Cabaleiro, J.C. (eds.) Euro-Par 2017. LNCS, vol. 10417, pp. 345–357. Springer, Cham (2017). https://doi.org/10.1007/978-3-319-64203-1_25

15. Koulouzis, S., et al.: Time-critical data management in clouds: challenges and a Dynamic Real-time Infrastructure Planner (DRIP) solution. Concurr. Comput. Pract. Exp. (2019). https://doi.org/10.1002/cpe.5269

16. de la Hidalga, A.N., et al.: The ENVRI Reference Model (ENVRI RM) version 2.2 (2017). http://doi.org/10.5281/zenodo.1050349

Reference Model Guided System Design and Development

The ENVRI Reference Model

Abraham Nieva de la Hidalga[1]($\boxtimes$) (iD), Alex Hardisty[1] (iD), Paul Martin[2] (iD),
Barbara Magagna[3] (iD), and Zhiming Zhao[2] (iD)

[1] Cardiff University, Cardiff, UK
{nievadelahidalgaa,hardistyar}@cardiff.ac.uk
[2] Multiscale Networked Systems, University of Amsterdam,
1098XH Amsterdam, The Netherlands
pwmartin.research@gmail.com, z.zhao@uva.nl
[3] Environment Agency Austria, Vienna, Austria
barbara.magagna@umweltbundesamt.at

Abstract. Advances in automation, communication, sensing and computation
enable experimental scientific processes to generate data at increasingly great
speeds and volumes. Research infrastructures are devised to take advantage of
these data, providing advanced capabilities for acquisition, sharing, processing,
and analysis; enabling advanced research and playing an ever-increasing role in the
environmental and Earth science research domain. The ENVRI community identified several recurring requirements in the development of environmental research
infrastructures such as i) duplication of efforts to solve similar problems; ii) lack
of standards to harmonise and accelerate development, and bring about interoperability; iii) a large number of data models and data information systems within the
domain, and iv) a steep learning curve for integration complex research infrastructure systems. To address these challenges, the ENVRI community has developed
and refined the Environmental Research Infrastructures Reference Model (ENVRI
Reference Model or ENVRI RM), a modelling framework encoding this knowledge. The proposed modelling framework encompasses a language and a notation
to describe the research domain, its systems and the requirements and challenges
faced when implementing those systems. By adopting ENVRI RM as an integra-
tive approach, the environmental research community can secure interoperability
between infrastructures, enable reuse, share resources, experiences and common
language, reduce unnecessary duplication of effort, and speed up the understanding of research infrastructure systems. This chapter provides a short introduction
to the ENVRI RM.

Keywords: Reference Model · Research infrastructure · System modelling ·
Design framework

1 Motivation

The construction of a Research Infrastructure (RI) is often iterative, e.g. from simple
functionality to more rich set of features, or from small scale to large scale. A large

© The Author(s) 2020
Z. Zhao and M. Hellström (Eds.): Towards Interoperable Research
Infrastructures for Environmental and Earth Sciences, LNCS 12003, pp. 61–81, 2020.
https://doi.org/10.1007/978-3-030-52829-4_4

RI is often an evolution of many iterations and can be typically characterised in terms of phases of concept development, design, preparation, implementation, operation and termination[1]. The RIs in the ENVRIplus[2] project were in different phases when they joined the project. It is thus very challenging to develop those diverse RIs and make them interoperable.

During the past few years, interoperability between infrastructures has been extensively studied, e.g. between scientific models, workflow, metadata, semantics, middleware and infrastructure [1]. To enable interoperability among different systems, a common vocabulary for design descriptions is essential. The aim of the Environmental Research Infrastructures Reference Model (ENVRI RM) is to provide a framework for specifying and building the data management services required by environmental and Earth sciences research infrastructures.

The current version of the ENVRI RM[3] was published in November 2017, following more than six years of work within the ENVRI [2] and ENVRIplus projects [3, 4]. These projects documented common practices and architectures supporting environmental research infrastructures, derived from the Reference Model for Open Distributed Processing (RM-ODP) [5–8].

The ENVRI RM provides the documentation of the basic concepts, the architectural model, and different examples of use with diagrams. The users of the ENVRI RM can be designers of RIs, but it is also intended to help people who build services to support RI activities, or who produce standards to capture best practice and reusable mechanisms. The ENVRI RM gives the designer a way of thinking about the system, and structuring its specification, but does not constrain the order in which the design steps should be carried out. The ENVRI RM can be used along with any type of design/development processes.

Since the design of an RI requires large collaborative efforts, it is likely that the actual process will be iterative, filling in detail in different parts of the specification as ideas evolve and requirements are better understood. The design of a new RI may follow a classical top-down, waterfall-style pattern, while the maintenance of an existing RI will start by capturing existing constraints. The development of services can follow an agile or rapid prototyping development model, stressing modularization and fine-grained iteration. The ideas for structuring specifications presented here can be applied within any of these methodologies. They remain valid if the design approach changes and provide a common framework and vocabulary for collaboration between designers using different processes.

Many competing architectural frameworks have recently been proposed; however, the ENVRI RM offers a set of distinguishing features that make it particularly relevant for the specification of an Environmental RI. First, it has the stability derived from continuous development during two successful European funded projects (ENVRI, and ENVRIplus) spanning more than six years (2011–2019) [1–4]; during this period the ENVRI RM has been reviewed and evaluated internally and externally by design experts

---

[1] http://roadmap2018.esfri.eu/strategy-report/the-esfri-methodology/.

[2] EU H2020 ENVRIplus project http://www.envriplus.eu.

[3] The ENVRI RM and its associated derivatives are published are available online www.envri.eu/rm.

and by the research community. Second, it documents common requirements of environmental research infrastructures and best practices for fulfilling those requirements. Third, there has been an extended campaign of validation and refinement which used the ENVRI RM, analysing different infrastructures and services. And fourth, the discoveries have been formalised in the Open Information Linking for Environmental Research Infrastructures (OIL-E), an ontology framework designed to facilitate analysis, classification, and validation of RI designs; supporting the documentation of crosscutting requirements; and facilitating metadata exchange.

In this chapter, we will discuss the development of the reference model. The main aspects discussed include the context for the development of the ENVRI RM (Sect. 4.1), the main concepts supporting the modelling of environmental research infrastructure systems (Sect. 4.2), the modelling process (Sect. 4.3), and the outlook for the ENVRI RM and links to further chapters (Sect. 4.4).

2 Background of the ENVRI RM

Research Infrastructures are often complex distributed systems. Describing their structure and external properties is required to understand and manage these systems. When the system description concentrates on the distillation of general principles, it is called architecture. However, if the description is presented in a way that is useful for the derivation of a whole family of systems, it is called a framework. Hence, when describing a system supporting a broad range of applications, it is common to talk of an architectural framework. In this sense, the ENVRI RM is an architectural framework for the design of a distributed system for environmental research infrastructures.

The ENVRI RM was developed as a research infrastructure architecture framework based on the Reference Model for Open Distributed Processing (RM-ODP) [5–8]. The following sections describe the three concepts required for understanding the RM-ODP modelling paradigm: the object model, design viewpoints, and correspondences.

2.1 Object Model

RM-ODP system specifications are expressed in terms of objects. Objects are representations of the entities to be modelled. The specification and design of complex systems following the object paradigm makes use of two important object properties abstraction and encapsulation [9]. Abstraction allows highlighting aspects of the system relevant from a given perspective while hiding those of no relevance. Encapsulation is the property by which the information contained in an object is accessible only through interactions at the interfaces supported by the object [9]. In the ENVRI RM, objects are used to represent abstract entities (measurements, data sets, metadata, systems, services), physical entities (sensors, servers, networks) and social entities (institution, research group, researcher).

2.2 Viewpoint Specification

The definition of objects is distributed in viewpoint specifications. The idea behind viewpoints is to break down a complex specification into a set of individual specifications which consistently support and complement each other [10, 11]. The design of RM-ODP aimed at serving different stakeholders by introducing the idea of a set of linked viewpoints to maintain flexibility and avoid the difficulties associated with constructing and maintaining a single large system description. RM-ODP defines five viewpoints, as shown in Fig. 1, designed to appeal to different user groups [9].

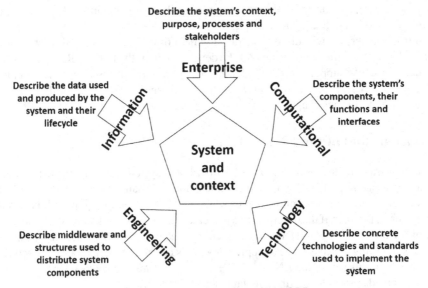

Fig. 1. The five viewpoints of RM ODP.

In the ENVRI RM, to better align the definition of viewpoints to the research domain, the *Enterprise Viewpoint* is renamed as the *Science Viewpoint*. The name change aims to acknowledge that the main type of systems modelled are intended for supporting scientific research. However, apart from this, the definition of the ENVRI RM Science Viewpoint respects the rationale, elements and structure of the RM-ODP Enterprise Viewpoint.

2.3 Correspondences

Dividing a system design in five viewpoint specifications facilitates the understanding of different groups of stakeholders. However, it is necessary to keep these specifications consistent with each other [9]. In RM-ODP, the consistency of the designs produced within each specification is maintained with the explicit mapping between elements defined in one viewpoint (e.g. objects, actions and constraints) to elements defined in other viewpoints. These mappings are formally defined as correspondence links between related elements. The correspondences can be one-to-one or one to many. A one-to-one correspondence allows mapping the representation of an element in one viewpoint to the representation of an element on another viewpoint. A one-to-many correspondence allows for an element representation in a viewpoint to be mapped to multiple elements in another viewpoint, providing a fine-grained description of that element (Fig. 2).

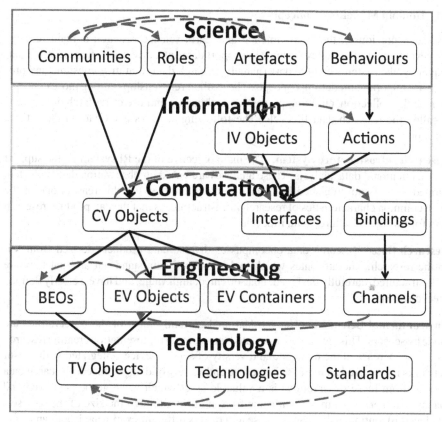

Fig. 2. Example of Viewpoint Correspondences in the ENVRI RM.

For the ENVRI RM, correspondences are formally defined in the Open Information Linking for Environmental Research Infrastructures (OIL-E) framework [12]. OIL-E is an ontology framework designed to facilitate analysis, classification, and validation of the design of a RI.

The three RM-ODP modelling mechanisms (objects, viewpoints, and correspondences) enable a complex system to be described as a set of interlinked viewpoint models. This set of models is equivalent to a single large and complex model with all viewpoints included; however, such a description is too complex to be useful. Instead, different groups of stakeholders will understand and use a subset of viewpoint specifications. A design team with members from all stakeholder groups is responsible for defining viewpoint correspondences when needed.

2.4 Domain Modelling Concepts

As stated previously, the environmental and Earth science research domain requires the development of complex systems to support data-intensive scientific research. Consequently, the systems and the data (namely research data) that they consume and produce are important modelling concepts. The explicit relationships among those concepts include the collection, curation, processing, publishing and use of research data, which is called the research data lifecycle. The following sections elaborate on these three concepts.

Research Infrastructure System. The main objective of the RI systems is the support of computational data analysis. These analyses are based on observation data collected, curated, stored and published by diverse research entities. For this reason, one of the main common characteristics of research infrastructures is that they all produce research data following a structured data lifecycle.

Research Data. Research data encompasses diverse data products derived from scientific research. The attributes which make research data stand out are that they are well-structured, carefully designed, goal-oriented, high value, and have a clearly defined lifecycle [13].

Environmental Science is observational, and currently most of the observations are made by sensors. This data is then translated into a digital representation creating research data. The increase in the number and diversity of sensor devices integrated with sensor networks has spurred an increase in the size and variety of data produced. Research data derived from these observations is a valuable asset which needs to be preserved and managed to derive the maximum value from it [13]. Although the size of the data sets produced is continuously growing, research data is different from what is known as big data. Big data usually includes data sets with sizes beyond the ability of commonly used software tools to capture, curate, manage, and process data within a tolerable elapsed time [14]. Big data encompasses unstructured, semi-structured and structured data, but the main focus is on unstructured data [15]. This difference comes from the processes that influence the creation of research data. In fact, research data are the product of carefully designed research projects. Moreover, taking advantage of big data requires the existence of well-structured datasets provided by research data (also called smart data) [13].

Research Data Lifecycle is the model of a process that covers the lifespan of research data products, from design to collection, curation, processing, publishing and reuse. Several data lifecycle models have been proposed in line with the importance assigned to research data products (for instance the data lifecycle models of the UK Data Service [16], Digital Curation Centre [17], and DataONE [18]). Inspired by these models and trying to find the most suitable for a wide range of cases presented by the institutions represented in the ENVRI consortium, the designers of the ENVRI RM looked at the commonalities of these models and produced a lightweight model of five stages (Fig. 3). The proposed lifecycle was designed to follow the main state changes to data (and metadata) as they are processed by RIs (acquired, curated, processed, published and used).

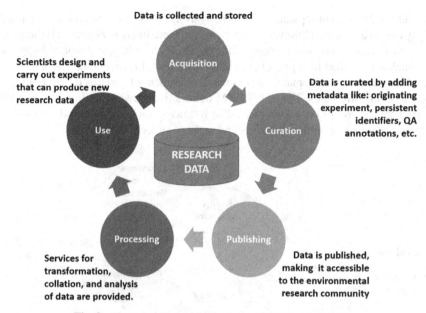

Fig. 3. The research data lifecycle model of the ENVRI RM.

The research data lifecycle model was refined with the analysis of the processes and practices for the management of research data of 26 research infrastructures (RIs) from four environmental areas (biosphere, lithosphere, atmosphere and hydrosphere) [1–4]. These analyses observed that the applications, services and software tools can be categorised following the five phases of the data lifecycle: acquiring data, storing and preserving data, making the data publicly available, providing services for further data processing, and using the data to derive other data products. The data lifecycle model was cross validated with an extended research campaign which visited seven research infrastructures. During these visits, it was observed that all the research infrastructures analysed exhibit behaviour that aligns with its phases. Furthermore, the campaign also served to validate structuring the ENVRI RM in line with the five phases of the data lifecycle.

3   The ENVRI Reference Model (ENVRI RM)

This section presents the ENVRI RM as the set of viewpoints, showing the main objects within each viewpoint, their structuring in line with the research data lifecycle and the correspondences to objects defined in other viewpoints. The ENVRI RM uses UML diagrams to produce the models of each viewpoint. UML4ODP [19] is the recommended notation for RM-ODP; however, this is not mandatory and different alternative notations can be used for each viewpoint as long as they can express equivalent concepts.

The viewpoint models proposed by the ENVRI RM aim to be as loosely coupled as possible, allowing parallel design and development among different teams. This approach allows some parts of the specification to reach a level of stability and maturity before others. The idea of separating concerns by using a set of viewpoints can be applied to many design activities. However, components are more likely to be reused if the same set of viewpoints is accepted by many different teams. The largest possible degree of commonality is needed to support the creation of a useful architectural framework to cover a large and diverse domain, such as the development of systems for environmental research infrastructures. The ENVRI RM defines five viewpoints (Fig. 4), intended to appeal to five groups of stakeholders. The following subsections introduce the five viewpoints, describing on the objectives and areas of concern they cover.

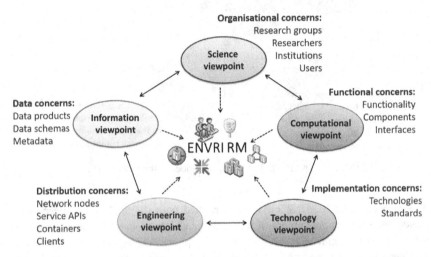

Fig. 4. The five viewpoints of the ENVRI RM

3.1 Science Viewpoint

The science viewpoint focuses on the institutional and social context of the domain in which the designed systems are intended to operate. This viewpoint concentrates on the objectives, processes, assets and policies that need to be supported by the system being modelled. The stakeholders to be satisfied are the research groups that promote the research processes, the managers making possible the operation of such processes, and the sponsors responsible for funding the research project. The emphasis is on the organisations, the research groups, their objectives, and on the environment within which the system operates.

The science viewpoint is intended to cover a wide range of operational setting; the target area can be whatever the designers are asked to describe. It can be a single experiment and its users, a research group, a larger institution, or a consortium with several partners.

The main modelling concepts of the Science Viewpoint are communities, roles, actions and artefacts. The main modelling concepts of the Science Viewpoint are *communities, roles, actions* and *artefacts*:

- *Roles* are fulfilled by objects defined in a *community*, which represents the different system stakeholders, scientists, scientific institutions, evaluation and certification agencies, as well as the information systems that provide the supporting IT services.
- *Actions* describe how the roles interact.
- *Artefacts* represent the information exchanged among them.

The diagram in Fig. 5 is a UML activity diagram. This type of diagram represents the relationships of the objects as containment (communities contain roles, roles contain behaviour and artefacts), sequencing ('take reading' precedes 'collect data'), and delegation ('acquisition system' performs 'collect data' producing a '[raw] data set').

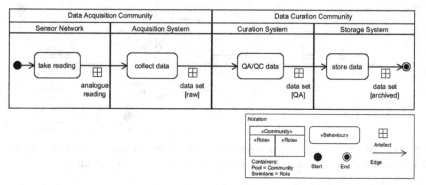

Fig. 5. The four main objects used to create science viewpoint specifications in the ENVRI RM: communities (outer container), roles (inner container), behaviours (rounded corner rectangles) and artefacts (small squares under the edge connectors (arrows)).

The science viewpoint specification enables the clear and concise representation of data processes at a high level. This specification is intended to be understood and shared by all the research infrastructure stakeholders.

3.2 Information Viewpoint

The information viewpoint specification enables the clear and concise representation of the data assets consumed and produced by the processes in the research infrastructure. The information viewpoint concentrates on modelling the data manipulated within the research infrastructure. Providing a common model that can be referenced from throughout a complete design specification assures that the same interpretation of information is applied at all levels. Common understanding about data and its interpretation minimises divergence and incomplete information among design, development and implementation teams.

The aim of the ENVRI RM information viewpoint is to achieve a shared model for the design activity, given that it can cover a federation of systems (possibly independently developed) and integration of legacy (preexisting) systems.

The main modelling concepts of the Information Viewpoint are *information objects* and *information activities*. The diagram in Fig. 6 is a UML activity diagram. This type of diagram represents the relationships of the objects as sequencing of actions ('take reading' precedes 'collect data'), and information objects ('analogue reading' precedes '[raw] data set'). In the information viewpoint the emphasis is on the data, their evolution (change) and the activities which enable that evolution. In the information viewpoint, the artefacts specified at a high level in the science viewpoint are refined, providing a clear specification of the types, states and relationships between different data products. In addition to the activity diagrams, the specifications at this level also include class diagrams to specify the hierarchy of data assets (Fig. 7).

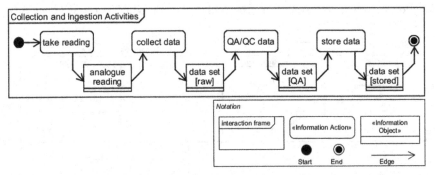

Fig. 6. The two main objects used to create information viewpoint specifications in the ENVRI RM: information objects (rectangles) and information actions (rectangles with rounded corners).

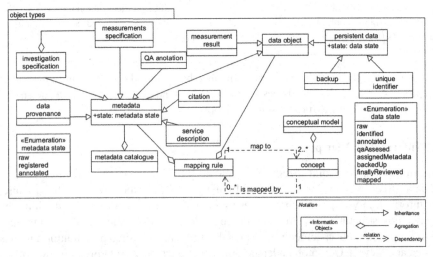

Fig. 7. A hierarchy of information objects. The class diagram emphasises the relationships of information objects such as composition, aggregation, generalisation, and multiplicity.

The correspondences between information viewpoint and science viewpoint objects can be seen directly by comparing this diagram with the one in Fig. 5. Artefacts in Fig. 5 correspond to information objects in Fig. 6 and behaviour in Fig. 5 can be mapped to information actions in Fig. 6.

3.3 Computational Viewpoint

The computational viewpoint specification models the units that provide different functionalities for processing data assets. The computational viewpoint is concerned with the development of the high-level design of the processes and applications supporting the RI research activities. This viewpoint expresses models in terms of objects with strong encapsulation boundaries, interacting at typed interfaces by performing a sequence of operations (or passing continuous streams of information). The computational viewpoint specification refers to the information viewpoint for the definitions of data objects and their behavioural constraints.

The main modelling concepts of the Computational Viewpoint are computing objects, their passive and active interfaces, and the relevant configurations in which objects are integrated to provide their services. The diagram in Fig. 8 is a UML component diagram. This type of diagram represents the relationships of the components as containment (nested subcomponents), and sequencing ('take reading' precedes 'collect data').

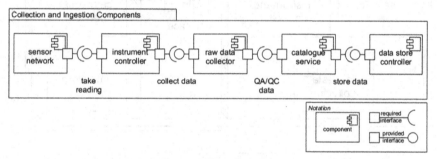

Fig. 8. Component objects and their interfaces. Component diagrams like this are used to create computational viewpoint specifications in the ENVRI RM.

3.4 Engineering Viewpoint

The main goal of the engineering viewpoint is to represent the distribution of components among different hardware and software systems. For instance, containers representing subsystem can be nested inside containers representing hardware platforms (servers and/or networks). The engineering viewpoint tackles the problem of diversity in infrastructure provision, and it gives the prescriptions for supporting the necessary abstract computational interactions in a range of different situations. It thereby offers a way to avoid lock-in to specific platforms or infrastructure mechanisms. An interaction may involve communication between subsystems, or between objects hosted in various servers, and accordingly different engineering solutions will be used.

The engineering viewpoint is also concerned with providing a set of guarantees (called transparency) to the designer. Providing a transparency involves taking responsibility for a distribution problem, so that the computational design does not need to worry

72   A. N. de la Hidalga et al.

about it. The transparency mechanisms needed are provided in the form of standard middleware or web services components, simplifying the engineering specification, since it can reference the existing solutions and merely state how they are combined to meet the infrastructure needs of the system.

The main modelling concepts of the Engineering Viewpoint are engineering objects, containers and channels. In Fig. 9, the diagram represents two subsystems (acquisition and curation) which in turn contain (host) different basic engineering objects. The objects in one subsystem can communicate with other objects using standard interfaces (e.g. APIs).

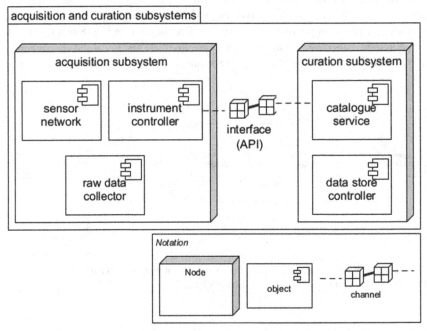

Fig. 9. Deployment diagrams are used to create engineering viewpoint specifications in the ENVRI RM. This type of diagram represents the relationships of the engineering objects as containment (nested node containers), and interfaces (communication channels).

3.5 Technology Viewpoint

Technology Viewpoint specifications are intended to represent the concrete dependencies between design and implementation. The technology viewpoint is concerned with managing real-world constraints, such as restrictions on the hardware available to implement the system within budget, or the existing application platforms on which the applications must run. The designer never really has the luxury of starting with a green-field, and this viewpoint brings together information about the existing environment, current procurement policies and configuration issues. It is concerned with selection of ubiquitous standards to be used in the system, and the allocation and configuration of real resources. It represents the hardware and software components of the implemented system, and the communication technology that provides links between these components. Bringing all these factors together, it expresses how the specifications for an ODP system are to be implemented.

This viewpoint also has an important role in the management of testing conformance to the overall specification because it specifies the information required from implementers to support this testing. The main modelling concepts of the Technology Viewpoint are *conformance points* and *standards*. In Fig. 10, the diagram represents a system component (catalogue service) and the technology constraints which condition its operation. The diagram shows three conformance points each paired with a corresponding standard or implementation constraint. For instance, the catalogue service API is a conformance point to be provided as part of the service, and its corresponding constraint indicates that the corresponding API definition should use a standard such as Open API.

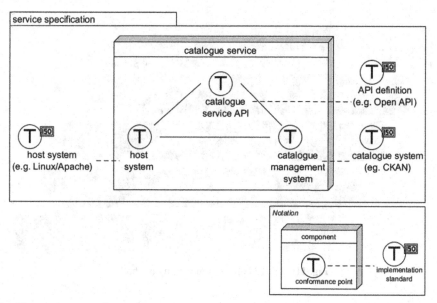

Fig. 10. Deployment diagrams are used for technology viewpoint specifications in ENVRI RM. This type of diagram represents the relationships of the objects and their implementation constraints as relationships to requirements, system configurations and services.

4   The Modelling Process

Diagrams can help understand part of the operation of a RI. However, a single diagram without context can invite many interpretations and needs to be complemented with further information when presented to different stakeholders. Different stakeholder groups can be interested in issues such as standards, data and metadata formats, chains of responsibility, communication protocols, software and hardware dependencies and many other issues which are hard to convey on a single representation. Moreover, it is expected to find multiple sources describing how many of those concerns are addressed.

During the period from April 2017 to January 2018, the ENVRI Reference Model development team, consulted with nine environmental research infrastructures from different domains about their status and development plans[4]. The interactions during those consultations served to define a structured modelling method [20].

The proposed modelling method is recursive and consists of five steps: identification, modelling, refinement, review-revision, and mapping (Fig. 11). In this method, the designer is free to select a starting viewpoint, model the characteristics of interest within that viewpoint and then model additional details by mapping the specification to other viewpoints. The advantage of modelling using the ENVRI RM in this way is that the designer can add detail to the models while keeping consistency at different levels of abstraction. The following sections will elaborate on each of the modelling steps illustrating them with an example.

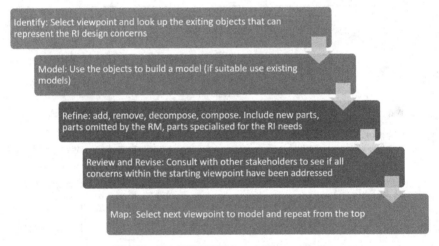

Fig. 11. The ENVRI RM modelling method.

4.1 Identify

The identification step requires gathering existing RI documentations and use it to determine the viewpoint from which to start modelling. The main representation of a system coincides with the main interest of the system designers. For instance, if the system must provide data with well-established formats, the information viewpoint might be the best described specification of the system. Similarly, if the main challenge is the integration

---

[4] As part of the ENVRI visits [19].

of processing components, then a computational specification that describes the operations to be supported might contain the most complete description of the system. In
this scenario, the recommendation is to identify the most complete specification of the
system and start by mapping it to one of the existing viewpoints. This will help in further
understanding the systems and discovering which attributes of the system are common
(shared with other RIs, domain independent) and which are special (unique, domain
dependent).

In the case of EPOS, the main model describes the architecture of the RI systems
using a block diagram (Fig. 12). This description is complemented with the definition
of the functions of each of the components [21]. The description of components, their
functionalities, and integration matches the concepts described by the computational
viewpoint of the ENVRI RM, which is designated as the starting viewpoint to model.
After deciding to start with the computational viewpoint, the viewpoint objects are
revised to select the ones that can be used to represent the concepts of the initial model.
Figure 13 shows how computational viewpoint components can be used to build a model
equivalent to the EPOS architecture. The mapping is not one to one, there are components
which cannot be mapped to existing computational viewpoint components, such as the
Thematic Core Services and Workspace Connector, these are addressed by creating
custom models, as explained in the next section.

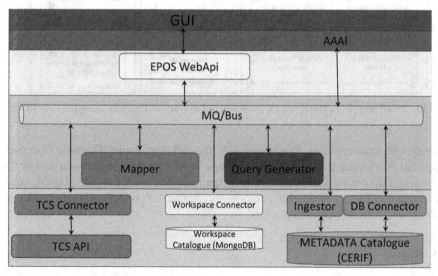

Fig. 12. EPOS Integrated Core Services Layered Architecture [21].

4.2 Model

The ENVRI RM is not expected to cover all possible cases, consequently some of the
entities described in the infrastructure design will not have equivalent viewpoint object
representations. In these cases, new objects can be defined and modelled to implement

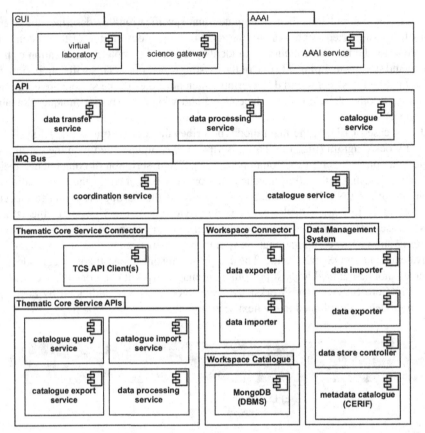

Fig. 13. Initial mapping of Integrated Core Services Layered Architecture using the ENVRI RM.

the required functionalities. Continuing with the EPOS example, Thematic Core Services and Workspace Connector are two cases in which components described in the architecture do not map one-to-one to existing reference model objects. For instance, the diagram in Fig. 14 shows the components required to provide the functionality of the temathic core services components.

4.3 Refine

The refinement of the models requires integrating the components in different configurations to provide additional functionalities. Continuing with the example, ENVRI RM components can be composed as shown in Fig. 15. The diagrams show the composition of the catalogue export service. Notice that the model is built using existing ENVRI RM components.

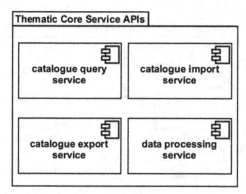

Fig. 14. ENVRI RM model components selected providing the functionality of Thematic Core Services (TCS). TCS require components for cataloguing and data processing (four services).

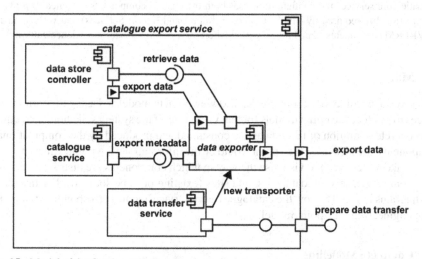

Fig. 15. Model of the Catalogue Export Service component, required for implementing the export data functionalities required by the Thematic Cores Services of the EPOS Architecture. The model is a refinement of the component specified in Fig. 14.

4.4 Review

In the review step, the models and compositions are discussed with the relevant stakeholders to determine if the models are complete and represent the entities considered in the original RI representation. To facilitate the discussion, further configuration diagrams can be produced, to show how the components are supposed to interact. For example, Fig. 16 shows a configuration describing how the components can be integrated to support importing data from different thematic core services for the EPOS case example.

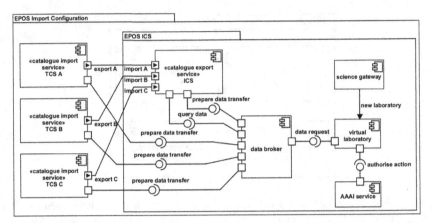

Fig. 16. Model of the configuration of components to support importing data from different thematic core services the configuration uses both the custom components designed to provide the functionality required by EPOS (Catalogue Import and Export Services) and with standard ENVRI RM components (data broker, virtual laboratory, AAAI service, and science gateway).

4.5 Map

The next stage requires determining the next viewpoint to model and using the correspondences to produce the initial models for that viewpoint. If the system stakeholders require to a concrete definition of the data assets consumed and produced by the computational components, the ideal next viewpoint would be the information viewpoint. Alternatively, if the stakeholders need to visualise the way in which components are distributed across the resources i.e. servers, databases, and sites (existing or to be sourced). For instance, the diagrams in Fig. 16 show the catalogue query service and its corresponding mapping to an engineering viewpoint model.

4.6 Complete Modelling

The basic modelling process (identify, model, refine, review, map) can be repeated several times to obtain models covering complementary design concerns. The point at which the process should stop varies according to the intended use of the models (documentation, reporting, validation, etc.). The modellers should evaluate the benefits of creating models for each viewpoint with the rest of the stakeholders and stop the modelling process once a sufficiently fit for the purpose set of models has been obtained (Fig. 17).

5 Outlook

The ENVRI RM was designed and developed to support understanding emerging and established research infrastructures, and their operation environments (processes, systems and assets). The main goals of this research effort were to (1) discover common operations, (2) describe the systems and services which they provide and depend-on, and

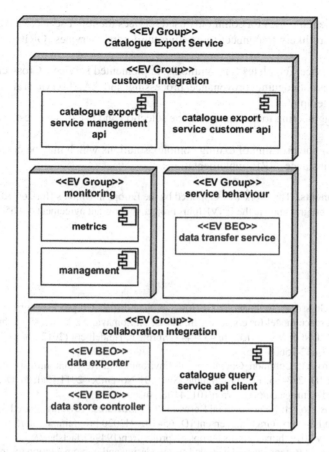

Fig. 17. Engineering Viewpoint Model of the Catalogue Export Service. This model includes the three components used in the corresponding computational model shown in Fig. 13.

(3) identify the requirements and challenges of integrating (required services, standards, and coordination).

The recommendation for the engineering viewpoint follows a microservice architecture model which allows the definition API interfaces that support flexible integration of services and systems. The recommendation for the Technology Viewpoint allows the use of templates for defining conformance points to verify the suitability of technologies and standards.

The ENVRI RM serves as a reference architecture for the evolution of the services offered and consumed by different research infrastructures into a coherent software product line. During the past years, ENVRI RM has not only been used by the RIs within ENVRIplus projects, but also application outside, e.g. for a Chinese agricultural data management infrastructure [22]. This software product line can facilitate:

- Creating client libraries for commonly used services Identifier services are a good use case, they are likely to connect to existing third-party Services (ORICID, DOI and ePIC.);
- Creating service Templates for commonly implemented services. Cross-cutting services such as cataloguing, provenance, processing, and AAAI services are candidates for service templates;
- Creating engineering tools supporting the selection and use of services;

Facilitating the profiling of exiting complex solutions which may be considered for adoption, for instance, VRE implementations.

Acknowledgements. This work was supported by the European Union's Horizon 2020 research and innovation programme via the ENVRIplus project under grant agreement No 654182.

References

1. Zhao, Z., Grosso, P., de Laat, C.: OEIRM: an open distributed processing based interoperability reference model for e-science. In: Park, J.J., Zomaya, A., Yeo, S.-S., Sahni, S. (eds.) NPC 2012. LNCS, vol. 7513, pp. 437–444. Springer, Heidelberg (2012). https://doi.org/10.1007/978-3-642-35606-3_52
2. ENVRI Project: Common Operations of Environmental Research Infrastructures. Grant agreement ID: 283465. FP7 programme. Start: 01 November 2011 End: 31 October 2014. https://cordis.europa.eu/project/rcn/101244/factsheet/en
3. ENVRIplus Project: Environmental Research Infrastructures Providing Shared Solutions for Science and Society. Grant agreement ID: 654182. H2020 programme. Start: 01 May 2015 End: 31 July 2019. https://cordis.europa.eu/project/rcn/194947/factsheet/en
4. Zhao, Z., et al.: Reference model guided system design and implementation for interoperable environmental research infrastructures. In: 2015 IEEE 11th International Conference on e-Science, Munich, Germany, pp. 551–556. IEEE (2015). https://doi.org/10.1109/eScience.2015.41
5. IISO/IEC 10746-1:1998 Information technology – Open Distributed Processing – Reference model: Overview
6. ISO/IEC 10746-2:2009 Information technology – Open Distributed Processing – Reference model: Foundations
7. ISO/IEC 10746-3:2009 Information technology – Open Distributed Processing – Reference model: Architecture
8. ISO/IEC 10746-4:1998 Information technology – Open Distributed Processing – Reference model: Architecture Semantics
9. Linington, P.F., Milosevic, Z., Tanaka, A., Vallecillo, A.: Building Enterprise Systems with ODP. CRC Press, Boca Raton (2012)
10. IEEE, 1471-2000, IEEE Recommended Practice for Architectural Description of Software Intensive Systems (2000)
11. ISO/IEC/IEEE Standard 42010:2011 Systems and Software Engineering – Architecture description
12. Martin, P., et al.: Open information linking for environmental research infrastructures. In: 2015 IEEE 11th International Conference on e-Science, Munich, Germany, pp. 513–520. IEEE (2015). https://doi.org/10.1109/eScience.2015.66

13. Guo, Y.K.: Assimilated learning-bridging the gap between big data and smart data. In IoTBDS, p. 7 (2017)
14. Snijders, C., Matzat, U., Reips, U.D.: Big data: big gaps of knowledge in the field of internet science. Int. J. Internet Sci. 7(1), 1–5 (2012)
15. Dedić, N., Stanier, C.: Towards differentiating business intelligence, big data, data analytics and knowledge discovery. In: Piazolo, F., Geist, V., Brehm, L., Schmidt, R. (eds.) ERP Future 2016. LNBIP, vol. 285, pp. 114–122. Springer, Cham (2017). https://doi.org/10.1007/978-3-319-58801-8_10
16. UK Data Service: Research Data Lifecycle. https://www.ukdataservice.ac.uk/manage-data/lifecycle.aspx. Accessed 28 Apr 2019
17. Digital Curation Centre: Curation Lifecycle Model. http://www.dcc.ac.uk/resources/curation-lifecycle-model. Accessed 28 Apr 2019
18. Data One: Data Life Cycle. https://www.dataone.org/data-life-cycle. Accessed 28 Apr 2019
19. Rec. ITU-T X.906 (10/2014) I ISO/IEC 19793 ITU-T: 2015 Information technology – Open Distributed Processing – Use of UML for ODP system specifications
20. Nieva de la Hidalga, A., Hardisty, A., Magagna, B., Martin, P., Zhao, Z.: Use of the ENVRI Reference Model to Support the Design of Environmental Research Infrastructures. Geophysical Research Abstracts, vol. 20, EGU2018-18552, 2018 EGU General Assembly (2018)
21. EPOS: Integrated Core Services Architecture. https://www.epos-ip.org/data-services/ict-architecture/ics-architecture. Accessed 28 Apr 2019
22. Zhao, Z., Liao, X., Wang, X., Ruan, C., Zhu, Y., Feng, D.: A Reference Model approach for developing agriculture big data infrastructures. J. East China Normal Univ. (Nat. Sci.) (2) 77–96 (2019). http://xblk.ecnu.edu.cn/EN/10.3969/j.issn.1000-5641.2019.02.009

Reference Model Guided Engineering

Zhiming Zhao[1](✉) ⓘ and Keith Jeffery[2] ⓘ

[1] Multiscale Networked Systems, University of Amsterdam,
1098XH Amsterdam, The Netherlands
z.zhao@uva.nl
[2] Keith G Jeffery Consultants, Faringdon, UK
keith.jeffery@keithgjefferyconsultants.co.uk

Abstract. Environmental research infrastructures (RIs) support their respective research communities by integrating large-scale sensor/observation networks with data curation and management services, analytical tools and common operational policies. These RIs are developed as service pillars for intra- and interdisciplinary research; however, comprehension of the complex, interconnected aspects of the Earth's ecosystem increasingly requires that researchers conduct their experiments across infrastructure boundaries. Consequently, almost all data-related activities within these infrastructures, from data capture to data usage, need to be designed to be broadly interoperable in order to enable real interdisciplinary innovation and to improve service offerings through the development of common services. To address these interoperability challenges as they relate to the design, implementation and operation of environmental RIs, a Reference Model guided engineering approach was proposed and has been used in the context of the ENVRI cluster of RIs. In this chapter, we will discuss how the approach combines the ENVRI Reference Model with the practices of Agile systems development to design common data management services and to tackle the dynamic requirements of research infrastructures.

Keywords: Research infrastructure · Reference Model · Interoperability · Agile

1 Introduction

Many key problems in environmental science are intrinsically interdisciplinary; the study of climate change, for example, involves the study of the atmosphere, but also earth processes, the oceans and the biosphere. Modelling these processes individually is difficult enough, but modelling their interactions is another order of complexity entirely. Scientists are challenged to collaborate across conventional disciplinary boundaries, but must first discover, extract and understand data dispersed across many different sources and formats.

Data-centric research differs from classical approaches for analytical modelling or computer simulation insofar as new theories are measured first and foremost against huge quantities of observations, measurements, documents and other data sources culled from

© The Author(s) 2020
Z. Zhao and M. Hellström (Eds.): Towards Interoperable Research
Infrastructures for Environmental and Earth Sciences, LNCS 12003, pp. 82–99, 2020.
https://doi.org/10.1007/978-3-030-52829-4_5

a range of possible sources. To enable such science, the underlying research infrastructure must provide not only the necessary tools for data discovery, access and manipulation but also facilities to enhance collaboration between scientists of different backgrounds.

Environmental research infrastructures (RIs) support user communities by providing federated data curation, discovery and access services, analytical tools and common operational policies integrated around large-scale sensor/observer networks, often deployed on a continental scale. Examples in Europe include LifeWatch[1] (concerned with biodiversity), EPOS[2] (solid Earth science), Euro-Argo[3] and EMSO[4] (ocean monitoring), as well as ICOS[5] and the new EISCAT_3D system (atmosphere)[6]. These infrastructures are developing into important pillars for their respective user communities, but are also intended to support interdisciplinary research as well as more specific research data aggregators such as Copernicus[7] within the context of GEOSS[8]. As such, it is very important that data-related activities are well integrated in order to enable data-driven system-level science [2]. This requires standard policies, models and e-infrastructure to improve technology reuse and ensure coordination, harmonization, integration and interoperability of data, applications and other services. However, the complex nature of environmental science seems to result in the development of environmental RIs that meet only the requirements and needs of their own specific domains, with very limited interoperability of data, services, and operation policies among infrastructures.

It is thus important to identify technical and organizational commonalities for the cluster of research infrastructures in environmental and Earth sciences and provide a unified data discovery and access services to the whole RI activity cycle. This chapter presents the engineering model developed in the EU H2020 projects ENVRI, ENVRIplus and ENVRI-FAIR [3] for 1) combining both domain-specific characteristics and common abstractions; 2) harmonising RI-specific requirements with common operations; and 3) accounting for both existing generic e-infrastructures already adopted by existing RIs. The chapter is an extension of the earlier publication in IEEE eScience 2015 [1].

2 Engineering Challenges in Environmental RIs

Environmental RIs collectively play an important role in environmental and Earth science research in Europe, as shown in Fig. 1, with more than half of them, prioritised in the roadmap of the European Strategy Forum on Research Infrastructures (ESFRI) [4].

The RIs are in one or across multiple environmental domains: atmosphere; bio- or ecological; aquatic; and solid earth. There is considerable variation in their states of development.

---

[1] http://www.lifewatch.eu/.

[2] http://www.epos-eu.org/.

[3] http://www.euro-argo.eu/.

[4] http://www.emso-eu.org/.

[5] http://www.icos-infrastructure.eu/.

[6] https://www.eiscat3d.se/.

[7] http://copernicus.eu/.

[8] http://www.earthobservations.org/geoss.php.

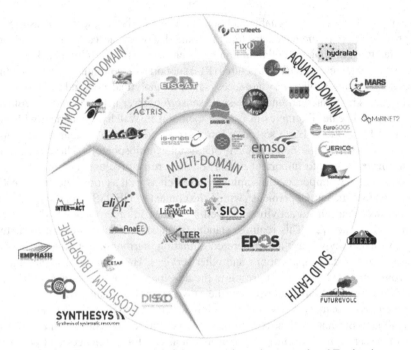

Fig. 1. Key European research infrastructures in environmental and Earth sciences.

2.1 Interoperability Challenges

In the earlier chapters, we discussed that one of the key missions in the cluster project of ENVRI is to provide reusable solutions to common problems these research infrastructures face and promote their interoperability for future system level of sciences [3].

In the ENVRI project, we reviewed existing interoperability solutions [5] from different specific aspects: infrastructure, middleware, and workflow. Typically, these solutions are realised iteratively, building adapters or connectors between two components and then deriving new service layer models for standardization via a community effort. Such a process of iteration can gradually promote the evolution of new standards for both infrastructures and the service layers above them, but will not completely solve all interoperability problems while the diversity between infrastructures and the gaps between standards remain significant [6]. White et al. [7] argued that an interoperability reference model is needed to complement models of application and infrastructure.

For those environmental RIs that are currently under construction or in preparation, it, therefore, becomes urgent to guide their development so that they can be immediately interoperable once operational.

2.2 Challenges for Enabling System-Level Science

To perform system-level environmental science, scientists face challenges with respect to data accessing, processing and publication:

1. *Obtaining and harmonizing data from different sources.* Data are often in different formats, annotated using different metadata, and retrieved via catalogues with different interfaces.
2. *Identifying different levels of data from the same instruments and experiment.* Data, being quality controlled and processed, are labelled as being of different levels during the data lifecycle, for example, raw input data (level 0) versus derived datasets (levels 1 or higher). Identifying different levels of data from the same instruments is crucial for precisely understanding their meaning.
3. *Selecting and combining data processing models from different domains.* Data processing models are often represented as workflows of services with attached datasets in different languages and require different execution engines to realise.
4. *Selecting optimal infrastructure upon which to execute applications.* Infrastructures often provide different scheduling and monitoring tools.
5. *Publishing data objects in different research infrastructures.* Data objects should be both identifiable and citable.

Environmental RIs provide the tools to help with this, but only if their services are sufficiently interoperable. To enable interdisciplinary research across RIs from different sub-domains of environmental science, there are a number of principles that any interoperable services and their supporting infrastructure should adhere to:

- *Simple but effective.* Scientists should be able to use, analyse, compose and store data from distributed sources in an easy but effective way, with appropriate metadata generated at all stages in order to trace data provenance.
- *Formal syntax.* the datasets should possess (a) a well-defined schema to describe attributes, types and permitted values (for validation); (b) referential integrity to avoid any updating problem; (c) functional integrity so that each attribute has no dependencies other than the object being described in order to ensure correct representation of the world of interest. Software services should have defined functionality through formally-defined APIs with parameter lists and defined non-functional properties covering performance and trust, security, privacy.
- *Bridgeable semantics.* A certain degree of semantic mapping is required to bridge the diverse complex knowledge organizing systems needed by different scientific and technical domains, but all the tools and resources need to be documented in a principled, formal way first. For datasets, the semantics of attribute values must be defined and for services the semantics of the parameters in the API must be defined. In both cases the semantics of descriptions and keywords in the catalogue require definition.
- *Extensible and robust.* Available resources change and user demands fluctuate; core RI services must be elastic and fault-tolerant, and provide programmatic interfaces for service composition.
- *Open yet secure.* Although most research data is open, there is a need to protect the privacy of researchers, attribute credit to individuals and organizations, embargo new research prior to publication and preserve authority and accountability constraints when transferring data between different technical and political domains.

In order to meet these rather wide-ranging principles, the ENVRIplus solutions build upon the results of earlier projects, the expertise of individual RIs, and the services of e-infrastructure initiatives. Filling in the gaps, the ENVRI community continues to work to:

1. Optimise data processing and develop common models, rules and guidelines for research data workflow documentation.
2. Facilitate data discovery and (re-)use following the FAIR principles[9], and provide integrated end-user information technology to access heterogeneous data sources.
3. Make data citable by building upon existing approaches with practical examples, exchanges of expertise, and agreements with publishers.
4. Facilitate the discovery of software services and their possible compositions.
5. Characterise users and build a community on top of existing RI communities.
6. Characterise ICT resources (including sensors and detectors) to allow virtualisation of the environment (for instance onto the grid- or cloud-based platforms) such that data and information management and analysis is optimised in terms of resource and energy expenditure.
7. Facilitate the connection of users, composed software services, appropriate data and necessary resources in order to meet end-user requirements.

2.3 Engineering Challenges

The development of Research Infrastructures in environmental Earth sciences has to consider not only the requirements discussed in Sect. 2.2, but also the status of the existing work, e.g. types of legacy assets, the maturity of available services, and usage of standards. Figure 1 shows a clear diversity among the research infrastructures in the cluster of environmental and Earth sciences:

1. *At different levels of development:* some infrastructures are partially or wholly in operation (e.g. Euro-Argo), while others are still under preparation or development when the ENVRIplus project starts.
2. *Having different development roadmaps:* infrastructures in the ESFRI roadmap (e.g. EMSO and LifeWatch) have a clear timeline with an established funding stream, while some other infrastructures are still funded under specific projects.
3. *Using different standards for specifying metadata information:* standards such as CERIF, used in EPOS, and ISO 19115, used in SeaDataNet, are also used for creating data catalogues.
4. *Providing different support for end-users to perform scientific experiments:* for example, ICOS provides a web-based environment, the Carbon Portal[10], to allow scientists to discover data, visualise its content, and perform customised data processing workflow, while LifeWatch provides specific deployments of software environments (virtual laboratories) to its users.

---

[9] https://www.force11.org/group/fairgroup/fairprinciples.
[10] https://www.icos-cp.eu/.

To be interoperable, the data or services from different RIs need to be discovered, accessed and integrated across their boundaries. It is important to identify the common problems faced by the RIs, and provide reusable solutions to those problems. To effectively deal with such issues, RI development faces a number of challenges:

1. How to effectively deal with the diversities, so that developers can identify and model the common problems faced by the RIs?
2. How to design reusable solutions to their common problems, so that each individual RI can effectively take the solution and customise it in their own software stacks?
3. How to effectively handle new requirements from each RI, e.g. demands from user communities?
4. How to effectively select technology and standards for prototyping the solutions to those common problems?

Based on those challenges, the ENVRI community proposed a reference guided approach, which we discuss in Sect. 4.

3 The State of the Art: Software Architecture and Development Models

In this section, we shall briefly review the software engineering technologies and methodologies from the perspectives of engineering model, software architecture, and reference model guidance.

3.1 Software Architecture

The architecture of a software system models the high-level structure of the system; the functional components and the logical relations among those components have been modelled using different orientations [10] e.g. of objects, components, software agents and services. Since 2000, service-oriented architecture has been widely adopted in the software industry for automating the cross-organization of business processes, hiding complexity in software delivery, and simplifying software reuse [11, 12]. In this context, a number of trends can be highlighted as arising during recent decades:

1. When running on virtualised infrastructure, loosely coupled distributed architectures are more scalable than the monolithic architectures in which all components reside in one integrated system;
2. Service-oriented architectures (SOA) are playing an increasingly important role in enterprise computing, and internet applications.
3. Web services can be deployed on remote hosts and can be invoked by remote clients via standardised internet-based protocols (e.g. HTTP). They can be implemented using Remote Procedure Call (RPC) based technologies, e.g. XML RPC or Simple Object Access Protocol (SOAP), or using Representational State Transfer (RESTful) mechanisms.

4. Microservices design the services in "suitable" granularity [13] with atomic functionality, which can be better reusable and scalable. The concept of microservice is typically driven by elastic computing in Cloud, where the required service function can be flexibly scaled out by adding more instances to overcome performance bottlenecks.

3.2 Reference Model and Architecture in System Development

Reference models or architecture have been widely in the IT industry to standardise the abstraction of certain new technologies, e.g. the OSI reference model for network development [8] and workflow management reference model [9] for business process management. A reference model for a computational system provides an ontological framework for involved parties to clearly communicate.

In both the ENVRI and ENVRIplus projects, a reference model has been recognised as a promising contribution for realising interoperability for diverse environmental RIs. In this section, we will first review the work of the ENVRI Reference Model, and then summarise the lessons learned. Afterwards, we will discuss the approach for the ENVRIplus Reference Model.

In the ENVRI project, the development of the Reference Model (ENVRI-RM) was based on an analysis of six RIs involved in the project: ICOS, Euro-Argo, EISCAT_3D, LifeWatch, EPOS, and EMSO. By interviewing specialists from each of these RIs, and examining the requirements, design documents, and use cases collected, we abstracted some common operations and design patterns. This analysis had to cope with different viewpoints and varying vocabularies between (and even within) RIs.

The methodology for developing ENVRI-RM was to decompose system descriptions based on viewpoints. Open Distributed Processing (ODP) [14] provides five viewpoints from which to describe systems: *enterprise* (about system scenarios, involved communities and roles), *computation* (about system interfaces and bindings between system components), *information* (about data objects and schemas of the system), *engineering* (about system middleware and engineering principles) and *technology* (technology standards and decisions). This decomposition of complex systems by viewpoint is a useful technique for managing complexity and providing information tailored to different kinds of stakeholders. ENVRI-RM employs these viewpoints to model the characteristics of environmental research infrastructures, but we replace the Enterprise viewpoint with a "Science" Viewpoint to align the ODP with the RI view of the world. The current version is available online[11] (Fig. 2).

ENVRI-RM focused on the design of a small set of RIs and was produced at a time when most of them were in their preparatory phase of development. Since ENVRI began, many of them have made significant progress in their development, to some extent exceeding the expressiveness of ENVRI-RM. As such, a number of lessons can be learned:

---

[11] www.envri.eu/rm.

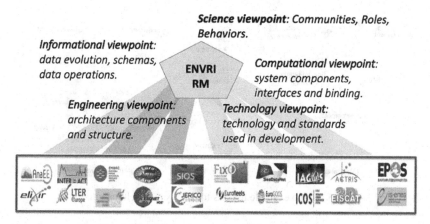

Fig. 2. The basic idea of ENVRI reference model.

1. The rapid evolution of environmental RIs was not sufficiently taken into account when building ENVRI-RM; there are mismatches between new insights in the RIs and what was encoded in the model.
2. Supporting use cases are required for validation and demonstration; the lack of these in ENVRI led to drifting requirements and difficulty explaining the model to potential users, although this was improved in ENVRIplus.
3. The development of the model did not involve enough domain-aware ICT specialists from the RIs themselves. This was partly due to the early development state of the RIs, but meant that the model was not really applied to that development.

3.3 Software Development Models

To efficiently manage the activities in the lifecycle of software development, different engineering models have been proposed and applied during recent decades. The water-fall model is a typical example, where requirement analysis, system design, software development, testing and integration, and delivery are organised sequentially. The development team focuses on a specific task at each stage. When the application problem is well understood and there is sufficient engineering time, the waterfall model is easy to apply in practice. However, when an application is difficult to describe precisely in the very beginning, or the time for delivery is fixed and urgent, e.g. when driven by specific market needs, the waterfall model exhibits a number of weaknesses: i) high cost in incorporating changing requirements or correcting mistakes, and ii) high risks in managing time because the project commonly is delayed if any mistakes are made at an earlier phase. The waterfall model has been adapted in different ways to overcome these issues:

1. The V model [15], in which the software testing and validation are performed against system design, architecture and requirements, as shown in Fig. 3-a.
2. The Iterative model [16], in which all phases in the lifecycle can provide feedback to the previous phase, and make corrections where necessary, as shown in Fig. 3-b;

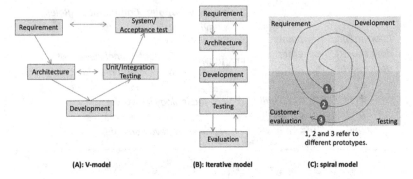

(A): V-model (A): V-model   (B): Iterative model   (C): spiral model

Fig. 3. Some example models for software development.

3. The Spiral model [17], in which the lifecycle is organised as a number of continuous phases, and each phase is a loop of all steps as defined in the waterfall model. The spiral model can reduce the risks of unbalanced time allocation and partial or inaccurate requirements analysis, as shown in Fig. 3-c.

In this evolution of software development models, we can clearly see several highlights: i) developers do not just execute engineering tasks sequentially and in a single round, ii) developers can flexibly switch engineering tasks forward or backward, and iii) the duration of the customer evaluation is also getting shorter. For applications which have clear time boundary and delivery constraints, a method called Agile development has emerged during the past decade, where the development team focuses on the prioritised tasks requested by the customer, and efficiently perform the development with well-controlled progress reviews. Highsmith [18] highlighted the key difference between classical waterfall model and the Agile model by using the relationships between **Feature(s), Cost** and **Time**. In the classic model, the set of **features** are derived from the requirements and commonly are fixed; the **timeline** and project **cost** often have to be adapted based on the original plan and the actual progress [21]. The Agile model is the opposite: the set of **features** has to be adaptable to meet the fixed **cost** and **timeline** of the project (Fig. 4).

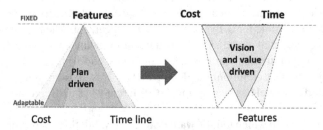

Fig. 4. Agile management: from plan driven to vision and value driven [18].

3.4 Summary

Targeting at the interoperability of more than 20 research infrastructures in the cluster of environmental and Earth science, the ENVRIplus data for science theme has to simultaneously interact with the development teams in each RI [23]. Within the period of the project, the theme developers had to continuously:

1. collect and analyse requirements from each RI,
2. tackle common challenges, and
3. deliver useful solutions to the development teams of the RIs, even while each RI clearly has its own development roadmap and timeline.

To effectively manage the development process of the theme team, and the interaction with individual RIs, the engineering approaches we reviewed above needed to be carefully selected and applied. A reference model guided approach was thus proposed.

4 The Reference Model Guided Approach

The **ENVRIplus reference model guided engineering model** builds upon abstracted concepts derived from analysing common operations of a selected set of RIs and subsequently defines an ontological reference model for all environmental RIs. Figure 5 shows the basic idea of the reference model guided approach.

Fig. 5. The basic idea of the reference guided approach.

The proposed approach uses the ENVRI-RM as the common ontological framework to:

1. formulate requirement collection questionnaires;
2. align the input acquired from different research infrastructures;

3. analyse the requirements from different viewpoints;
4. design and validate the solution using the architectural patterns provided by the reference model.

The development teams carried out the development tasks of the designed solution in an iterative way. In the meantime, a number of small parallel use case teams were dynamically established based on the demands and the priority of each solution development team. The use case projects were managed using the agile approach: via a dynamically maintained task list, the project teams aimed to deliver a rapid prototype or technical validation in a timely way. The successful results from the use case teams were curated and included regular development task teams interaction.

A high-level steering committee was established to control the selection of successful results and establish a portfolio for the entire theme.

In the rest of the chapter, we will discuss this approach in more detail.

4.1 Reference Model Guided: Requirement Collection, Technology Review and Gap Analysis

Based on the requirements collected from each of the four main environmental science domains and their respective RIs, we **identified and developed common operations**, by characterising RIs' individual current solutions with consideration given to underlying common technologies and engineering challenges. These individual operations will be characterised in terms of the engineering model, which will then be used in the design and implementation of common operations. The common operations are of two kinds: (a) those needed by any RI for data management, cataloguing, curation, provenance, analytics, visualisation; (b) those required for interoperation across RIs.

To benefit from existing technologies, we reviewed early results from specific RIs and interacted with computational e-infrastructures (such as EGI), data infrastructures (such as EUDAT[12]), and other initiatives (such as D4Science[13]) that work on related issues. We reviewed other interoperation technologies including CERIF [19] from EPOS for describing datasets, users, software, facilities, services and resources, and DCAT[14] for high-level exposure of basic dataset information.

This approach was used to (a) reduce risk; (b) maximise utilization of e-infrastructures in individual RIs developed with EC or other public funding; (c) provide an opportunity for convergence of ideas among the RIs without discarding work already done; and (d) maximise the chances of successful interoperation between environmental RIs, both technically and socially.

4.2 Identifying Common Data Management Services Using the ENVRI-RM

The ENVRI-RM assists in defining commonalities in the operations of environmental RIs, e.g. common services that support a particular subdomain of environment research,

---

[12] http://www.eudat.eu/.

[13] http://www.d4science.org/.

[14] http://www.w3.org/TR/vocab-dcat/.

or set of such sub-domains. ENVRIplus is not concerned with the unique services of a specific RI. The focus is on common services that are useful for significant subsets of environmental RIs.

We have identified six common concerns based on the demands of the RIs involved in ENVRIplus, which we will work to provide solutions for.

1. **Data identification and citation** requires the implementation of a common policy model for handling persistent identifiers for publishing and citing data. Moreover, services for assigning and handling identifiers and for retrieving data based on identifiers should also be provided.

2. **Interoperable data processing, monitoring and diagnosis** services make it significantly easier for scientists to aggregate data from multiple sources and to conduct a range of experiments and analyses upon those data. Expanding upon the data processing workflow modelled in ENVRI, this service focuses on the engineering aspects of managing the entire lifecycle of computing tasks and application workflows for efficient utilization of underlying e-infrastructure. In particular, the service enables scientists to enrich the data processing environment by injecting new algorithms to be reused by others.

3. **Performance optimization for big data science** is increasingly required in environmental science. ENVRIplus focused on high-level, generically-applicable optimization mechanisms for making decisions on resources, services, data sources and potential execution infrastructures, and on scheduling the execution of big data applications [22].

4. Data quality control and annotation were modelled as basic curation services in ENVRI-RM, although they have different (but related) requirements. **Self-adaptable data curation for system-level science** covers different levels of data. The service provided by ENVRIplus complies with data and metadata standards such as OASIS[15] and INSPIRE[16] and provides rich, interoperable metadata for geospatial semantic annotation. The quality of user experience, when checking the quality of data and when annotating different data using the aforementioned metadata standards, is explicitly modelled and considered in the development of curation services.

5. To perform complex data-driven experiments, scientists want simple but effective mechanisms to discover data recorded in catalogues and to integrate data into computing processes. An **interoperable data cataloguing** service provides interoperable solutions for accessing, retrieving and integrating data from different catalogues. The service extended the open search tools developed in the ENVRI project by reusing the latest technologies. It investigated key issues in interoperable cataloguing and metadata harmonization with consideration of other ongoing initiatives.

6. Higher-level data products provided by RIs have to be clearly reproducible. Therefore, provenance services that record the evolution of data by tracking each operation processed have to be further developed and integrated within existing RIs. A **cross-RI data provenance** service provides tracing services for data manipulation between

---

[15] https://www.oasis-open.org/.

[16] http://inspire.ec.europa.eu/.

different infrastructures. Standardised interfaces for querying, accessing and integrating provenance data will be realised, building on current standardization efforts such as W3C-PROV[17] or natively in CERIF as used in EPOS.

4.3 Reference Model Guided System Design

The architectural patterns defined in the ENVRI-RM provides an abstraction for designers to design a data management service. The current ENVRI-RM provides the following information:

1. Science viewpoint: different roles involved in the service, and the interaction among those roles via the service;
2. Information viewpoint: the data evolution in the service including data schemas and data objects, and the actions that modify those data objects;
3. Computational viewpoint: the binding among components in the service, including key computational objects, and the artefacts transferred among those objects;
4. Technology viewpoints: the standards and technologies to be employed in the service;
5. Engineering viewpoint: the architecture of the service. Currently, microservice-based architectures are highly recommended by the RM.

Using the patterns provided by the ENVRI-RM, a developer can model the basic interface of the data management service and identify the key internal components. Figure 6 presents a typical design scenario for infrastructure optimization service.

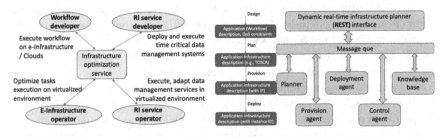

Fig. 6. Example of reference model guided design for infrastructure optimization service.

4.4 Agile Use Case Teams for Technology Investigation and Validation

The third step is **validation and service deployment**, deploying the implemented common operations within generic e-infrastructures (such as EGI or EUDAT), and operating them in the service of specific RIs. This approach aligns with ongoing work and trends in the provision of e-Infrastructure, especially grid-based (e.g. EGI), cloud-based (e.g. HELIX-Nebula) and data-centric projects (e.g. EUDAT), as well as the developments being proposed (and implemented) under the umbrella of Research Data Alliance

---

[17] http://www.w3.org/TR/prov-overview/.

(RDA)[18]. To enable the final usage of developed common services, the results will be tested and deployed in RIs, possibly via computing and data infrastructures such as EGI and EUDAT.

To engage the users of those data services in the loop in time, the requirements need to be formulated as "stories" and further elaborated as cases for the development teams. Based on the complexity of the cases, we identified three different levels: implementation cases, test cases, and science cases [20].

1. Implementation cases are relatively simple and can be finished in a relatively short time period. An implementation case often focuses on a specific feature of data management.
2. Test cases are those focusing on problem scenarios which require features from different services. Test cases are often bigger than implementation cases and need more time.
3. Science cases are often based on research problems which require data and services from different RIs. A science case can drive a number of test cases.

In a large project like ENVRIplus, more than 20 RIs participated in joint development activities. Cases were continuously collected and reviewed; the development teams of the specific data management services actively participated in the use cases, and established a use case project team, based on the Agile methodology, as explained in the next subsection (Fig. 7).

	Atmosphere			Biosphere					Solid earth	Marine							Muti domain				
	ACTRIS	IAGOS	EISCAT-3D	ANAEE	ELIXIR	INTERACT	LIFEWATCH	LTER	EPOS	EUROFLEET	EUROARGO	EUROGOOS	ESONET	FIXO3	JERICO	SEADATANET	EMBRC	EMSO	IS_ENES	ICOS	SIOS
SC-3							X														
TC-2											X							X		X	
TC-4								X			X							X		X	
TC-16														X	X	X	X				
IC-1	X	X	X					X												X	
IC-2								X													
IC-3			X																		
IC-8		X	X					X								X				X	
IC-10								X													
IC-11								X	X												
IC-12								X													
IC-13																				X	
IC-14				X				X			X					X					X

Fig. 7. The relation between use cases and specific domains. The complete list of use cases can be found in [20].

---

[18] https://rd-alliance.org/.

4.5 Coordinated Team Collaboration

In the ENVRIplus project, the development efforts were structured via different teams:

1. The developers for each common data service working for all RIs, rather than for one single RI, working on services for identification, processing, infrastructure optimization, curation, cataloguing and provenance.
2. The developers from each RI were identified. In many cases, these developers were distributed, due to the complexity of the infrastructure. These developers were responsible for developing and maintaining services in individual RIs.
3. Developers focusing on specific agile use case projects, which are created based on the dynamic needs of the RI communities.

Figure 8 depicts the interactions among the different teams.

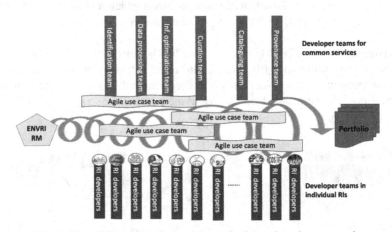

Fig. 8. How different developers interact in the engineering approach.

4.6 Portfolio Management

A service portfolio is a core repository that manages the evolution of the service and software assets that a company or organization delivers. It is an important strategy for the software industry to bridge the gaps among customer needs, development teams and the delivered software products (services). The portfolio is often broader than the service catalogue that an organization provides to the customer; it often contains the services to be developed, and inactive services after being replaced.

In the ENVRIplus project, the data for science theme adopts this strategy to manage the development plan of reusable solutions and use cases while interacting with the research infrastructures from different subdomains. We follow the practice from FITSM[19], based on the best practices from the e-Infrastructures EGI.

---

[19] https://www.fitsm.eu/.

In the ENVRIplus project, we organise the service portfolio in the data for science theme as four parts: 1) reference model related services and tools, 2) reusable solutions to common problems, 3) reusable solutions from use cases, and 4) testbeds. Figure 9 shows the basic idea.

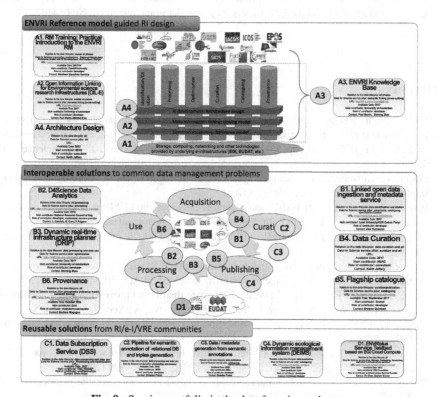

Fig. 9. Service portfolio in the data for science theme.

Figure 9 depicts the structure of the snapshot of the service portfolio in 2018.

5 Summary

Conducting system-level environmental science research requires advanced systems for collecting, curating and providing access to scientific data products. Various environmental research infrastructures (RIs) are being constructed to address this requirement; however, there is no coherent standard approach to constructing interoperable RIs that would permit the kind of interdisciplinary research needed to fully exploit the data now being made available.

In this chapter, we discussed the reference model guided approach adopted in the ENVRIplus project. This approach provided a uniform way of characterising existing RIs to permit the definition of required common and cross-cutting (interoperation) services.

However, building the reference model for ENVRIplus was labour-intensive and there is an ongoing discussion of the cost-benefit. In the rest of the book, we will discuss more details of how this approach is applied in the context of different development teams.

Acknowledgements. This work was supported by the European Union's Horizon 2020 research and innovation programme via the ENVRIplus project under grant agreement No 654182.

References

1. Zhao, Z., et al.: Reference model guided system design and implementation for interoperable environmental research infrastructures. In: 2015 IEEE 11th International Conference on e-Science, Munich, Germany, pp. 551–556. IEEE (2015). https://doi.org/10.1109/eScience.2015.41
2. Foster, I., Kesselman, C.: Scaling system-level science: scientific exploration and IT implications. Computer **39**, 31–39 (2006)
3. Petzold, A., et al.: ENVRI-FAIR - interoperable environmental FAIR data and services for society, innovation and research. In: 15th IEEE International Conference on e-Science, San Diego, US (2019). https://doi.org/10.1109/eScience.2019.00038. https://zenodo.org/record/3462816
4. European Strategic Forum of Research Infrastructure (ESFRI) Roadmap. http://ec.europa.eu/research/infrastructures/index_en.cfm?pg=esfri. Accessed 03 Apr 2019
5. Zhao, Z., Grosso, P., de Laat, C.: OEIRM: an open distributed processing based interoperability reference model for e-science. In: Park, J.J., Zomaya, A., Yeo, S.-S., Sahni, S. (eds.) NPC 2012. LNCS, vol. 7513, pp. 437–444. Springer, Heidelberg (2012). https://doi.org/10.1007/978-3-642-35606-3_52
6. Riedel, M., Laure, E., Soddemann, T., Field, L., et al.: Interoperation of world-wide production e-Science infrastructures. Concurr. Comput. Pract. Exp. **21**, 961–990 (2009)
7. White, L., et al.: Understanding interoperable systems: challenges for the maintenance of SOA applications. In: 2012 45th Hawaii International Conference on System Sciences, Maui, HI, USA, pp. 2199–2206. IEEE (2012)
8. ISO/IEC 7498-4:1989 – Information technology – Open Systems Interconnection – Basic Reference Model: Naming and addressing (1989)
9. Workflow reference model. http://www.wfmc.org/2-uncategorised/53-reference-model. Accessed 03 Apr 2019
10. Oussalah, M.C. (ed.): Software Architecture 1: Oussalah/Software Architecture 1. Wiley, Chichester (2014)
11. Carey, M.J.: SOA What? Computer **41**, 92–94 (2008)
12. Belqasmi, F., Singh, J., Bani Melhem, S.Y., Glitho, R.H.: SOAP-based vs. RESTful web services: a case study for multimedia conferencing. IEEE Internet Comput. **16**, 54–63 (2012)
13. Di Francesco, P., Lago, P., Malavolta, I.: Migrating towards microservice architectures: an industrial survey. In: 2018 IEEE International Conference on Software Architecture (ICSA), Seattle, WA, pp. 29–2909. IEEE (2018). https://doi.org/10.1109/ICSA.2018.00012
14. Linington, P.F. (ed.): Building Enterprise Systems with ODP: an Introduction to Open Distributed Processing. CRC Press, Boca Raton (2012)
15. Mathur, S., Malik, S.: Advancements in the V-Model. Int. J. Comput. Appl. (0975–8887), **1**(12), 29–34 (2010)

16. Choetkiertikul, M., Dam, H.K., Tran, T., Ghose, A., Grundy, J.: Predicting delivery capability in iterative software development. IEEE Trans. Softw. Eng. **44**, 551–573 (2018)
17. Boehm, B., Brown, W., Turner, R.: Spiral development of software-intensive systems of systems. In: Proceedings of the 27th International Conference on Software Engineering, ICSE 2005, St. Louis, MO, USA, pp. 706–707. IEEE (2005)
18. Highsmith, J.: Agile Project Management: Creating Innovative Products. Addison Wesley, Upper Saddle River (2010)
19. Jeffery, K., Houssos, N., Jörg, B., Asserson, A.: Research information management: the CERIF approach. Int. J. Metadata Semant. Ontol. **9**, 5 (2014)
20. Chen, Y., et al.: D 9.1. http://www.envriplus.eu/wp-content/uploads/2015/08/D9.1-Service-deployment-in-computing-and-internal-e-Infrastructures.pdf. Accessed 03 Apr 2019
21. Casale, G., et al.: Current and future challenges of software engineering for services and applications. CloudForward (2016). https://doi.org/10.1016/j.procs.2016.08.278
22. Koulouzis, S., et al.: Time-critical data management in clouds: challenges and a Dynamic Real-time Infrastructure Planner (DRIP) solution. Concurr. Comput. Pract. Exp. (2019). https://doi.org/10.1002/cpe.5269
23. Martin, P., Remy, L., Theodoridou, M., Jeffery, K., Zhao, Z.: Mapping heterogeneous research infrastructure metadata into a unified catalogue for use in a generic virtual research environment. Future Gener. Comput. Syst. **101**, 1–13 (2019). https://doi.org/10.1016/j.future.2019.05.076

Semantic and Knowledge Engineering Using ENVRI RM

Paul Martin[1] , Xiaofeng Liao[1] , Barbara Magagna[2] , Markus Stocker[3,4] ,
and Zhiming Zhao[1(✉)]

[1] Multiscale Networked Systems, University of Amsterdam,
1098XH Amsterdam, The Netherlands
pwmartin.research@gmail.com, {x.liao,z.zhao}@uva.nl
[2] Environment Agency Austria, Vienna, Austria
barbara.magagna@umweltbundesamt.at
[3] TIB Leibniz Information Centre for Science and Technology, Hannover, Germany
markus.stocker@tib.eu
[4] MARUM Center for Marine Environmental Sciences, PANGAEA Data
Publisher for Earth & Environmental Science, Leobener Strasse 8, 28359 Bremen, Germany

Abstract. The ENVRI Reference Model provides architects and engineers with
the means to describe the architecture and operational behaviour of environmental
and Earth science research infrastructures (RIs) in a standardised way using the
standard terminology. This terminology and the relationships between specific
classes of concept can be used as the basis for the machine-actionable specification
of RIs or RI subsystems.

Open Information Linking for Environmental RIs (OIL-E) is a framework for
capturing architectural and design knowledge about environmental and Earth science RIs intended to help harmonise vocabulary, promote collaboration and identify common standards and technologies across different research infrastructure
initiatives. At its heart is an ontology derived from the ENVRI Reference Model.
Using this ontology, RI descriptions can be published as linked data, allowing discovery, querying and comparison using established Semantic Web technologies.
It can also be used as an upper ontology by which to connect descriptions of RI
entities (whether they be datasets, equipment, processes, etc.) that use other, more
specific terminologies.

The ENVRI Knowledge Base uses OIL-E to capture information about environmental and Earth science RIs in the ENVRI community for query and comparison. The Knowledge Base can be used to identify the technologies and standards used for particular activities and services and as a basis for evaluating
research infrastructure subsystems and behaviours against certain criteria, such
as compliance with the FAIR data principles.

Keywords: Ontology · Knowledge base · Research infrastructure · Reference
model

© The Author(s) 2020
Z. Zhao and M. Hellström (Eds.): Towards Interoperable Research
Infrastructures for Environmental and Earth Sciences, LNCS 12003, pp. 100–119, 2020.
https://doi.org/10.1007/978-3-030-52829-4_6

1 Introduction

The ENVRI Reference Model[1] (ENVRI RM) provides a standard set of stereotypes for the different classes of actor, information object, behaviour, etc. found within environmental and Earth science research infrastructures (RIs) [1, 2]. These stereotypes were derived from the study of the RIs participating in the ENVRI community cluster for environmental RIs in Europe[2]. ENVRI RM places all of these stereotypes in the context of the research data lifecycle, identifying the critical elements needed to facilitate data acquisition, curation, publishing, processing and use by a community of researchers in the environmental and Earth sciences, though many stereotypes are applicable more broadly to research infrastructure in general. By referring to the model, RI architects can identify the elements that are most important to them, determine any gaps within their own (planned) infrastructure, and compare against other RI specifications—in particular allowing them to look at how other RIs solved the same problems, and what technologies and standards they used to do so. Given the instantiation of ENVRI RM for a particular RI however, there is still the question of how the resulting information can be published in a way that is useful to as broad an audience as possible. For example, in addition to published documentation describing the modelling of a particular RI (with in-depth textual explanations and diagrams of major subsystems and their organisation and construction), it would also be convenient to be able to translate those documents into a form that can be programmatically queried and compared against other RI models in a systematic way.

This chapter describes how ENVRI RM was used as a basis to create Open Information Linking for Environmental Research Infrastructures (OIL-E) [3], a multi-view machine-readable ontology for describing RIs based on ENVRI RM that can act as an upper ontology for describing different entities and activities attributable to environmental and Earth science RIs. OIL-E was intended to:

1. Capture the terminology of ENVRI RM as a controlled resource for use in the annotation of RI documentation and other semantic enrichment activities.
2. Permit the translation of specific RI models produced using ENVRI RM into machine-readable RDF data that can be stored in a suitable knowledge base.
3. Assist in the association of other semantic descriptions for data, services and other RI elements with one another by acting as a 'connective ontology' for environmental and Earth science RI entity specifications.

As part of the second objective, in particular, an ENVRI Knowledge Base[3] has been under development to serve as an online information corpus about the ENVRI cluster of environmental science RIs. The ENVRI Knowledge Base gathers information collected about RI design and RI resources, structured according to the OIL-E ontology (and ENVRI RM) and provides access based on established Semantic Web technologies. It thus serves as a practical demonstrator of the kind of semantic search and query that OIL-E can facilitate.

---

[1] http://envri.eu/rm/.

[2] https://envri.eu/.

[3] http://kb.oil-e.net/.

In Sect. 2, we examine more closely the background and motivation behind using ENVRI RM to develop semantic and knowledge resources for the environmental and Earth science RI community. We describe the methodology applied in developing the OIL-E ontology (Sect. 3), and how we applied it to the modelling of RIs in the ENVRI cluster (Sect. 4). We then move on to discuss the ENVRI Knowledge Base (Sect. 5). Finally, we discuss where further development is needed or desired (Sect. 6) before drawing our conclusions (Sect. 7).

2 Background and Motivation

Environmental research increasingly depends on the collection and analysis of large volumes of data gathered from various sources including field observations, sensor networks, laboratory experiments and simulations based on expert models. Societal challenges facing the world today like climate change, food security and disaster prediction/response can only be addressed by making optimal use of such data, which also requires scientists to collaborate across disciplinary boundaries, as these challenges are intrinsically transdisciplinary in nature. Environmental and Earth science RIs support researchers in their interactions with a host of different data sources and analytical tools by providing access to combined corpora of curated research datasets via unified services and data portals, but no one RI fully encompasses the full research ecosystem [4], each typically serving a specific environmental domain or catering for a specific class of data. The challenge, therefore, is to functionally integrate existing environmental RIs to permit researchers to freely and effectively interact with the full range of research assets potentially available to them, allowing them to collaborate and conduct innovative interdisciplinary research regardless of the particular research community to which they belong. Realising this ideal requires a broad understanding of the fundamental commonalities of environmental science research infrastructure services, however: in terms of concepts, in terms of processes, in terms of data and services, and in terms of technology adoption. The process of achieving this understanding can be expedited by the use of a standard reference model (e.g. ENVRI RM), which can be used to construct formal descriptions of RIs and their major component elements.

ENVRI RM was constructed using the Reference Model for Open Distributed Processing (ODP) [5] for modelling complex distributed systems. ODP requires the modelling of a system from five different viewpoints (enterprise, information, computation, engineering and technology), with the correspondences between the five resulting views ensuring their mutual validity. This viewpoint-based approach provides clarity to each 'facet' of the end model by reducing the number of competing elements to only those that match a particular set of concerns (such as the flow of information through the system), while still retaining the aggregate complexity needed to model any substantive distributed system. ENVRI RM provides the five views prescribed by ODP (renaming the enterprise view as the *science* view in light of its subject) specialised for the common elements of environmental and Earth science RIs, as revealed in the study of participating RIs in Europe:

- The **science** viewpoint, which considers the main behaviours facilitated by a RI and the communities and resources involved in those behaviours.

- The **information** viewpoint, which identifies the information objects handled by a RI and their various states throughout the operation of the RI.
- The **computational** viewpoint, which identifies the logical computational elements that interact to support various RI operations.
- The **engineering** viewpoint, which describes how computational elements are distributed in an infrastructure, and the communication channels between infrastructural nodes.
- The **technology** viewpoint, which identifies the software, hardware and standards used to implement data and computational entities in a RI.

ENVRI RM uses three of the five views prescribed by ODP to capture the generic aspects common across all RIs (those being the science, information and computational views), and then uses the engineering and technology viewpoints to explore the more specific solutions and design patterns observed as being used by current RIs for the generic components prescribed in the three former views. Each view has its own concerns, and parts of those concerns may correspond to concerns in other views (for example information in one view may be used by computational elements in another); each view is thus able to describe particular key RI activities. For example, in Fig. 1, we show the components prescribed for raw data collection in the computational view as a UML component diagram. A *data transfer service* provides a *raw data collector* which brokers the streaming of data from an instrument (represented by an *instrument controller* computational object) to a data store (represented by a *data store controller*), with a *persistent identifier service* invoked to acquire an identifier for the resulting dataset and that dataset's existence registered with a *catalogue service*. For any given RI, these components are expected to be present in some form; perhaps the data collector is not a distinct component from the data transfer service, and perhaps the PID service is only invoked for certain types of data, but most actual cases of raw data collection should be describable in terms of this interaction template.

The use of methodologies such as ODP [6] helps guide the software engineering process by recognising the existence of different kinds of stakeholder in system development with different primary concerns and providing a multi-faceted modelling context that addresses each while maintaining an overall coherent specification. This benefits all parties by providing distinct specifications of each facet of the system that is sufficiently revealing the key characteristics of the system from one perspective. Simultaneously, these specifications can ignore details that are less relevant to that perspective, as long as those details are made evident in at least one of the other views so that they are not neglected by the combined specification. A similar benefit can be obtained in the design of ontologies and other formal models, where simple decompositions of systems with a particular perspective in mind often produce the most useful and easy to apply models. Conversely, trying to do 'too much' within the framework of a single ontology can make it more difficult to use and more likely to contain errors or controversies. We can instead create a linked set of interconnected ontologies (or partition a larger ontology into parallel-connected sub-ontologies with independent hierarchies); each sub-ontology then represents a different viewpoint, but with links to corresponding concepts in the other ontologies. This allows for more complex systems to be modelled while retaining the clarity of a simple yet serviceable ontology for each viewpoint. Such an approach also provides the option to focus on a case-by-case basis on modelling those specific views deemed most useful, ignoring the other viewpoints not applicable to modellers' immediate concerns.

OIL-E is intended to provide such a multi-view framework for the modelling of environmental and Earth science RIs. The OIL-E ontology captures all stereotypes defined by ENVRI RM along with their essential relations and distributes them across the ENVRI RM viewpoints while also adding more cross-view relations to better facilitate classification and validation of RI models described using the ontology. For example, OIL-E allows for technologies (including both software and standards) used within a RI to be linked directly to the information objects and computational services that implement or use them. Using the stereotypes of ENVRI RM to produce a high-level, 'connective' ontology for RI specifications, OIL-E can provide a means for describing and maintaining constellations of loosely-coupled views on the same RI system, where the correspondences between concepts in different views might be difficult to express with complete precision, making the conception of a single canonical representation that integrates the full scope of all views difficult or intractable.

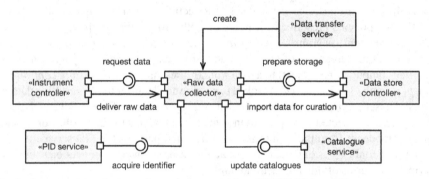

Fig. 1. A computational view of raw data acquisition: ENVRI RM specifies components and activities using UML (in this case, a component diagram).

3   Methodology

The ENVRI semantic linking framework was developed based on ENVRI RM. Open Information Linking for Environmental RIs (OIL-E) was designed to provide an upper ontology for RI descriptions based on ENVRI RM that can be used to contextualise different kinds of RI assets from an architectural or operational perspective. This is in contrast to being a general-purpose ontology for describing scientific phenomena like ENVO [7] or BFO [8]; OIL-E has more in common with conceptual models such as CERIF [9] that focus on the products and tools of research rather than on scientific classification itself, albeit more concerned with providing a controlled vocabulary for environmental science RIs in particular.

The multi-viewpoint approach intrinsic to ENVRI RM and inherited from ODP informs the design of OIL-E in many ways. Most notably, each viewpoint essentially provides its own micro-ontology, with instances of the concepts defined that can then be related to concepts in other views via correspondences. Correspondences, as defined by ODP, describe relationships between entities existing in different views, and are used to anchor the different views with one another to ensure a coherent description of the same system. This allows OIL-E to operate as a 'hub' ontology, whereby specifications created in one view (e.g. information) can be used to dictate requirements on another view (e.g. computation). For example, given the specification of an information action to produce newly processed data from a persistent dataset, there must be an accompanying computational operation to carry out that action. Likewise, given a behaviour by which a researcher processes such data, there must be a computational service on which that operation can be invoked. It is also possible to extend each view using other, more specific ontologies (e.g. for describing datasets in the information view), which then inherit the relationships with concepts in the other views.

As a Semantic Web [10] ontology, OIL-E is written in OWL 2.0 [11] and published online[4], with the ontology itself split into two parts. The full ontology:

- Captures notions of research infrastructure from multiple perspectives: social infrastructure, physical research infrastructure (i.e. sites, observatories and devices) and computational infrastructure being the most evident.
- Clearly separates these different views on infrastructure, and then establishes their correspondences.
- Captures the most significant interactions between different actors and resources, and the information that is produced by such interactions.
- Helps establish the relationships between other existing standards and vocabularies in terms of the facets of infrastructure, infrastructure assets and infrastructure activity to which they apply.

The foundation of OIL-E is the **oil-base** base ontology, which provides a set of abstract concept classes derived from the most common elements observed in the ENVRI RM and distributed across the five standard ODP views. The purpose of **oil-base** is to capture the generic concepts not specific to environmental science RIs, and to act as a simple upper ontology for all further OIL-E extensions. Despite its application to research infrastructure, **oil-base** is not a general-purpose upper ontology for describing scientific phenomena, but rather is a means to gather architectural and procedural concepts used in a complex system, distribute them across the most appropriate views, and then model the correspondences across those views.

Figure 2 illustrates the core concept hierarchy and its subdivision into the top-level concepts for each ODP view. These concepts generally refer either to *objects* of discourse, *activities* involving such objects, or *attributes* of objects and activities. This simple categorisation is used as the basis for defining exclusivity and restrictions on object properties, as well as allowing certain concepts to exist in multiple views and many generic properties to be defined for use in multiple views (or across views). The separation of specific concepts to specific views is then done via inference using *classifier* concepts

---

[4] http://www.oil-e.net/.

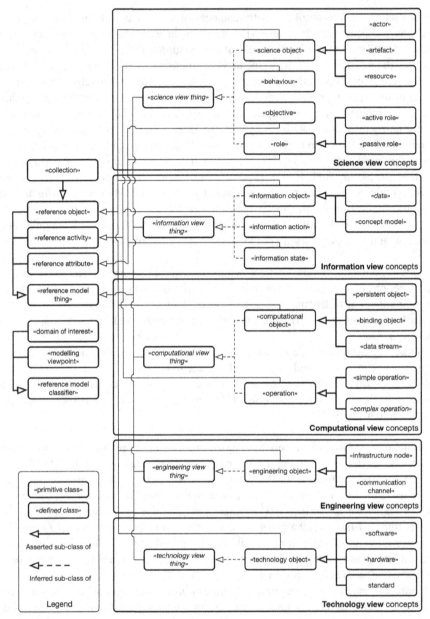

Fig. 2. The top-level concept hierarchy of each viewpoint in **oil-base**. Some sub-concepts have been omitted for brevity.

for which there are default definitions for each of the five ODP viewpoints. This has been done to make it easier to specify alternative viewpoints (e.g. a virtualisation viewpoint or a privacy viewpoint) should the original five ODP viewpoints be deemed insufficient to future modelling needs without requiring a substantial restructuring of the ontology. This approach also minimises the number of concept classes that are derived from multiple parent classes, in line with standard ontology design best practices.

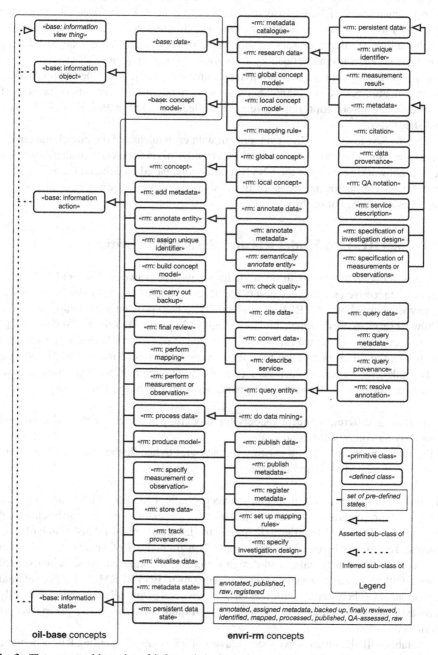

Fig. 3. The concept hierarchy of information viewpoint concepts defined by **envri-rm**. Some defined classes (e.g. for persistent data objects with specific states) have been omitted for brevity.

Defined from the ENVRI RM specification, the **envri-rm** ontology is the primary extension of **oil-base**. This ontology takes the sets of archetypes in each view defined by ENVRI RM, being the classes of object and process considered common across environmental science RIs, and extends OIL-E with concept classes (over 250 at the time of writing) for all of them, allowing better contextualised classification of RI entities and the ability to infer necessary relationships between them. As an example, Fig. 3 shows the set of concepts defined by ENVRI RM for the *information viewpoint*; the concepts are mainly split across sub-classes of information object (e.g. persistent dataset) and information action (e.g. annotate data), with some added information states used to differentiate information objects (e.g. annotated, assigned metadata, backed up or published).

ENVRI RM is an on-going development; with each release of the model, the **envri-rm** ontology must be updated accordingly. Currently, this is done via consultation within the relevant working group in the ENVRI community, based on demand for new stereotypes for RI entities or activities, or discussion regarding the correctness of specific properties or other relationships.

4   Using OIL-E to Model RIs and Research Activities

OIL-E is intended to assist with semantic harmonisation between different RIs by providing a connective ontology for describing RI components and activities based on the archetypes defined by ENVRI RM—essentially to help provide a landscape overview of how RI services are designed and implemented. In particular, OIL-E was designed to be a framework by which we can study how different metadata schemes and controlled vocabularies are used in practice to describe various entities of interest to RIs. Such a study, in the correct context, can be used to expedite alignment and transformation of formal specifications in the service of greater RI interoperability. This entails:

- Comparing different concept models for modelling research assets and data, and identifying commonalities and gaps.
- Building generic tools using existing technologies to handle the search and mapping of models related to RI architecture and specification.

The linking component of OIL-E glues concepts both inside ENVRI RM and between ENVRI RM and external vocabularies. In the latter case, external models can be classified in terms of ENVRI RM in order to help map the landscape of RI-related standards and models. The **envri-rm** ontology only contains a limited set of vocabularies derived from common RI functionality and design patterns, so linking **envri-rm** with external models will also enable domain-specific extensions to ENVRI RM itself. The internal correspondences between the different OIL-E views can potentially be used to indirectly draw associations between concept models with quite different foci (e.g. data versus services).

Notably, OIL-E conflates two major classes of information regarding RIs: schematic information, about the general 'kinds' of the element found in a given RI; and instance information, about actual services, datasets, technologies currently found in a RI. Take for example the Integrated Carbon Observation System (ICOS)[5]. Modelling this RI, we can assert that "ICOS Level 1 data" concerns a general class of dataset found in the ICOS Carbon Portal, the properties of which apply to all instances of such datasets, while there may also be individual examples of Level 1 data product in ICOS that we also model. A description of the former is schematic information, while a description of the latter is instance information. In practice, most OIL-E data so far produced is a mix of schematic information and instance data about invariant parts of RIs. For example, the "ICOS Carbon Portal" is a specific component of the ICOS RI rather than a class of component and thus is instance data, as is the metadata standard "ISO 19139" used for metadata produced by many RIs (though there may also be a class of "ISO 19139 compliant metadata records"). Whether schematic or instance information, the combination of this data provides a description for a RI that can be used to classify not only persistent RI entities such as datasets and services, but also transient events, which (for example) allows such extensions of OIL-E to be used to classify or validate provenance traces.

Information specific to individual RIs is created by providing specific instances of RM archetypes implemented by the RI as well as extending **envri-rm** with concepts particular to the RI, for example as shown in Fig. 4. In this case (which for brevity has been simplified from reality to serve as an exemplar), we extend **envri-rm** for the AnaEE[6] RI (for Analysis and Experimentation in Ecosystems) and show a few of the concepts for AnaEE-specific processes involving the AnaEE metadata catalogue across three views (science, information and computational views). We also show a couple of the specific entities that must be instantiated to support these processes—for example, the AnaEE discovery catalogue service that must invoke all updates to the metadata catalogue in the RI.

RI-specific concepts may apply to any of the views defined by ENVRI RM, with OIL-E providing the vocabulary necessary to relate concepts within and between views. The technology viewpoint of OIL-E, in particular, allows for the identification of specific technologies (i.e. software, hardware and standards) to be linked to particular types and instances of RI datasets and services, which can then be mapped out as knowledge graphs to show how technologies are used and how they relate to various RIs and RI systems, for example as shown in Fig. 5. We can also identify the context in which such technologies are used (e.g. for what kind of dataset or to implement what service) and provide information about where such technologies can be acquired.

It is also possible to extend OIL-E for a specific kind of process rather than a specific RI. Creating a taxonomic model for data quality control (QC) processes is an example of an extension to the base OIL-E ontology that elaborates upon a specific part of RI design. OIL-E defines a class of RI behaviour, *'quality checking behaviour'*, which can be used to classify QC behaviours performed by RIs; an extension for QC processes can enhance that concept by providing a greater range of relations between QC behaviours and other entities representing terms of a taxonomy for QC processes.

---

[5] https://www.icos-ri.eu/.

[6] https://www.anaee.com/.

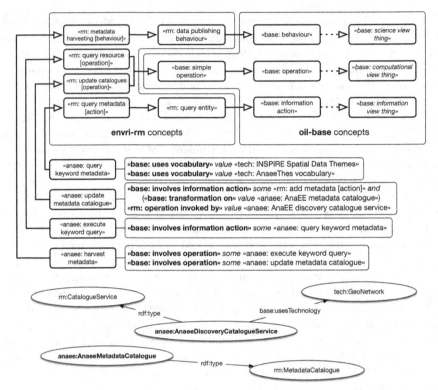

Fig. 4. Extending OIL-E to model components and activities of AnaEE: a (simplified) example of activities involved in the harvesting of metadata for the AnaEE metadata catalogue.

In Fig. 6, a QC process is defined for the EISCAT_3D[7] RI. EISCAT_3D is concerned with using radar observations and the incoherent scatter technique for studies of the atmosphere and near-Earth space above the Arctic. Once fully operational, it will provide a considerable volume of data, in real-time, from its sensor arrays deployed in Norway, Sweden and Finland. These data need to be checked for possible errors or anomalies. As for many research data streams, there need to be multiple phases of quality control to ensure the quality of data reaching researchers. The first of these is described in RDF in Fig. 6. It is performed shortly upon acquisition of new data, is a semi-automated process conducted in real-time by a human technician, performed within the RI itself (since EISCAT_3D acquires the data directly, rather than via intermediaries) and involves a set of activities: statistical checks, corrective measures, technical checks and data enhancements.

There are many other processes defined by ENVRI RM for different parts of the research data lifecycle such as data acquisition or publication. For every such process, OIL-E provides a base stereotype, often with additional requirements for e.g. the actors involved in the process, which can be easily extended with additional controlled vocabulary and sub-concepts using standard Semantic Web technology and techniques.

---

[7] https://www.eiscat.se/eiscat3d/.

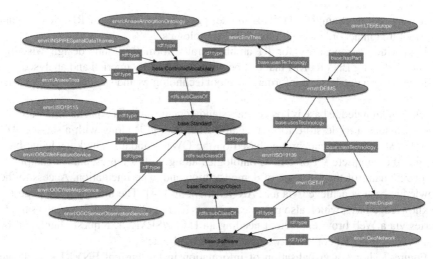

Fig. 5. Linking technologies and standards: the use of different technologies by different RIs can be explored via the knowledge graph generated using RI data in OIL-E.

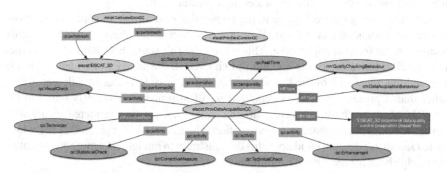

Fig. 6. Modelling quality control processes in OIL-E: example of provisional real-time data quality control on newly-acquired data in EISCAT_3D.

5   The ENVRI Knowledge Base

A key outcome in ENVRIplus that naturally resulted from the creation of OIL-E is the creation of a knowledge base to collect together information about the RIs in the ENVRI community and their activities [12]. The need for such a knowledge base was motivated by the need to better map the semantic landscape of environmental science RIs in Europe, and in particular to gather information about the different metadata schemes, ontologies, thesauri and other controlled vocabularies used by RIs specifically in terms of their application in RI subsystems (as opposed to simply providing another ontology portal).

The ENVRI Knowledge Base in its first iteration as a product of the ENVRIplus project serves three basic purposes:

1. It provides an example of OIL-E in use, providing examples of RI-oriented data structured in accordance with the OIL-E ontologies.
2. It provides a repository for RI architectural information and 'design wisdom' encoded using ENVRI RM that can be programmatically queried and analysed.
3. It serves as a database of information about technologies and standards used by RIs.

The current knowledge base is hosted via a standalone instance of Apache Jena Fuseki[8], which provides a triple store for aggregated RDF data [13] along with a service API and internal reasoning capabilities based on the OWL standard. The knowledge base contains the complete set of OIL-E ontologies along with a representative sample of RI-specific data for the purposes of demonstration and experimentation. Access to the knowledge base is achieved via a SPARQL endpoint [14]. The main landing page for the knowledge base, which also provides a means to try out and modify various sample queries via a Web browser without needing an HTTP/SPARQL request client, can be found via the ENVRI community site at: http://kb.oil-e.net[9].

Figure 7 shows a visualisation of information in the current ENVRI knowledge base as can be viewed by visiting the above landing page. Nodes are colour coded to distinguish concept classes from instance data and data properties, with additional information accessible by directly selecting individual nodes.

When resolving queries sent to it, the knowledge base is able to apply the relations and classifications defined by OIL-E in order to infer results beyond those explicitly asserted in the internal triple store. This allows the full set of ENVRI RM archetypes to be used to guide the discovery and search over all the RI data provided. It should be noted that the scope of the knowledge base as of writing is that of a demonstrator, rather than a production-level system, and so all information about RIs found in the ENVRI knowledge base is provisional, and should not yet be considered an accurate representation of the infrastructures in question. The ENVRI-FAIR project[10] [26], which started January 2019, will build upon this demonstrator to implement a more authoritative knowledge base for the ENVRI community.

For the ENVRI Knowledge Base, we identified four key knowledge capabilities that we wanted to support:

A Survey of the Technical Landscape. The web of knowledge created by semantic linking should help us understand what technologies (including software, standards and vocabularies) are being used by environmental science RIs.

Comparative Solution Analysis. It should be possible to compare solutions developed by environmental science RIs—specifically, given the knowledge of how technologies are used in their proper context, we should be able to compare developments in equivalent contexts.

Gap Analysis and Component Recommendation. Given a reference model for environmental science RIs (i.e. ENVRI RM), it should be possible to identify what is missing in the current development state of a given RI, and based on both that model

---

[8] https://jena.apache.org/documentation/fuseki2/.

[9] In the ENVRI-FAIR project, it is planned to be deployed using the ENVRI community domain.

[10] https://www.envri.eu/envri-fair/.

and the solutions developed by other RIs, it should be possible to then make certain recommendations.

Linked Open Research Infrastructure. The web of knowledge created by semantic linking should itself be publicly accessible, machine-navigable, and provide a gateway to the services and data held by the RIs. It should include (where available) data provenance and resource catalogues, and it should (where appropriate) make use of other ENVRI services such as the catalogue service for cross-RI search.

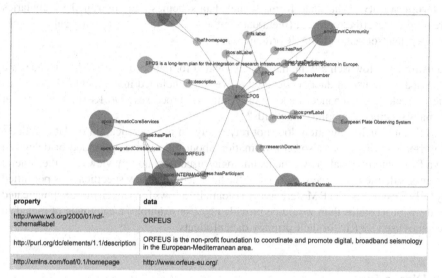

property	data
http://www.w3.org/2000/01/rdf-schema#label	ORFEUS
http://purl.org/dc/elements/1.1/description	ORFEUS is the non-profit foundation to coordinate and promote digital, broadband seismology in the European-Mediterranean area.
http://xmlns.com/foaf/0.1/homepage	http://www.orfeus-eu.org/

Fig. 7. Visualising information in the ENVRI Knowledge Base; showing entities associated with the EPOS research infrastructure.

Given a sufficiently-detailed corpus of information regarding environmental and Earth science RIs backed by a 'standard model' for how such RIs are constructed (i.e. ENVRI RM), it is possible to evaluate individual RIs or RI subsystems in terms of how they compare with similar RIs or against some kind of base criteria. For example, it is possible to compare data quality control processes applied by RIs in the same domain (e.g. marine science) against one another, or against a specific prescribed methodology (e.g. the quality control methodology used for oceanographic data collected by GODAE [15]). Another possibility is to provide tools for RI designers to evaluate their own RIs in terms of compliance with the FAIR data principles [16]. Wilkinson et al. define a number of guidelines for providing Findable, Accessible, Interoperable and Reproducible data that would better support open data science. There are a number of possible approaches that could be taken to evaluate some of these guidelines based on the content of an OIL-E-structured knowledge base:

Findability. Which published data products include globally unique persistent identifiers? Are those identifiers included in the product metadata? What other core metadata does each published product include (or not include)? Does the RI provide an index or registry for search and discovery of data products? Does it contribute to any external registries?

Accessibility. Can data product metadata be retrieved by a standard, open and free communication protocol, and if so, which one? Does the RI define an authentication and authorisation process for accessing data, and does it use standard, open mechanisms? Are metadata accessible via some means even if the data product described is no longer available?

Interoperability. What data formats, metadata schemes and controlled vocabularies are used to describe/represent (meta)data in the RI? Do those terminological resources comply themselves to the FAIR principles?

Reusability. How rich are the metadata provided for data products? What licences are assigned to the use of data? Is detailed provenance included in the metadata, and does the RI include provenance tracking in its internal processes? Do RI (meta)data meet domain-specific community standards?

Notably, such evaluation does not rely solely on the specification of data products (information view), but also on information about the services provided or delegated by a RI (computational view), the technologies used (technology view) and the general processes defined (science view). Thus, the holistic multi-view specifications permitted by OIL-E using ENVRI RM stereotypes potentially allows for a much more sophisticated analysis of RI status than would be provided by (for example) a catalogue of metadata schemes used by RIs for their primary data products.

6 Discussion

The knowledge base and OIL-E are both the basis for more tools with which to support several useful functions. We can envisage a number of avenues of further development (or in most cases, alignment with existing developments for mutual benefit). These include:

Cross-RI Search and Discovery. OIL-E provides a standard taxonomy for various entities and activities related to RIs, which can be used to classify different kinds of resources as part of a faceted search pipeline. An OIL-E knowledge base can hypothetically act directly as a catalogue service for multiple RIs, but this is not necessarily the best possible approach, as OIL-E is optimised for describing RI design and contextualising RIs' component parts, rather than providing a more traditional metadata scheme for describing RI resources. A knowledge base can be the basis however for a discovery service for heterogeneous research assets (including other catalogues) based on its internal network of relationships based on ENVRI RM, which could conceivably be used to direct queries dispatched to a common search portal to the correct RI resources.

Faster RI Specification Using ENVRI RM. Detailed descriptions of RIs in terms of their architecture, core data products and processes allow for more in-depth investigations and comparisons of RI solutions to various technical problems. ENVRI RM provides the basis for such descriptions, but requires specialist expertise to use effectively, and has previously been used manually, resulting in the creation of a body of documentation for each RI modelled. OIL-E captures all the key concepts defined by ENVRI RM, and thus a tool based on OIL-E that allows RI architects to more easily specify their RIs using ENVRI RM templates would accelerate the creation of RI data; this data can then be directly inserted into the ENVRI Knowledge Base and used in comparative analyses. The application of standards such as the Shapes Constraint Language (SHACL) [17], to validate data entry into such templates in a way that complements the basic classification capabilities of the OIL-E OWL ontology, would be particularly helpful.

Requirements Recommendation. Using tools such as OIL-E and the ENVRI Knowledge Base, it is possible to do a comparative analysis of the solutions provided by RIs in terms of technology and processes to address various common problems regarding the handling of research data (and other things). This requires a certain degree of constructive analysis of a number of queries. Tools which can interact with the knowledge base on behalf of a user, constructing and interpreting queries behind a more friendly interface, could be very useful for taking full advantage of the corpus of knowledge built up from RI modelling [12].

Provenance Exploration. There are two notable ways in which OIL-E data can interact with provenance data, especially data encoded to the W3C PROV standard [18]. The first is as linking data to various provenance repositories, contextualising the role of the repositories and providing a reference to where the provenance is and how it can be extracted. The second is as a validation framework; given descriptions of RI processes encoded in OIL-E, provenance traces can be checked against those descriptions by mapping agents, entities and activities to the correct OIL-E concepts and then checking whether the relationships described in the provenance trace match those of prescribed by the process model.

Natural Language-Based Document Analysis and Annotation. A significant corpus of existing information about RIs exists in the form of written documentation produced by RI architects and developers. The ability to apply a framework such as OIL-E to annotate uploaded documents, identifying possible references to concepts defined in ENVRI RM in the text, for example, would be useful both to contextualise documents automatically and provide initial descriptions for the RIs and RI components described by the documents. Such descriptions can be verified and extended by human experts, and also used as training data for producing better annotations in future, or perhaps even to identify possible extensions (e.g. new concepts or alternative synonyms for existing concepts) to ENVRI RM. Machine learning tools would thus provide a valuable additional source of data for the knowledge base, or to validate existing models of RIs [27].

The Semantic Web relies on a number of foundational technologies for representing and associating semantics to information, from RDF to OWL and SKOS [19], along with standards for interacting with semantic information (e.g. for search and discovery) such as SPARQL. Considerable attention has been given to the openness, extensibility and computability of such standards, with different options for controlled vocabulary specification depending on the circumstances (e.g. the choice of SKOS over OWL for many vocabulary specification cases [20, 21]). While RI designs could be specified using something other than Semantic Web technologies (for example based on traditional relational database models), the openness and extensibility of the Semantic Web fit well with the heterogeneity of RI designs and the varying levels of detail in which specific aspects of RI design may or may not be modelled. It should also be noted that RI models are not themselves particularly large in terms of data volume, being constructed of relatively 'high-level' propositions that nonetheless need to be very carefully structured; this also fits the Semantic Web knowledge graph meta-model.

OIL-E's use of RM-ODP (via ENVRI RM) is not wholly new; RM-ODP has been expressed in ontology form as early as 2001 [22]. Applications of ODP have been studied extensively [23], and ODP has been applied to the design of various kinds of infrastructure, including in the Internet of Things (IoT) and Smart Cities contexts [24]. The applicability of ODP, a standard that was developed in the 1990s, to modern concepts of service-oriented architecture and Cloud have been discussed before in research literature [25]. Certainly, the advancement and wide-scale adoption of virtualisation and programmable infrastructure mean that the separation of concerns between the computational and engineering viewpoints (for example) are less clear than they perhaps were originally; modelling a system deployed on a virtual infrastructure and modelling the virtual infrastructure service itself, for instance, would each result in a very different assignment of concepts between the two views. On the other hand, ODP supports the notion of transparencies, the selection of aspects of system design (such as authentication and migration of components) to not be explicitly modelled in specifications so as to reduce confusion, clutter or repetition in design documents. In this light, the explicit acknowledgement that the resources and channels described in the engineering view of a RI specification happen to be virtualised becomes simply another transparency option. Certainly, regardless of whether ODP can be considered to be a sufficiently contemporary specification for the modelling of modern distributed systems, the notion of specifying systems across multiple views is still well-regarded in software engineering research literature.

In many cases, it has been found that most queries on the state of the modelled RI systems focus on a single view, but defining correspondences between views can still be very useful for validating consistency between views of the same RI. Conceivably, different views on the same RI might be maintained by different authorities; the OIL-E multi-view ontology helps keep the different views consistent by identifying expected links between concepts in different views, which RI architects can then evaluate and try to align, either in the description of the RI or, where deficiencies in the subject are identified, in the design of the RI itself. One additional possible extension to OIL-E now being investigated is the integration of SHACL functions into OIL-E. SHACL is a constraint language used to validate RDF graphs and is a refinement of prior de facto standards such as SPIN and SWRL. Unlike OWL it performs closed world validation rather than open-world classification, and also makes the unique name assumption that OWL explicitly does not. SHACL can be used to embed SPARQL queries into RDF graphs as part of rules or functions that can be applied on the content of the graph, providing a means for RI service developers to publish instructions for building (for example) parameterised HTTP requests to their services that other actors can retrieve from the knowledge fabric. Such an approach allows interaction logic to be defined (and updated) in one place (e.g. the knowledge base or a successor system that may be distributed over several nodes perhaps directly curated by RIs). It also admits the possibility that other information in the linked knowledge graph can be used in a dynamic fashion to introduce some additional interstitial intelligence into the logic.

7 Conclusion

In this chapter, we described how the ENVRI Reference Model (ENVRI RM) was used as the basis for a formal ontology for describing research infrastructure, RI subsystems and RI processes. This ontology, called Open Information Linking for Environmental Research Infrastructures (OIL-E), preserves the multi-view approach of ENVRI RM to provide a flexible framework for RI modelling that can be tailored to the particular interests of different stakeholders in RI design and development; for example to focus on the behaviours of the main actors in a RI, the computational services provided by the RI, or the main datasets curated by the RI. We described the main design principles of OIL-E and how we captured the extensive lexicon of ENVRI RM in a logical concept hierarchy in a way that could be easily applied or extended for specific RIs or particular kinds of activity.

The use of ontologies to capture the vocabulary and relations between entities is a useful means to model information artefacts used in research infrastructure and other knowledge-based systems, but the balance of expressivity and computability poses a continuing challenge. Ontologies are a single tool alongside other forms of schema and rubrics for the capturing of design knowledge and architectural details of the infrastructure. To better explore how best to capture RI design information in a way that is machine-readable, programmatically queryable, and above all *useful*, we applied OIL-E as the underlying structure for a knowledge base of European environmental and Earth science RIs participating in the ENVRI community. This knowledge base was created to demonstrate how an information corpus for RIs might be used to analyse and compare RI designs, as well as to document the technologies, software and standards used by RIs in their appropriate operational contexts. We reviewed the current state of the knowledge base, and discussed a number of ways in which it can be improved to permit the handling of a greater range of queries of possible interest to RI designers.

The next major objective of the ENVRI community is to facilitate the adoption of the FAIR data principles for research data gathered in the atmospheric, marine, solid earth and biodiversity domains, and to develop sustainable FAIR data services for research communities as part of the broader push towards better open data science and more seamless interoperability between different data providers. Further development of the ENVRI Knowledge Base would greatly support this effort by providing a clear means for RI developers to evaluate their RIs' progress towards greater 'FAIRness' and to explore the technology choices made by their fellow developers in other scientific domains. Such development has been committed to as part of the next phase of ENVRI, the ENVRI-FAIR project, which began work in January 2019. The main priorities in knowledge base development will thus be to take the lessons and data prototypes developed in the previous ENVRIplus project in order to create a more extensive, complete, 'production-level' knowledge resource.

Acknowledgements. This work was supported by the European Union's Horizon 2020 research and innovation programme via the ENVRIplus project under grant agreement No 654182.

References

1. Zhao, Z., et al.: Reference model guided system design and implementation for interoperable environmental research infrastructures. In: 2015 IEEE 11th International Conference on e-Science, pp. 551–556. IEEE, Munich (2015). https://doi.org/10.1109/eScience.2015.41
2. Nieva de la Hidalga, A., et al.: The ENVRI Reference Model (ENVRI RM) version 2.2, November 2017. https://doi.org/10.5281/zenodo.1050349
3. Martin, P., et al.: Open information linking for environmental research infrastructures. In: 2015 IEEE 11th International Conference on e-Science (e-Science), pp. 513–520. IEEE (2015). http://dx.doi.org/10.1109/eScience.2015.66
4. Martin, P., Chen, Y., Hardisty, A., Jeffery, K., Zhao, Z.: Computational challenges in global environmental research infrastructures (Chap. 12). In: Chabbi, A., Loescher, H.W. (eds.) Terrestrial Ecosystem Research Infrastructures: Challenges and Opportunities, pp. 305–340. CRC Press, Boca Raton (2017). https://zenodo.org/record/3361569
5. ISO 10746-1: Information technology—Open Distributed Processing—Reference Model: Overview. ISO/IEC standard. International Organization for Standardization (1998)
6. Linington, P.F., Milosevic, Z., Tanaka, A., Vallecillo, A.: Building Enterprise Systems with ODP: An Introduction to Open Distributed Processing. CRC Press, Boca Raton (2011)
7. Buttigieg, P.L., Morrison, N., Smith, B., Mungall, C.J., Lewis, S.E.: The environment ontology: contextualising biological and biomedical entities. J. Biomed. Semant. 4(1), 43 (2013)
8. Arp, R., Smith, B., Spear, A.D.: Building Ontologies with Basic Formal Ontology. The MIT Press, Cambridge (2015)
9. Jörg, B.: CERIF: the common European research information format model. Data Sci. J. **9**, 24–31 (2010)
10. Berners-Lee, T., Hendler, J., Lassila, O., et al.: The semantic web. Sci. Am. **284**(5), 28–37 (2001)
11. W3C OWL Working Group: OWL 2 web ontology language. W3C recommendation, W3C (2012). https://www.w3.org/TR/2012/REC-owl2-overview-20121211/
12. Zhao, Z., et al.: Knowledge-as-a-service: a community knowledge base for research infrastructures in environmental and earth sciences. In: 2019 IEEE World Congress on Services (SERVICES), pp. 127–132. IEEE, Milan (2019). https://doi.org/10.1109/SERVICES.2019.00041
13. Wood, D., Cyganiak, R., Lanthaler, M.: RDF 1.1 concepts and abstract syntax. W3C recommendation, W3C (2014), http://www.w3.org/TR/2014/REC-rdf11-concepts-20140225/
14. W3C SPARQL Working Group: SPARQL overview. W3C recommendation, W3C. http://www.w3.org/TR/2013/REC-sparql11-overview-20130321/
15. Cummings, J.A.: Ocean data quality control. In: Schiller, A., Brassington, G. (eds.) Operational Oceanography in the 21st Century, pp. 91–121. Springer, Dordrecht (2011)
16. Wilkinson, M.D., et al.: The FAIR guiding principles for scientific data management and stewardship. Sci. Data **3**, 1–9 (2016). https://doi.org/10.1038/sdata.2016.18
17. Kontokostas, D., Knublauch, H.: Shapes constraint language (SHACL). W3C recommendation, W3C, July 2017. https://www.w3.org/TR/2017/REC-shacl-20170720/
18. Groth, P., Moreau, L.: PROV-overview. W3C note, W3C (2013). http://www.w3.org/TR/2013/NOTE-prov-overview-20130430/

19. Bechhofer, S., Miles, A.: SKOS simple knowledge organization system reference. W3C recommendation, W3C (2009). http://www.w3.org/TR/2009/REC-skos-reference-20090818/
20. Stellato, A.: Dictionary, thesaurus or ontology? Disentangling our choices in the semantic web jungle. J. Integr. Agric. **11**(5), 710–719 (2012)
21. Baker, T., Bechhofer, S., Isaac, A., Miles, A., Schreiber, G., Summers, E.: Key choices in the design of simple knowledge organization system (SKOS). Web Semant.: Sci. Serv. Agents World Wide Web **20**, 35–49 (2013)
22. Wegmann, A., Naumenko, A.: Conceptual modelling of complex systems using an RM-ODP based ontology. In: 2001 Proceedings of the Fifth IEEE International Enterprise Distributed Object Computing Conference. EDOC 2001, pp. 200–211. IEEE (2001)
23. Kilov, H., Linington, P.F., Romero, J.R., Tanaka, A., Vallecillo, A.: The reference model of open distributed processing: foundations, experience and applications. Comput. Stand. Interfaces **35**(3), 247–256 (2013)
24. Román, I., Madinabeitia, G., Jimenez, L., Molina, G., Ternero, J.: Experiences applying RM-ODP principles and techniques to intelligent transportation system architectures. Comput. Stand. Interfaces **35**(3), 338–347 (2013)
25. Jebbar, M., Sekkaki, A., Benamar, O.: Integration of SOA and cloud computing in RM-ODP. In: 2012 6th International Conference on Sciences of Electronics, Technologies of Information and Telecommunications (SETIT), pp. 97–105. IEEE (2012)
26. Petzold, A., et al.: ENVRI-FAIR - interoperable environmental FAIR data and services for society, innovation and research. In: 15th IEEE International Conference on e-Science, San Diego, US (2019). https://doi.org/10.1109/escience.2019.00038, https://zenodo.org/record/3462816
27. Liao, X., Zhao, Z.: Unsupervised approaches for textual semantic annotation, a survey. ACM Comput. Surv. **52**(4), 45 (2019). https://doi.org/10.1145/3324473. Article 66

Common Data Management Services in Environmental RIs

Data Curation and Preservation

Keith Jeffery(✉) 🔟

Keith G Jeffery Consultants, 71 Gilligans Way, Faringdon SN7 7FX, UK
keith.jeffery@keithgjefferyconsultants.co.uk

Abstract. Data is a valuable resource. In some scientific disciplines, experiments can be redone to reproduce the data. In environmental sciences, the observations and measurements of the earth and its surroundings commonly can be made only once: each time point records uniquely the state of the many earth processes. This demands that environmental data - structured to information - is preserved in such a way that it may be reused. Phenomena like the ozone hole, biodiversity and climate change depend on data curated over a long period of time. However, it is not just the data that must be curated. The software used to process and analyse the data - or more accurately an executable specification of the software - must be preserved along with associated libraries and computing operational environment. Information on the equipment and sensors used must be preserved since this affects the relevance and quality of the data for future use. Equally challenging is the decision to discard data - for reasons of costs of storage (although that is reducing rapidly) or cost of curation. Curation is blended inextricably with cataloguing and provenance and the core requirement is for rich metadata to characterise the digital asset for all three purposes.

Keywords: Data · Information · Preservation · Curation · Storage · Metadata · Cataloguing, provenance

1 Introduction, Context and Scope

"Digital curation is the selection, preservation, maintenance, collection and archiving of digital assets. Digital curation establishes, maintains and adds value to repositories of digital data for present and future use. This is often accomplished by archivists, librarians, scientists, historians, and scholars" (Wikipedia)[1].

Cataloguing, Curation and Provenance are commonly grouped together since the metadata, workflow, processes and legal issues associated with each have a high degree of intersection in recorded metadata attribute values and therefore rather than generating independent systems a common approach is preferable. Moreover, there are strong interdependencies with identification and citation, with AAAI (Authentication, Authorisation, Accounting Infrastructure), with processing, with optimisation, with modelling and with architecture.

---

[1] https://en.wikipedia.org/wiki/Digital_curation.

© The Author(s) 2020
Z. Zhao and M. Hellström (Eds.): Towards Interoperable Research
Infrastructures for Environmental and Earth Sciences, LNCS 12003, pp. 123–139, 2020.
https://doi.org/10.1007/978-3-030-52829-4_7

A key aspect of curation is the interplay between governance and technology. Finding technological solutions to satisfy the principles of governance is not always easy. The increased acceptance of the Data Curation Lifecycle and the increasing use of Data Management Plans (DMPs) evidences this. Another key aspect is involving the researchers in the decision making of what to keep and what to discard; this provides motivation for the process of curation, including the provision of appropriate metadata.

2   Curation Within ENVRIplus

The ENVRI community observes and analyses many aspects of Earth's changing phenomena. Observations and analyses today may be needed or reviewed in ways that are impossible to predict. Consequently, preparing the platform for future researchers as best we can by investing in curation has to be a key element of the ENVRI research culture with broad support by Research Infrastructures (RIs) and researchers. This requires leadership, education and collaborative development.

The ideal curation culture will ensure – via an appropriate IT system including both technological and governance aspects - the availability of digital assets through media migration to ensure physical readability, redundant copies to ensure availability, appropriate security and privacy measures to ensure reliability and appropriate metadata to allow discovery, contextualisation (for relevance and quality) and use, including information on provenance and rights.

At the curation stage of the lifecycle we record metadata concerning quality. Such metadata is – by its nature – domain specific and to some extent subjective. The required quality of the asset described by the metadata depends heavily on the purpose to which it is to be put. Decisions that are of broad scope and/or urgent may require only summary quality metadata whereas decisions relating to critical and detailed information such as in reproducibility of research may need detailed technical quantitative parameters recorded in the metadata. Thus, the end-user has to decide – based on the metadata available, guidelines established by governance and training to develop the skills – whether the asset is of appropriate quality for the intended purpose and whether – based on cost-benefit analysis - it should be curated. Clearly, the richer and more comprehensive the metadata providing context, the better judgement on quality can be made. The quality processes for some RIs in the environmental sciences have been studied in [1] and both a quality taxonomy and potential improvements recommended.

There has been significant progress over the period of the ENVRplus project: (1) the RIs appreciate the curation lifecycle; (2) the RIs generally have developed DMPs usually using the DCC (Digital Curation Centre) template appropriate for H2020 (EC Horizon 2020) projects; (3) the RIs appreciate the interplay between curation and both cataloguing and provenance; (4) the RIs understand the requirements for rich metadata to effect curation (and also cataloguing and provenance); (5) some RIs are planning future evolution utilising these principles.

3   Current Curation Activity

In the ENVRI community, there is a curation activity [12, 13]. Starting from a relatively low base at the beginning of the ENVRIplus project, curation activity has risen steadily encouraged by the presentations at ENVRI meetings and by the collection of information on curation associated with requirements collection.

3.1 Curation Lifecycle

The desirable lifecycle is represented by a DCC diagram, as shown in Fig. 1. The DCC in the UK is responsible for advising researchers and others on digital curation. The lifecycle model emphasises the steps in curation, the information required and the decisions to be taken at each step.

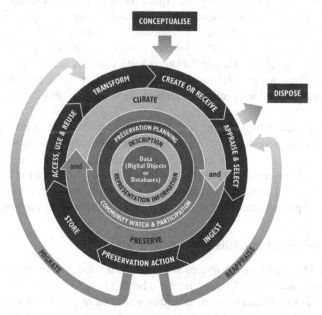

Fig. 1. The curation lifecycle model ("The DCC Curation Lifecycle Model", JISC/DCC, http://dcc.ac.uk/sites/default/files/documents/publications/DCCLifecycle.pdf) from DCC.

3.2 Data Management Plan

A Data Management Plan is defined as "A data management plan or DMP is a formal document that outlines how you will handle your data both during your research, and after the project is completed" (Wikipedia)[2].

The ENVRIplus RIs now generally have DMPs and utilise these as a basis for internal policymaking, road mapping, technological planning and governance of asset management, the latter within the framework of governance established by the RI e.g. the governance of a consortium through a consortium agreement. They are also used in the context of external agreements on transnational access.

_____

[2] https://en.wikipedia.org/wiki/Data_management_plan.

3.3 OAIS Reference Model

The Open Archival Information Systems (OAIS) Reference Model - ISO 14721:2002 [2] - provides a generic conceptual framework for building a complete archival repository and identifies the responsibilities and interactions of Producers, Consumers and Managers of both paper and digital records. The standard defines the processes required for effective long-term preservation and access to information objects while establishing a common language to describe these. It does not specify an implementation but provides a framework to make a successful implementation possible, through describing the basic functionality required for a preservation archive. It identifies mandatory responsibilities, and provides standardised methods to describe a repository's functionality by providing detailed models of archival information and archival functions [3]. Some RIs have considered OAIS as a framework but none has implemented it fully, although the concepts of Submission Information Package (SIP): which is the information sent from the producer to the archive, Archival Information Package (AIP): which is the information stored by the archive and Dissemination Information Package (DIP): which is the information sent to a user when requested have influenced the curation work in environmental and Earth science RIs.

In order to populate such a framework, a rich metadata element set is required. Much work has been done investigating various metadata standards to assess their suitability for curation (as well as for cataloguing and provenance). Within the work of the RDA (Research Data Alliance) MIG (Metadata Interest Group) – of which the chapter author is co-chair – a set of metadata elements in a structure for the purposes of curation, cataloguing and provenance according to the FAIR principles[3] has been proposed[4].

3.4 RDA (Research Data Alliance)

The Research Data Alliance has groups working on curation, provenance and catalogue metadata as well as citation. Clearly there is a benefit to ENVRIplus in alignment with the evolving RDA metadata recommendations which assist greatly not only in curation but also cataloguing, provenance and citation leading to improved discovery, contextualisation (for relevance and quality), interoperability, scientific reproducibility, and general governance of research assets. However, the RDA work is brought together with that of other groups in the specification of metadata[5]. The other groups are either domain-specific (e.g. in agriculture) or cross-cutting (e.g. in citation).

RDA proposed some metadata principles which are now generally accepted in that community:

- The only difference between metadata and data is a mode of use;
- Metadata is not just for data, it is also for users, software services, computing resources;
- Metadata is not just for description and discovery; it is also for contextualisation (e.g. relevance, quality, restrictions (rights and costs)) and for coupling users, software and computing resources to data (to provide a VRE);

---

[3] https://www.force11.org/group/fairgroup/fairprinciples.

[4] https://drive.google.com/drive/folders/0B8FnM3PsoL2dd2RnYVBmcjRMYXc.

[5] https://www.rd-alliance.org/groups/metadata-ig.html.

- Metadata must be machine-understandable as well as human-understandable for autonomicity (formalism);
- Management (meta)data is also relevant (e.g. research proposal, funding, project information, research outputs, outcomes and impact); and furthermore, a metadata element set that covers all the uses of metadata (not just curation):

  - Unique Identifier (for later use including citation);
  - Location (URL);
  - Description;
  - Keywords (terms);
  - Temporal coordinates;
  - Spatial coordinates;
  - Originator (organisation(s)/person(s));
  - Project;
  - Facility /equipment;
  - Quality; Availability (licence and persistence) including curation duration;
  - Provenance;
  - Citations;
  - Related publications (white or grey);
  - Related software;
  - Schema;
  - Medium/format;

It should be noted that many elements within this set have an internal structure (syntax) and semantics (meaning – usually represented by an ontological structure with term explanation and relationships) and so are not simple attributes with values. The RDA groups continue working on 'unpacking' the elements to a form suitable for discovery, contextualisation and action by both humans and computers.

4   Problems to Be Overcome for Curation in ENVRI

4.1   Current State

Some important problems associated with curation were discovered during requirements collection:

> *"ENVRI research communities will expect an integrated and seamless curation service that supports their routine work well and that opens paths for innovative research. This will require engagement from the practising domain scientists to help the ICT experts deliver relevant curation systems"* [4].

The incremental progress achieved for each problem is documented below:

Motivation

Problem to Be Overcome: There is little motivation for researchers to curate their digital assets. At present curation activity obtains no 'reward' such as career preferment based on data citations. In some organisations curation of digital assets is regarded as a librarian function but without the detailed knowledge of the researcher the associated metadata is likely to be substandard. Increasingly funding agencies are demanding curation of digital assets produced by publicly funded research.

Progress Achieved: Motivation has increased significantly – but not sufficiently yet. Use cases that provided significant scientific results dependent on curation are well-known and have provided motivation. The requirement by funding agencies for DMPs has also caused increased interest in and compliance with curation principles. Finally, the increasing availability to researchers of 'data stewards' and curators are improving curation.

Business Model

Problem to Be Overcome: Curation involves deciding what assets to curate and of those, for how long they should be kept. Determining an appropriate duration of retention for a digital asset is a problem; economics and business models do not manage well the concept of infinite time. First a business justification is needed in that (a) the asset cannot be collected again (i.e. it is a unique observation, experiment); (b) the cost of collecting again (by the same or another researcher) is greater than the cost of curation.

Progress Achieved: Awareness of the data curation lifecycle (within the research lifecycle) has increased leading to better governance and improved curation decisions.

The economic problem remains but decreasing costs of both storage and processing argue for increased curation by improving the cost/benefit ratio.

The major cost of curation is in expert staff providing guidelines and protocols and also – ideally – associated software tools. Increasing automation and autonomicity of curation processes will further reduce costs leading to an acceptable economic model in time.

Metadata

Problem to Be Overcome: Metadata collection is expensive unless it is automated or at least partially automated during the data lifecycle by re-using information already collected. Commonly, metadata is generated separately for discovery, contextualisation, curation and provenance when much of the metadata content is shared across these functions. A comprehensive but incrementally completed metadata element set is required that covers the required functions of the lifecycle. It needs sufficient application domain data that other specialists in that domain will be able to find and correctly interpret the associated data. Making the metadata handling facilities and tools that use them, such as workflows and data management, available to practical researchers to help them in their daily work, encourages them to invest in metadata, improves the quality of domain metadata and therefore facilitates the later curation processes [5].

Progress Achieved: Awareness of the need for - and benefits to be derived from – rich metadata is increasing substantially in the RIs as they evolve. This evolution is driven by researcher aspirations and requirements and is supported by improving technology. At present there are many metadata standards - international (e.g. ISO19115), local variants of international (e.g. INSPIRE[6] or APs (Application profiles) of DCAT[7], OpenSearch geoxtension[8]) and community/local all used in RIs within ENVRI. However, interconversion among all n of them requires $n(n-1)$ converters. Using a common canonical rich metadata schema as the 'switchboard' for interoperation between RIs reduces this to n convertors.

The co-development of rich metadata cataloguing, curation and provenance is a journey taking the RIs from a processing and governance environment where much human effort is required to re-use the assets with poor metadata to an automated environment with much re-use of the assets leveraged by rich metadata.

The cost of metadata creation is high. However, increasingly it is collected incrementally along the research workflow so reducing the perceived cost at each step. With rich metadata used for cataloguing, curation and provenance functions the scientific benefit increases relative to the costs of collection.

The utilisation of CERIF[9] additionally to CKAN[10] as the metadata standard for interoperation in ENVRIplus will improve the situation even further because of its much richer syntax and semantics (providing a superset canonical standard for interoperation) and its provision of referential and functional integrity.

Process

Problem to Be Overcome: The lifecycle of digital research entities is well understood and it needs process support. The incremental metadata collection aspect is critically important for success. Workflow models – if adapted to such an incremental metadata collection with appropriate validation –are likely to be valuable here [6].

Progress Achieved: Within some RIs we see increasingly the use of workflows (and, indeed, in some, automation of workflow deployment across multi-cloud or multiple processing environments managed by rich metadata). This allows for incremental metadata collection as predicted (with consequent benefits) but also highlights the need for rich metadata if automated processing – and thus reduction of human costs in research - is to be achieved. This was demonstrated in the PaaSage project[11] where the chapter author was scientific coordinator.

Curation of Data

Problem to Be Overcome: It may be considered that curation of data is straightforward –but it is not. First the dataset may not be static (by analogy with a type-specimen in a museum); both streamed data and updateable databases are dynamic thus leaving management decisions to be made on frequency of curation and management of versions with obvious links to provenance. Issues related to security and privacy change with

---

[6] https://inspire.ec.europa.eu/metadata/6541.

[7] https://www.w3.org/TR/vocab-dcat-2/.

[8] https://www.opengeospatial.org/standards/opensearchgeo.

[9] https://www.eurocris.org/cerif/main-features-cerif.

[10] https://ckan.org/portfolio/metadata/.

[11] https://paasage.ercim.eu/.

time and the various licences for data use each have different complexities. The data may change ownership or stewardship. Copies may be made and distributed to ensure availability but then have to be managed in systems such as LOCKSS[12]. Derivatives may be generated and require management including relationships with the original dataset and all its attendant metadata.

Progress Achieved: After the first half of the project, the RIs have increased their awareness – and appreciation – of this problem.

The relationship with provenance and cataloguing is clear – and the need for an integrated rich metadata catalog to cover all these processing and governance requirements in an integrated and consistent fashion is also becoming clear to the RIs.

The need for metadata covering not only description of the asset and its history, but also the persons and organisations - backed by funding – responsible is now understood.

Technology for the management of distributed copies – and their partitioning/replication/migration for processing efficiency overcoming latency – in a multi-cloud environment is being developed in the MELODIC project[13] where the chapter author is a consultant to the project.

The RDA Data Citation Working Group[14] has produced a recommendation for managing citation to parts of and versions of datasets. This relies on appropriate curation of the versions and services to define the partition.

Curation of Software

Problem to Be Overcome: Software written 50 years ago, is unlikely to compile (let alone compose with software libraries and execute) today. Indeed, many items of software, such as the workflows behind a scientific method, will either not run or give different results, six months later. Since many research propositions are based on the combination of the software (algorithm) and dataset(s) then the preservation and curation of the software become very important. It is likely that in future it will be necessary to curate not only the software but also a specification of the software in a canonical representation so that the same software process or algorithm can be reconstructed (and ideally generated) from the specification. This leaves the question of whether associated software libraries are considered part of the software to be curated or part of the operating environment (see below). Very often software contains many years-worth of intellectual investment by collaborating experts. It is not unusual for the software to encode the 'scientific method' used by the researcher which may be less well (or less formally) documented elsewhere (e.g. scholarly publications). This makes software very valuable and hard to replace. Taking good care of such assets will be a requirement for most research communities.

Progress Achieved: The issue was novel to most RIs when introduced in ENVRIplus Task T8.1 and recorded in Deliverable D8.1 [7]. The requirement is now appreciated but the metadata systems in use in most RIs are incapable of providing a technological

---

[12] https://www.lockss.org/.

[13] http://melodic.cloud/.

[14] https://www.rd-alliance.org/groups/data-citation-wg.html.

solution. It is further complicated because many developers – including those in some RIs – use GitHub[15] and related (or similar) technology to manage software development including versions, copies, compositions and deployments.

There is – as yet – no generally accepted way of managing this from both the technological and governance points of view. From an ENVRIplus perspective, the best we can do is to use rich metadata to catalogue the software and its evolution and monitor work elsewhere that will provide appropriate solutions.

Curation of Operational Environments

Problem to Be Overcome: It is necessary to record the operational environment of the software and dataset(s). The hardware used – whether instrumentation for collection or computation devices – has characteristics relating to the accuracy, precision, operational speed, capacity and many more. The operating system has defined characteristics and includes device drivers – i.e., a software library used by the application. It is a moot point whether software libraries belong to the application software or to the operational environment for the purposes of curation. Finally, the management ethos of the operational environment normally represented as policies requires curation.

Progress Achieved: The issue was novel to most RIs when introduced. The requirement is now appreciated but the metadata systems in use in most RIs are incapable of providing a technological solution.

There appears to be no generally accepted solution available. The best we can do in ENVRIplus is to collect rich metadata covering the operational environments and monitor external developments to find solutions as they are developed.

Increasingly, there appears to be a partial solution in containerisation using e.g. Docker[16] or Kubernetes[17]. Unlike Virtual Machines (VMs) (which have the contents of the container plus the operating system and are thus heavier on resources) containers include just the application and associated libraries and runtime environment and thus can be moved from one operating system environment to another, utilising the operating system kernel read-only and permitting writing to the container through its own 'mount' (access to the container).

Curation of 'Raw' Data Collected by Sensors or Instruments

Problem to Be Overcome: This is a special class of operational environment of importance to RIs in environmental science. The problems are manifold due to the data collection volume, velocity, variety, veracity and value and the difficulties of analytics, simulation and visualisation of streamed data.

Progress Achieved: While the requirements collected early in the project concentrated on the curation of validated or part-processed data, some RIs require curation of (at least some) raw data to allow subsequent reprocessing in calibration for precision and accuracy. Some examples illustrating the variety of practice are given below. EMSO

---

[15] https://github.com/.

[16] https://www.docker.com/.

[17] https://kubernetes.io/.

has distributed observatories with differing policies. In contrast Euro-Argo centralises quality control and curation. IAGOS validates the raw data manually or automatically before curation. ICOS stores (a kind of curation) raw sensor data collected at the stations and curates validated data. LTER does ingestion and quality control (curation) at individual sites. SeaDataNet relies on local centres curating quality-controlled data. An aspect particularly relevant increasingly to ENVRI communities is semi-automated curation of metadata which can be achieved if instrument metadata is available (SensorML[18] or SSNO[19]) and e.g. linked by PIDs with the incoming data stream[20].

4.2 A Longer-Term Horizon

There is some cause for optimism. Work within the ENVRIplus project has increased knowledge and understanding among the RIs and has exposed the issues and challenges to be addressed. A list of reasons for the optimism is:

1. Media costs are decreasing – so more can be preserved for less (and the cost reduction hopefully matches the expansion of volume). Media costs have decreased even more in the last 24 months and the trend shows no sign of changing;
2. Awareness of the need for curation is increasing; partly through policies of funding organisations and partly through increased responsibility of some researchers. The awareness has increased substantially not only through the efforts of ENVRIplus but also international efforts such as RDA and the FAIR initiative. The link with open science (i.e. open access to scholarly publications and datasets) is an effective driver.
3. Research projects in ICT are starting to produce autonomic systems that could be used to assist with curation. In particular MELODIC (mentioned above) is offering solutions combining curation and deployment.
4. Increasing standardisation of metadata and approaches to curation based on rich metadata are emerging and it is to be expected that this will continue producing richer and more effective curation services.

The cost of collecting metadata for curation remains a problem. Reducing storage costs mean that more data (even raw data to allow later re-processing before interpretation) can be stored. However, the major cost is that of creating appropriate metadata for the purposes of curation and subsequent discovery, contextualisation (including provenance) and action on the asset. The relative cost against benefit is reduced considerably by collecting the metadata once and using it for curation, cataloguing and provenance. Incremental collection along the workflow with re-use of existing information has been shown to decrease costs – but particularly to decrease researcher resistance to providing metadata - further. Improving techniques of automated metadata extraction from digital objects offer a further possibility of cost-reduction. There was some hope that they

---

[18] https://www.opengeospatial.org/standards/sensorml.

[19] https://www.w3.org/TR/vocab-ssn/.

[20] https://www.rd-alliance.org/group/persistent-identification-instruments/case-statement/persistent-identification-instruments.

may reach production status in the ENVRIplus time frame [8]. At present – although progress has been made – there are no appropriate systems although research indicates some cause for optimism.

4.3 Issues and Implications

Lack of Common Metadata Elements

Issues and Implications: Commonality of metadata elements across curation, provenance, cataloguing (and more) implies that a common core metadata scheme should be used for interoperability – possibly with extensions for particular domains where interoperability is not required.

Ongoing Work: The joint work especially with cataloguing - and following the recommendations from both the ENVRIplus cataloguing activity and the architecture – has led to the development of two catalogues, one using CKAN as in EUDAT B2FIND[21] and the other using CERIF as used in EPOS[22]. Experiments are underway to evaluate the two approaches for capability as the core metadata scheme.

Metadata Collection Expense

Issues and Implications: Metadata collection is expensive so incremental collection along the workflow is required: workflow systems should be evolved to accomplish this and scientific methods and data management working practices should be formalised using such workflows to reduce chores and risks of error as well as to gather the metadata required for curation;

Ongoing Work: There is evidence of increased use of workflows in the RIs although many are human-driven and not automated. Nonetheless, this provides the governance process to ensure incremental metadata collection to provide the required rich metadata.

Automated Metadata Extraction

Issues and Implications: Automated metadata extraction from digital objects shows promise but production system readiness is some years away. However, metadata provision from equipment-generated streamed data is available;

Ongoing Work: This has been monitored but the current systems are not yet at production status sufficient to be recommended to the RIs

DCC Recommendations

Issues and Implications: ENVRIplus should adopt the DCC recommendations;

Ongoing Work: Following acceptance by the RIs, this is achieved. However, implementation is incremental.

---

[21] https://www.eudat.eu/services/b2find.
[22] https://www.epos-ip.org/.

RDA Tracking and Involvement

Issues and Implications: ENVRIplus should track the relevant RDA groups and – ideally – participate.

Ongoing Work: Following acceptance by the RIs both tracking and participation have been pursued actively. Of particular relevance is the work on the RDA Metadata Element set which could be a candidate for a future common metadata scheme.

Education and Awareness

Issues and Implications: ENVRIplus should consider educational and practical steps to increase awareness of curation issues for all practitioners, particularly those concerned with curation organizational and technical strategy – collaboration and coordination could reduce the cost of this.

Ongoing Work: Curation has been presented at ENVRI meetings and elsewhere to raise awareness and encourage best practice in both governance and technical solutions.

The appreciation of the data curation lifecycle and the increasing use of DMPs is an achievement. The appreciation of the need for rich metadata for curation (alongside cataloguing and provenance) is also an achievement.

5   Architectural Design for Curation in ENVRI

5.1   Context

5.1.1   Initial State

At the beginning of the project, we asserted three aspects of the then-current state. Each, below, is supplemented by the work done during the project:

1. Technologies are available for curation but they may not be compatible with those for cataloguing and provenance. There has been a rapid and voluminous increase in understanding the need – for technological and governance reasons – to utilise one common metadata standard (in each RI and for interoperation across RIs) covering cataloguing, curation and provenance. Furthermore, it is widely understood and appreciated that this metadata standard has to be rich in syntax and semantics.
2. Governance principles for curation were lacking widely among the ENVRI community. The appreciation of the Data Lifecycle (within the research lifecycle) and the increasing use of DMPs has seen a marked improvement in governance.
3. Most RIs in the ENVRI community appreciate the importance of curation but are not practising it – partly because existing used metadata standards do not support it explicitly and/or can only be made to support it partially. All RIs appreciate the importance of curation and understand the rationale behind the WP8 work towards a rich metadata standard for curation (as well as cataloguing and provenance).

Further work on curation has considered also other, wider, aspects. In particular:

1. The use of personal data;
2. Fixity or preservation of state against possible data corruption.

The use of personal data – even in open science – is a contentious issue. The GDPR[23] (General Data Protection Regulation) of the EU makes provision for protecting personal data and its use. In open science, the name of a person, their institution, the equipment they use, their publications and their research assets are highly relevant to contextualisation (assessing relevance and quality for a new purpose). At present, there is no case law testing the limits of GDPR so this requires tracking and incorporating statements based on any judgements into the governance of RIs and their management (including curation) of data.

Environmental research data is the evidence base for some active political discussions, especially concerning climate change, utilisation of resources and pollution. Clearly, for environmental research, it is essential to have the observations made at a particular location and time preserved (possibly after assessment for accuracy, precision and/or any calibration corrections, smoothing or aggregation). This requires appropriate security to protect the integrity of the research product (asset) against 'tampering'.

5.1.2 Current State

It is clear that in the ENVRIplus project timespan the RIs have appreciated the need for a common rich metadata standard covering not only curation but also cataloguing and provenance (chapter 8 and 12). The requirement for protection of personal data and assurance of integrity (including fixity) underlines the need for rich metadata appropriate for enforcing access control. The ICT team has been working towards this and has been evaluating the solutions described in within the context of the requirements and architecture.

The final architectural solution for curation post-ENVRIplus will be decided as a result of that evaluation.

5.2 Architectural Design

5.2.1 Introduction

The initial design for curation was based not just on the state of the art and requirements for curation, but also for cataloguing and provenance (and also identification, citation and processing) for the reasons outlined above. The design consists of two components: the catalogue metadata and the curation processes. The final design confirms the initial design and adds detail.

5.2.2 Catalogue Metadata

The catalogue – for the purposes of curation – needs to describe the asset to be curated with rich metadata. The metadata must provide sufficient information for asset discovery, contextualization (for relevance and quality) and action. This is analogous to – but goes

---

[23] https://eugdpr.org/.

beyond in the area of action – the FAIR principles. In the case of curation, the action is to ensure an asset can be (a) made available when required; (b) is understandable to human and computer systems. The use of a logic representation provides advantages in deduction (facts from rules) and induction (rules from facts) which reduces potentially the metadata input burden and increases the validity of the metadata. Furthermore, because of versioning and the relationship to provenance, the metadata must include temporal information.

This system design aspect, therefore, depends on the cataloguing activity of ENVRIplus and to some extent on the Provenance activity, all within the overall architectural design.

However, the required metadata elements can be specified, derived from the use cases and requirements and the work of the Metadata Interest Group (and its sub-groups) of RDA (see above under 'State of the Art') which attempts to bring together experience and best practice from many international and national domain-specific efforts at standardising metadata for multiple uses, including curation. The base entities (objects) typically required (but note these may be complex with internal structure (syntax) and semantics) are:

Research Product (i.e. asset), Person, Organisation, Project, Research Publication, Citation, Facility, Equipment, Service, Geographic bounding box, Country, Postal address, Electronic address, Language, Currency, Indicator, Measurement, and Funding.

Of course, the entities appropriate to a particular DMP would be selected and used.

These entities need to be linked by linking entities to provide the role relationship (semantics) between base entities and the temporal duration of the truth of the assertion (the role linking the base entities). The linking entities can refer to instances within the same base entity (e.g. Research Product related to Research Product: with role 'derived' or Research Product related to Organisation: with role 'rightsholder'). Concepts such as availability are a relationship between the Research Product and e.g. Organisation with an appropriate role (e.g. manager) and a temporal duration. A similar relationship exists between a Research Product and an Organisation in the form of a licence (role) with temporal duration.

This structure gives great flexibility: the role relationships between Research Product and Person could be creator, reviewer, user…; those between Research Product and Facility, Equipment and service record the digital collection of the asset (Research Product). Indicators and measurement relate to quality when linked to Research Product. The address information may be linked to an organisation (such as the one owning the facility), the facility itself, the person or the organization employing the person (for the purpose of research).

The metadata structure outlined above has been encoded – partially - in the CKAN metadata of EUDAT B2FIND/B2SAVE and – using RDF – could be made compatible with the W3C PROV-O[24] standard for provenance (so linking curation and provenance). Additionally, the above conceptual structure has been encoded in CERIF (Common European Research Information Format; an EU recommendation to the Member States) which is used widely for research information management but also for the

---

[24] https://www.w3.org/TR/prov-o/.

EPOS project where it forms the catalogue. The ongoing ENVRIplus rich metadata catalogue (CERIF) involves harvesting from EPOS and conversion of CKAN records from the ENVRIplus CKAN catalogue harvested from other RIs. CERIF has been mapped to DC (Dublin Core)[25], DCAT (Data Catalogue Vocabulary), CKAN (Comprehensive Knowledge Archive Network which has its own metadata format based on DC) and ISO19115/INSPIRE (an EU directive). The initial mapping to/from PROV-O has been done in joint work between euroCRIS and CSIRO, Canberra [9]. CERIF provides a 'switchboard' for interoperability as a superset model compared with the others, capable of representing a fully connected graph and having declared semantics with crosswalk capability [10, 11].

However, the existing metadata standards used within the RIs do not reach this level of richness of representation. Convertors have been provided from within the project and from other projects e.g. VRE4EIC[26], but RIs need to provide additional information, supplementing that in their existing metadata, to achieve appropriate curation (and for that matter, provenance and cataloguing) especially for interoperation purposes. For example, typical provenance information in metadata standards such as DC, DCAT, ISO19115 and others is human-readable text and not machine-understandable.

The chapter on cataloguing (Chapter 8) describes the catalogue implementation using CKAN and CERIF as the canonical metadata standard and implements them as a prototype.

5.2.3 Curation Processes

The processes associated with curation are:

1. Store an asset (e.g. dataset) with metadata sufficient for curation purposes;
2. Discover an asset using the metadata – the richer the metadata and the more elaborate the query the greater the precision in discovering the required asset(s);
3. Copy an asset with its updated metadata (to have a distributed backup version);
4. Copy an asset with its updated metadata (media migration to ensure availability)
5. Move an asset with its updated metadata (to a distributed location if the original location is unable to manage curation);
6. Partition an asset and copy/move across distributed locations with its updated metadata (for performance, privacy and security);
7. Partition an asset and copy/move across distributed locations with its updated metadata (for performance including locality of e.g. data with software and processing power)

The processes were defined based on the requirements solicited [6]. All these processes could be applied to a set of assets as well as a single asset. These processes are all simple given rich metadata in the catalogue as outlined above. The processes are documented and specified in the ENVRI RM (Reference Model).

---

[25] https://www.dublincore.org/.
[26] https://www.vre4eic.eu/.

6 Conclusion

The final design of the curation functionality aims to maximise flexibility while retaining compatibility with provenance and the catalogue. The catalogue is central to the design and implementation. The choice of the metadata elements in the catalogue (including their syntax and semantics) is crucial for the processes not only of curation but also of provenance and catalogue management and utilisation. The metadata model of the catalogue has also to permit interoperation among RIs as well as the usual processes associated with metadata catalogues: discovery, contextualisation and action. This implies that the model must be a superset (in the representation of syntax and semantics) of the metadata models used or planned within the RIs.

The chapter on cataloguing (Chapter 8) covers the implementation of CKAN (as used in EUDAT) and CERIF for the metadata catalogue.

This curation work relates closely to other tasks: cataloguing and provenance but also Identification and citation and processing leading towards representation in the reference model and the overall architecture design and evaluation.

The choice of a metadata standard for the catalogue was a critical decision for the project and the ability of RIs to compare CKAN and CERIF for cataloguing (related to the cataloguing processes of discovery, contextualisation and action), curation and provenance has been instructive.

The work on curation has caused the RIs to increase their attention to – and effort on – curation. RIs will now – with their DMPs – decide which assets to keep and curate, and which to delete and lose. The result of positive action is archives of curated environmental data essential for later research especially comparing the state of the environmental domain at that (future) time with now and past states as recorded. Some RIs need to store raw data to allow subsequent reprocessing/validation before interpretation. Reducing storage costs make this feasible but the cost of metadata generation is high and needs to be weighed against the benefits. Some RIs may be engaged in global collaborations, e.g. Euro-Argo or operate under global coordination, e.g. for atmospheric observations that need to be recognised by the IPCC[27]. The RIs need to fit their curation plans into this larger context and may even draw on the resources provided by that context. If these commitments to compatibility for curation demand only metadata and processes that are a subset of those proposed here, then interoperability and compatibility are assured. This will be clarified via DMPs, so that ENVRIplus can more accurately judge the residual requirement.

Acknowledgements. This work was supported by the European Union's Horizon 2020 research and innovation programme via the ENVRIplus project under grant agreement No. 654182.

References

1. Zhao, Z., et al.: Knowledge-as-a-service: a community knowledge base for research infrastructures in environmental and earth sciences. In: 2019 IEEE World Congress on Services (SERVICES), pp. 127–132. IEEE, Milan (2019). https://doi.org/10.1109/SERVICES.2019.00041

---

[27] https://www.ipcc.ch/.

2. The Consultative Committee for Space Data Systems (CCSDS): Reference Model for an Open Archival Information System (OAIS), recommended practice CCSDS 650.0-M-2, June 2012 (2012). https://public.ccsds.org/Pubs/650x0m2.pdf. Accessed 01 Dec 2019
3. Using OAIS for Curation. DCC Briefing Papers: Introduction to Curation. Edinburgh: Digital Curation Centre. Handle: 1842/3354. http://www.dcc.ac.uk/resources/briefing-papers/introduction-curation. Accessed 01 Nov 2019
4. Atkinson, M., et al.: A consistent characterisation of existing and planned RIs. ENVRIplus Deliverable 5.1, submitted on 30 April 2016. http://www.envriplus.eu/wp-content/uploads/2016/06/A-consistent-characterisation-of-RIs.pdf. Accessed 01 Dec 2019
5. Myers, J., et al.: Towards sustainable curation and preservation: the SEAD project's data services approach. https://experts.illinois.edu/en/publications/towards-sustainable-curation-and-preservation-the-sead-projects-d. Accessed 01 Nov 2019
6. Jeffery, K., Asserson, A.: Supporting the research process with a CRIS. In: Asserson, A.G.S., Simons, E.J. (eds.) Enabling Interaction and Quality: Beyond the Hanseatic League; Proceedings 8th International Conference on Current Research Information Systems CRIS2006 Conference, Bergen, pp. 121–130 Leuven University Press (2006). ISBN 978 90 5867 536 1
7. Jeffery, K., et al.: Data curation in system level sciences: initial design. ENVRIplus deliverable report D8.1 (2017). http://www.envriplus.eu/wp-content/uploads/2015/08/D8.1-Data-Curation-in-System-Level-Sciences-Initial-Design.pdf
8. Dorbeva, M., Kim, Y., Ross, S.: Instalment on "Automated Metadata Generation". http://www.dcc.ac.uk/webfm_send/1513. Accessed 06 Jan 2020
9. Compton, M., Corsar, D., Taylor, K.: Sensor data provenance: SSNO and PROV-O together at last, In: Taylor, K., Gruetter, R. (eds.) Terra Cognita - Semantic Sensor Networks, TC-SSN 2014 - ISWC 2014. CEUR Workshop Proceedings, Trentino, Italy, pp. 67–82 (2014)
10. Martin, P., Remy, L., Theodoridou, M., Jeffery, K., Zhao, Z.: Mapping heterogeneous research infrastructure metadata into a unified catalogue for use in a generic virtual research environment. Future Gener. Comput. Syst. **101**, 1–13 (2019). https://doi.org/10.1016/j.future.2019.05.076
11. Remy, L., et al.: Building an integrated enhanced virtual research environment metadata catalogue. J. Electron. Libr. (2019). https://zenodo.org/record/3497056
12. Zhao, Z., et al.: Reference model guided system design and implementation for interoperable environmental research infrastructures. In: 2015 IEEE 11th International Conference on e-Science, pp. 551–556. IEEE, Munich (2015). https://doi.org/10.1109/eScience.2015.41
13. Chen, Y., et al.: A common reference model for environmental science research infrastructures. In: Proceedings of EnviroInfo 2013 (2013)

Data Cataloguing

Erwann Quimbert[1]([envelope]), Keith Jeffery[2] [ID], Claudia Martens[3], Paul Martin[4] [ID],
and Zhiming Zhao[4] [ID]

[1] Ifremer, BP 70, 29280 Plouzané, France
erwann.quimbert@ifremer.fr
[2] Keith G Jeffery Consultants, 71 Gilligans Way, Faringdon SN FX, UK
keith.jeffery@keithgjefferyconsultants.co.uk
[3] German Climate Computing Center [DKRZ], 20146 Hamburg, Germany
martens@dkrz.de
[4] Multiscale Networked Systems, University of Amsterdam,
1098XH Amsterdam, The Netherlands
paulmartin.research@google.com, z.zhao@uva.nl

Abstract. After a brief reminder on general concepts used in data cataloguing activities, this chapter provides information concerning the architecture and design recommendations for the implementation of catalogue systems for the ENVRIplus community. The main objective of this catalogue is to offer a unified discovery service allowing cross-disciplinary search and access to data collections coming from Research Infrastructures (RIs). This catalogue focuses on metadata with a coarse level of granularity. It was decided to offer metadata representing different types of dataset series. Only metadata for so-called flagship products (as defined by each community) are covered by the scope of this catalogue. The data collections remain within each RI. For RIs, the aim is to improve the visibility of their results beyond their traditional user communities.

Keywords: Catalogue · Metadata · Data · Interoperability · Standard · ISO · OGC · Format · Schema

1 Introduction

Data catalogues have been used in data management for a long time. Under the impetus of European regulations, the number of metadata catalogues has been growing steadily over the last decade, and more specifically thanks to the Inspire Directive [1], which has made it mandatory for public authorities to create metadata more easily and to share them more widely. Data catalogues provide information about data concerning one or many organizations, domains or communities. This information is described and synthesised through metadata records. Data catalogue centralised metadata is gathered in one location, usually accessible online through a dedicated interface. In this chapter, we will focus on data catalogues related to environmental sciences.

A common definition is that metadata is "data about data". Metadata provide information on the data they describe to specify who created the data, what it contains, when it

Z. Zhao and M. Hellström (Eds.): Towards Interoperable Research
Infrastructures for Environmental and Earth Sciences, LNCS 12003, pp. 140–161, 2020.
https://doi.org/10.1007/978-3-030-52829-4_8

was created, why it was created, and in which context. Metadata can be created automatically or manually and they are structured to allow easy and simple reading by end-users and by automated services.

As proposed by Riley [2], metadata can be classified into 3 categories:

1. **Descriptive metadata** give a precise idea about the content of a resource. Descriptive metadata may include a title, a description, keywords and one or many points of contact (creator, author, and editor). These metadata elements allow end-users to easily find a resource and to know if this resource fits their purpose and their research needs.
2. **Administrative metadata** include technical metadata (providing information about the format, file size, how they have been encoded, and software used), rights metadata (including user limitations, access rights, intellectual property rights and copyright constraints) and provenance metadata (lineage of the data, why this data has been created, by whom, and in which context).
3. **Structural metadata** provide information about the files that make up the resource and specify the relationships between them.

To complete this classification, it is often accepted that good metadata is metadata that is able to answer the 5 W's: *Who, What, Where, When* and *Why*.

RDA (Research Data Alliance) has developed agreed principles concerning metadata discussed in (Chapter 7) including the assertion that there is no difference between metadata and data except the use to which it is put. A library catalogue card used by a researcher to locate a scholarly paper is metadata when among other cards used by a librarian to count articles on river pollution it is data.

The purpose of data catalogues is multifold. One of its biggest benefits is to organise and centralise the metadata in one location which greatly facilitates data discovery for end-users and make data more accessible for different types of users (data consumers, data scientists or data stewards).

Data catalogues also avoid duplication of data.

Data catalogues exist to collect, create and maintain metadata. These records are indexed in a database and end-users should access the information through a user-friendly interface. This interface should offer common data search functionalities allowing users to narrow down their search according to different criteria: keywords (controlled vocabularies), geographic location, temporal and spatial resolution, and data sources.

Data catalogues have become an important pillar in the data management lifecycle. Indeed, almost every step of the data lifecycle is described in the metadata fields or accessible through the data catalogue online interface. Curated data are described by effective and structured metadata (cf. Riley's list above) providing information about data collection (e.g. metadata automatically produced about sensors/instruments) data processing (data lineage, software used, explanations of the different steps of data construction), data analysis (description of methods applied), data publishing (discovery metadata, policies for access, reuse and sharing) and data archiving (preserving data).

2 Metadata Standards and Interoperability Between Data Catalogues

2.1 Metadata Standards

"Metadata is only useful if it is understandable to the software applications and people that use it" [2]. We often speak about schema to illustrate the metadata structure. To facilitate this understanding metadata generally follow standardised schemas implementing recommendations from international organizations such as ISO[1] (International Organization for Standardization). There are several metadata standards widely used in the environmental science domain. It will not be possible to fully describe them in this chapter but a short description is given explaining in which community they are commonly used. To simplify integration within systems metadata, a machine-readable language is often used such as XML or RDF or even JSON-LD.

Metadata Standards versus Metadata Schemas
The terms 'schema' and 'standard' are used in an interchangeable way, but all refer to "the formal specification of the attributes (characteristics) employed for representing information resources" [3]. Yet another definition for 'metadata schema' is a "logical plan showing the relationships between metadata elements, normally through establishing rules for the use and management of metadata specifically as regards the semantics, the syntax and the optionality" [ISO/TC46, 2011] whereas 'syntax' describes the structure of a schema (language, rules to represent content) and 'semantics' describe the meaning of its elements, properties or attributes. Following Haslhofer and Klas [4] a metadata schema could be seen as a set of elements with a precise semantic definition and optionally rules how and what values can be assigned to these elements; a metadata standard then is a schema which is developed and maintained by an institution that is a standard-setting one. Hence a standard is a standard insofar as there is an institutional or organizational standardization unit developing and maintaining a standard - whereas all parties and persons involved agree this institution to be trustworthy and reliable. Some relevant standards are mentioned below.

ISO 19115 [5] is an internationally adopted schema for describing geospatial data. As indicated in their website "it provides information about the identification, the extent, the quality, the spatial and temporal aspects, the content, the spatial reference, the portrayal, distribution, and other properties of digital geographic data and services."

DataCite[2] [6] is an international consortium founded in 2009 with an emphasis to make explicitly *research data* citable, giving them a 'value' during the scientific process: "a persistent approach to access, identification, sharing and re-use of datasets" [6][3]. DataCite promotes the use of Persistent Identifiers for Digital Objects in order to unambiguously identify a digital resource, established as DOIs[4].

---

[1] https://www.iso.org/.

[2] https://schema.datacite.org/.

[3] https://schema.datacite.org/meta/kernel-4.1/doc/DataCite-MetadataKernel_v4.1.pdf.

[4] http://www.doi.org/index.html.

Dublin Core Metadata Initiative[5] [7] was founded in the aftermath of a World Wide Web conference during a workshop at the OCLC[6] (an organisation for a global digital library providing technology) headquartered in Dublin, Ohio (USA), aiming at achieving "consensus on a list of metadata elements that would yield simple descriptions of data in a wide range of subject areas for indexing and cataloguing on the Internet" [7]. Dublin Core was originally developed mainly by librarians, where 15 (initially 13 but extended when additional attributes were required) 'core' metadata elements[7] contain resource descriptions (contributor, coverage, creator, date, description, format, identifier, language, publisher, relations, rights, source, subject, title, and type). As these descriptions have been regarded as not sufficient, they were refined to 'qualified DC' by 55 'terms'[8]. DC has been represented progressively over time by text, HTML, XML and - recently - RDF. Only in this latter form does it approach the requirement for formal syntax and declared semantics.

CERIF[9] is a data model recommended by the European Union to the Member States for research information. It is described in some detail below.

DCAT [8] is a W3C recommendation 'data catalogue vocabulary' and has the advantage of being conceived natively with qualified relationships and use of RDF triples. It is currently undergoing revision by the DXWG (Data Exchange Working Group)[10].

Schema.org[11] is an initiative from Google and Microsoft now a community activity. It essentially provides a list of attributes, some with related vocabularies, for entities. In this way it is like CERIF: schema.org has entities for person and organisation, product and place for example. It may be encoded in RDF or JSON-LD.

All have some relevance to ENVRI. RIs are encouraged to choose a schema that has the capability to describe their 'world of interest'. Only rich metadata schemas (such as CERIF) can provide a unifying data model to which the others may be converted in a lossless manner.

Specification versus Interoperability
While Dublin Core and DataCite are generic metadata standards that aim to provide a minimum of metadata elements for describing a digital resource, ISO19115/19139 [8] is a standard especially for georeferenced data. The question is how to find an equilibrium between 'general' information that is sufficient to search and access research data across scientific disciplines on the one side and 'specific' information describing resources from certain research communities on the other side is not clearly answered yet (and maybe can´t be answered at all). RDA (Research Data Alliance) is working on a set of common metadata elements (each with syntax and semantics) linked by qualified references to act

---

[5] https://www.dublincore.org/.
[6] https://www.oclc.org/.
[7] http://dublincore.org/documents/dces/.
[8] http://dublincore.org/documents/dcmi-terms/.
[9] https://www.eurocris.org/cerif/main-features-cerif.
[10] https://www.w3.org/2017/dxwg/wiki/Main_Page.
[11] https://schema.org/.

as rich metadata set for FAIR (Findability, Accessibility, Interoperability, Reusability) [9] with the aim of overcoming this problem[12].

2.2 Data Catalogues Tools

There are many tools used by scientific communities to create data catalogues. Two example tools used by the environmental and Earth science research communities are GeoNetwork and CKAN.

GeoNetwork[13] is an open-source software allowing the creation of customised catalogue applications. This tool is mainly used for describing and publishing geographic datasets and is related to ISO 19115/19139.

CKAN[14] is an open-source Data Management System widely used in the world of open data. It uses essentially some Dublin Core metadata elements[15] but allows for an infinite extension of additional attributes thus making interoperation difficult. EUDAT B2FIND uses CKAN for its frontend.

Independently of the software used, protocols exist for sharing metadata between data catalogues, in particular OGC-CSW[16], OAI-PMH[17], SPARQL[18] and others.

3 Design for ENVRI

3.1 ENVRIplus Context

Data cataloguing is a key service in the data management lifecycle of ENVRIplus [18–20]. For ENVRIplus, an interoperable catalogue system aims at organizing the maintenance and access to descriptions of resources and outcomes of multiple Research Infrastructures in a framework which implements a number of functions on these descriptions. As defined in the ENVRI Reference Model (Chapter 4), maintenance of a catalogue is a strategic component of the curation process and the descriptions maintained in the catalogue support the acquisition, publication and re-use of data. The system must provide to users a function for the seamless discovery of the description of resources in the Research Infrastructures, encoded using a standardised format. The multi-Research Infrastructures context of ENVRIplus implies that, in addition to the descriptions usually available within each Research Infrastructure, resources may also have to be described at a higher granularity so to provide context.

The goal of the so-called Flagship catalogue is to expose and highlight products that best illustrate the content of Research Infrastructures catalogues. This demonstrator aims to provide a better overview to users of existing catalogues and resources, mostly data, indexed by these catalogues.

---

[12] https://drive.google.com/drive/folders/0B8FnM3PsoL2dd2RnYVBmcjRMYXc.

[13] https://geonetwork-opensource.org/.

[14] https://ckan.org/.

[15] https://ckan.org/portfolio/metadata/.

[16] https://www.opengeospatial.org/standards/cat.

[17] https://www.openarchives.org/pmh/.

[18] https://www.w3.org/TR/rdf-sparql-query/.

A Top-Down approach has been used with the aim of showcasing the products of the Research Infrastructures so that they reach new inter-disciplinary and data science usages. The homogeneous and qualified descriptions provided in a single seamless framework is a tool for stakeholders and decision makers to oversee and evaluate the outcome and complementarity of Research Infrastructure data products.

3.2   RIs Involved in the Flagship Catalogue

For a first version, the following Research Infrastructures have been targeted as first priority to have their resources described in the ENVRIplus catalogue system:

- AnaEE[19] (Analysis and Experimentation on Ecosystems) focuses on providing innovative and integrated experimentation services for research on continental ecosystems.
- Euro-Argo[20] is the European contribution to the Argo program. Argo is a global array of 3,800 free-drifting profiling floats that measures the temperature and salinity of the upper 2000 m of the ocean.
- EMBRC[21] is a pan-European Research Infrastructure for marine biology and ecology research.
- EPOS[22] (European Plate Observing System) is a long-term plan to facilitate integrated use of data, data products, and facilities from distributed research infrastructures for solid Earth science in Europe.
- IAGOS[23] (In-Service Aircraft for a Global Observing System) is a European Research Infrastructure for global observations of atmospheric composition using commercial aircraft.
- ICOS[24] is a pan-European research infrastructure for quantifying and understanding the greenhouse gas balance of Europe and its neighbouring regions.
- LTER[25] (Long Term Ecological Research) is an essential component of world-wide efforts to better understand ecosystems.
- SeaDataNet[26] is a pan-European infrastructure to ease the access to marine data measured by the countries bordering the European seas.
- Actris[27] is the European Research Infrastructure for the observation of Aerosol, Clouds, and Trace gases.

---

[19] https://www.anaee.com/.

[20] https://www.euro-argo.eu/.

[21] http://www.embrc.eu.

[22] https://www.epos-ip.org/.

[23] http://www.iagos-data.fr/.

[24] http://www.icos-ri-eu.

[25] http://www.lter-europe.net/.

[26] https://www.seadatanet.org/.

[27] https://www.actris.eu/.

Four kinds of users were identified for this flagship catalogue:

- Users outside a Research Infrastructure, researching data-driven science.
- Users inside a Research Infrastructure, such as data managers, coordinators, and operators as well as data scientists.
- Stakeholders, decision-makers and funders of the Research Infrastructures who need to have a broad picture of the Research Infrastructure resources in the European landscape to control their efficiency and complementarity.
- Policymakers, using ENV RI information for government policy and laws.

3.3 Proposed Architecture

At the beginning of the project, it was decided to not create a new metadata model. The requirements on product description were defined by adopting the metadata elements of the RDA metadata interest group[28]. We noticed that this schema gathers most of the common properties among different data models exposed above. The idea is to automatically map the metadata model from each Research Infrastructures to a canonical schema. We also encouraged the use of existing controlled vocabularies.

CERIF and CKAN frameworks are both chosen candidates for prototyping an ENVRIplus community catalogue for Research Infrastructures flagship data products.

To streamline the implementation of this flagship catalogue, it was decided to start with the EUDAT/B2FIND[29] demonstrator. The demonstrator on CERIF has also been developed jointly with EPOS and other relevant projects, e.g. VRE4EIC[30].

4 Cataloguing Using B2FIND

4.1 B2FIND Description and Workflow

B2FIND[31] is a discovery service for research data distributed within EOSC-hub and beyond. It is a basic service of the pan-European data infrastructure EUDAT CDI (Collaborative Data Infrastructure)[32] that currently consists of 26 partners, including the most renowned European data centres and research organisations. B2FIND is an essential service of the European Open Science Cloud[33] (EOSC) as it is the central indexing tool for the project that constitutes the EOSC (EOSC-Hub).

Therefore a comprehensive joint metadata catalogue was built up that includes metadata records for data that are stored in various data centres, using different meta/data formats on divergent granularity levels, representing all kinds of scientific output: from huge netCDF files of Climate Modelling outcome to small audio records of Swahili

---

[28] https://rd-alliance.org/groups/metadata-ig.html.

[29] http://b2find.eudat.eu/.

[30] https://www.vre4eic.eu/.

[31] http://b2find.eudat.eu/.

[32] https://www.eudat.eu/eudat-cdi.

[33] https://www.eosc-portal.eu/about/eosc.

syllables and phonemes; from immigrant panel data in the Netherlands to a paleoenvironment reconstruction from the Mozambique Channel and from an image of "Maison du Chirugien" in ancient Greek Pompeia to an xlsx for concentrations of Ca, Mg, K, and Na in throughfall, litterflow and soil in an Oriental beech forest.

In order to enable this interdisciplinary perspective, different metadata formats, schemas and standards are homogenised on the B2FIND metadata schema[34], which is based on the DataCite schema extended with the additional element <Disciplines>, allowing users to search and find research data across scientific disciplines and research areas. Good metadata management is guided by FAIR principles, including the establishment of common standards and guidelines for data providers. Hereby a close cooperation and coordination with scientific communities, Research Infrastructures and other initiatives dealing with metadata standardisation (OpenAire Advance, RDA interest and working groups and the EOSCpilot project to prepare the EOSC including a task on 'Data Interoperability'[35]) is essential in order to establish standards that are both reasonable for community-specific needs and usable for enhanced exchangeability. The main question still is how to find a balance between community-specific metadata that serve their needs on the one side and a metadata schema that is sufficiently generic to represent interdisciplinary research data but at the same time is specific enough to enable a useful search with satisfying search results.

Harvesting
Preferably B2FIND uses the Open Archives Initiative Protocol for Metadata Harvesting (OAI-PMH) to harvest metadata from data providers. OAI-PMH offers several options that make it a suitable protocol for harvesting: a) possibility to define diverse metadata prefixes (default is Dublin Core), b) possibility to create subsets for harvesting (useful for large amounts of records, resp. divergent records e.g. from different projects or sites or measurement stations) and c) the possibility to configure incrementally harvesting (which allows to harvest only new records). Nonetheless, other harvesting methods are supported as well, e.g. OGC-CSW, JSON-API or triples from SPARQL endpoints.

Mapping
The mapping process is twofold as it includes a format conversion as well as a semantic mapping based on standardised vocabularies (e.g. the field 'Language' is mapped on the ISO 639 library[36] and research 'Disciplines' are mapped on a standardised closed vocabulary). Therefore, entries from XML records are selected based on XPATH rules that depend on community-specific metadata formats and then parsed to assign them to the keys specified in the XPATH rules, i.e. fields of the B2FIND schema. Resulting key-value pairs are stored in JSON dictionaries and checked/validated before uploaded to the B2FIND repository. B2FIND supports generic metadata schemas as DataCite and Dublin Core. Community specific metadata schemas are supported as well, e.g.

[34] http://b2find.eudat.eu/guidelines/mapping.html.
[35] https://www.eoscpilot.eu/content/d69-final-report-data-interoperability.
[36] https://iso639-3.sil.org/code_tables/639/data.

ISO19115/19139 and Inspire for Environmental Research Communities or DDI[37] and CMDI[38] for Social Sciences.

Upload and Indexing

B2FIND's search portal and GUI is based on the open-source portal software CKAN, which comes with Apache Lucene SOLR Servlet allowing indexing of the mapped JSON records and performant faceted search functionalities. CKAN was created by the Open Knowledge Foundation (OKFN) and is a widely used data management system. CKAN has a very limited internal metadata schema[39] which has been enhanced for B2FIND while creating additional metadata elements as CKAN field "extra". B2FIND offers a full text search, results may be narrowed down using currently 11 facets (including spatial/temporal search and facets <Discipline>, <ResourceType>, <Publisher>, <Contributor>, <Language>, <Community>, <Tags> and <Creator>). "Community" here is the data provider where B2FIND harvests from.

4.2 B2FIND and FAIR Data Principles

FAIR data principles [9] are recommended guidelines to increase the impact of data in science generally by making them findable, accessible, interoperable and reusable. While these principles are increasingly recognised, specific elements need to be clarified: how to implement FAIR data principles during the data lifecycle? How to measure "FAIRness"? By whom? Currently, supporting FAIR data principles are done in varying ways with different methods[40]. The approach of B2FIND to these guidelines may be characterised as supporting 'Findability' by offering a discovery portal for research data based on a rich metadata catalogue, supporting 'Accessibility' by representing Persistent Identifiers for unique resolvability of data objects, supporting 'Interoperability' by implementing common standards, schemas and vocabularies and finally supporting 'Reusability' by offering licenses, provenance and domain-specific information. However, while FAIR principles refer to both data and metadata, B2FIND may manage only the metadata aspect.

4.3 Flagship Implementation

The implementation of ENVRIplus Flagship catalogue in B2FIND faced two main challenges: 1) how to integrate metadata records that are representing Research Infrastructures rather than Datasets, and 2) how to represent these RIs as part of ENVRIplus

---

[37] The Data Documentation Initiative (DDI) is an international standard for describing the data produced by surveys and other observational methods in the social, behavioural, economic, and health sciences. https://ddialliance.org/.

[38] The Component MetaData Infrastructure (CMDI) provides a framework to describe and reuse metadata blueprints. Description building blocks ("components", which include field definitions) can be grouped into a ready-made description format (a "profile"). https://www.clarin.eu/content/component-metadata.

[39] https://docs.ckan.org/en/ckan-1.7.4/domain-model.html#overview.

[40] GO FAIR initiative is a good example, therefore: one aim is to support 'Implementation Networks', whereas these networks define in how far they are FAIR. See therefore: https://www.go-fair.org/.

within the B2FIND architecture. These questions concerned both the technical level and content-related issues and are described below. The implementation process itself revealed challenges that may be seen as exemplary: how to deal with persistent identifiers and how to deal with granularity issues.

A. RI Dataproducts

As described above B2FIND is first and foremost a search portal for research data that should be findable across scientific disciplines. It is not primarily meant to be a search portal for other information as e.g. funding bodies, site information or research infrastructure descriptions. Concerning RIs that are part of ENVRIplus, most of them have their own search interface and some of them have already made their repositories harvestable. Thus, the flagship implementation started with harvesting already existing RI endpoints (DEIMS[41], NILU[42], EPOS, SeaDataNet, Euro-Argo, AnaEE, ICOS Carbon Portal[43]) and integrating them as "Communities" into a B2FIND testing machine[44], which means representing their data as e.g. "DEIMS". One challenge on B2FIND side was to develop the software stack[45] in order to be able to harvest from CSW endpoints. On the Data Provider side, the proper CSW configuration has been a task insofar as CSW does not yet allow the creation of Subsets (which would enable harvesting of just one subset for testing) and resumption token. Another issue concerned incrementally harvesting: OAI-PMH allows to exchange information of 'record status' and 'timestamp', which means that it is possible to harvest just those records that are not e.g. 'deleted' or those from a certain period of time (e.g. every week). CSW does not yet support these features. Creating a mapping for each "Community" has been relatively simple as all RIs use either DublinCore or ISO19139 as their metadata standard and usually XML as an exchange format. The only exception is ICOS that expose their metadata as triples. The decision to use the Flagship Catalogue for representing Data products (which means records that describe the *services* offered by the RIs rather than their *data*) compelled the RIs to create metadata records that fitted this purpose and expose them in a way that enabled B2FIND to ingest them.

B) B2FIND/Flagship architecture

Initially, the Flagship catalogue should have been visible in a way that would display both ENVRIplus as the main project and each RI as a part of it. CKAN allows to create "Groups" and "Subgroups"; however, B2FIND is constructed as CKAN "Group" and its "Communities" as CKAN "Subgroups" which means that a further distinction between ENVRIplus and RIs could not be implemented. In order to enable a search for RIs the decision was to create a 'Community' = ENVRIplus and use the metadata element

---

[41] https://deims.org/.

[42] https://www.nilu.no/en/.

[43] The data centre of ICOS, https://icos-cp.eu/.

[44] http://eudat7-ingest.dkrz.de/dataset.

[45] B2FIND uses CKAN only for GUI and search interface while the backend is developed B2FIND code, it´s Open Source on GitHub: https://github.com/EUDAT-B2FIND.

<Contributor> as a distinctive feature (Fig. 1). As the flagship implementation enforced B2FIND to enhance its metadata schema (to enable a faceted search via <Contributor>) it was implemented on a test machine at DKRZ[46]. The demonstrator may be seen here: http://eudat7-ingest.dkrz.de/dataset?groups=envriplus.

Fig. 1. Flagship catalogue in B2FIND: partial search result page.

As described above B2FIND links to a certain resource by using persistent identifiers (if offered within the metadata) in order to increase the reliability of a digital resource (Fig. 2). Therefore, an internal 'ranking' is used: if a DOI is provided it will be displayed, both as a link to the Landing page and additionally as a small icon on the single record

---

[46] https://www.dkrz.de/about-en.

entry page. If no DOI but another PID (e.g. a Handle) is offered this one will be shown, both as a link and as an icon. If none (DOI or PID) is given, B2FIND will represent any other URN or URL.

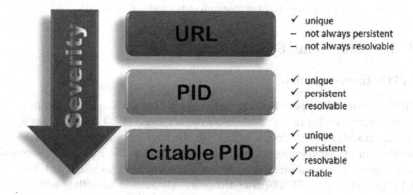

Fig. 2. Consistency of identifiers.

For the flagship implementation the RIs 'Dataproducts' did not all provide a DOI or PID (except for IAGOS, see Fig. 3) but an identifier that links to the described resource. Some effort was needed to define where the 'Source' information is given - some RIs presented internal identifier within the metadata element <identifier> (such as UUIDs) which are not automatically resolvable, sometimes this information was given in <alternateIdentifier> or <relatedIdentifier> attributes or within the header. To solve this problem a specific map file for each RI was created that defined the XPATH rules for each metadata element in order to map it onto the B2FIND schema.

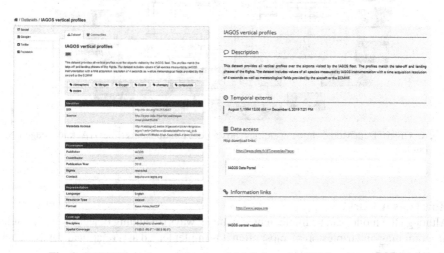

Fig. 3. B2FIND single record entry which links to IAGOS Landing page.

The effort spent on implementing the flagship product catalogue was useful as it initiated concrete technical developments on both sides (e.g. regarding CSW harvesting or enhanced B2FIND schema including <Contributor>). Nonetheless, it is questionable whether B2FIND is an adequate catalogue for ENVRIplus RI 'data products' as it is first and foremost a search portal for research data (and not services).

5   Cataloguing Using CERIF

5.1   EPOS Implementation

CERIF[47] (Common European Research Information Format) is an EU Recommendation to the Member States for research information since 1991. In 2000 CERIF was updated to a richer model, moving from a model like the later Dublin Core to the CERIF as used today: an extended-entity-relational-temporal model. The European Commission requested euroCRIS to maintain, develop and promote CERIF as a standard. It is a data model (Fig. 4) based on EERT (extended entity-relationship modelling with temporal aspects).

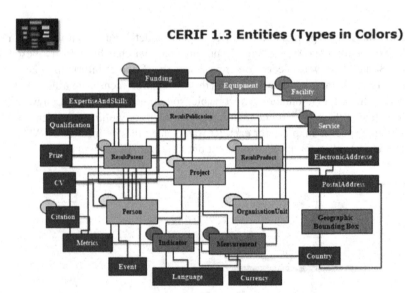

Fig. 4. CERIF Data Model showing entities (boxes) and relationships (lines) (Acknowledgement Brigitte Jörg).

How Does It Work?
Although the model can be implemented in many ways (including object-oriented, logic programming and triplestores), most often it is implemented as a relational database but

---

[47] An introductory presentation on CERIF: https://www.eurocris.org/cerif/main-features-cerif Tutorial: https://www.eurocris.org/community/taskgroups/cerif.

with a particular approach thus ensuring referential and functional integrity. CERIF has the concept of base entities representing real-world objects of interest and characterised by attributes. Examples are project, organization, research product (such as dataset, software), equipment and so on. The base entities are linked with relationship entities which describe the relationships between the base entities with a role (such as owner, manager, author) and date-time start and end so giving the temporal span of the relationship. In this way versioning and provenance are 'built-in'.

CERIF also has a semantic layer (ontologies). Using the same base entity/relationship entity structure it is possible to define relationships between (multilingual) terms in different ontologies. The terms are used not only in the 'role' attribute of linking relations (e.g. owner, manager and author) but also to manage controlled lists of attribute values (e.g. ISO country codes). CERIF provides for multiple classification schemes to be used – and related to each other.

Mappings have been done from many common metadata standards (DC, DCAT, ISO19115/19139, eGMS, DDI, CKAN(RDF), RIOXX and others) to/from CERIF, emphasizing its richness and flexibility.

Some Existing Use Cases

EPOS uses CERIF for its catalogue because of the richness for discovery, contextualisation and action and because of the built-in versioning and provenance, important for both curation and contextualisation. The architecture of the software associated with the catalogue (ICS: Integrated Core Services) is based on microservices (Fig. 5).

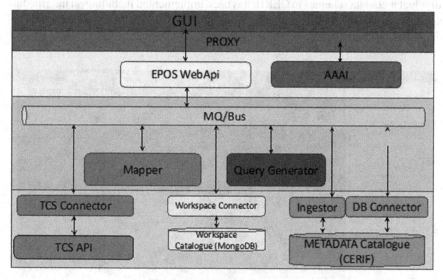

Fig. 5. EPOS ICS architecture.

The implementation uses PostgreSQL as the RDBMS and has been demonstrated on numerous occasions (Fig. 7). A mechanism for harvesting metadata from the various domain groups of EPOS (TCS: Thematic Core Services) and converting from their

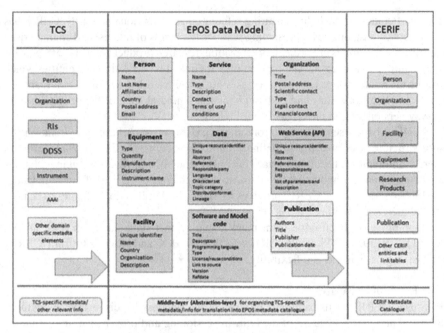

Fig. 6. EPOS metadata harvesting architecture.

individual metadata schemes to CERIF has been implemented including an intermediate stage using EPOS-DCAT-AP - a particular application profile of the DCAT standard [11]. (Figure 6).

Fig. 7. EPOS user interface.

CERIF thus provides EPOS users with a homogeneous view over heterogeneous assets allowing cross-disciplinary research as well as within-domain research.

The integration of metadata from different domains within EPOS is accomplished by a matching/mapping/harvesting/conversion process: to date 17 different metadata 'standards' from the RIs within EPOS have been mapped. The mapping uses 3 M technology[48] (from FORTH-ICS[49]) as used in the VRE4EIC project. The conversion is done in two steps, from the native metadata format of a particular domain to EPOS-DCAT-AP and thence to CERIF. This is to reduce the burden on the IT staff in the particular domains since their metadata standards are typically DC, ISO19115/19139, DCAT and so closer to DCAT than to CERIF. The onward conversion to CERIF not only permits richer discovery/contextualization/action but also provides versioning, provenance and curation capabilities while allowing metadata enrichment as the domains progressively provide richer metadata as needed for the processing they wish to accomplish.

euroCRIS also provide an XML linearization of CERIF for interoperation via web services, as well as scripts for the commonly-used RDBMS implementations.

The CERIF schema is documented[50] with a navigable model in TOAD[51].

CERIF has been used successfully within EPOS in the context of ENVRIplus. However, it is very widely used in research institutions and universities and in research funding organisations throughout Europe and indeed internationally. Of the 6 SMEs providing CERIF systems to the market, one has been taken over by Elsevier and one by Thomson-Reuters and thus incorporating CERIF in their products. OpenAIRE[52] uses the CERIF data model and it has influenced strongly the data model of ORCID[53].

The EPOS CERIF catalogue content has been loaded into an RDBMS at IFREMER which demonstrates portability and ease of set-up. The current work is to provide the user interface software to be used at that location. In parallel work proceeds on (a) converting CERIF to the metadata format based on DataCite and integrated with CKAN used at EUDAT for inclusion in the EUDAT B2FIND catalogue. Unfortunately, conversion from the B2FIND catalogue (based on CKAN) to CERIF is not possible because the records cannot be made available by the hosting organisation, largely due to resource limitations.

CERIF is natively FAIR since it supports all four aspects of the FAIR principles. Because of its referential and functional integrity, formal syntax and rich declared semantics CERIF is more machine-actionable than most metadata standards which usually require human intervention to interpret the metadata e.g. for the composition of workflows.

5.2 VRE4EIC and ENVRI

To prototype the use of CERIF as a joint catalogue service combining datasets from multiple RIs for use by a single VRE, a collaboration was established between ENVRIplus

---

[48] https://www.ics.forth.gr/isl/index_main.php?l=e&c=721.

[49] https://www.ics.forth.gr/.

[50] https://www.eurocris.org/Uploads/Web%20pages/CERIF-1.3/Specifications/CERIF1.3_FDM.pdf.

[51] https://www.eurocris.org/Uploads/Web%20pages/CERIF-1.5/MInfo.html.

[52] https://www.openaire.eu/.

[53] https://orcid.org/.

and the VRE4EICproject. VRE4EIC concerned itself with the development of a standard reference architecture for virtual research environments, as well as the prototyping of exemplar building blocks as prescribed by that reference architecture. In particular, the project consortium developed VRE4EIC Metadata Service to demonstrate how data from multiple RIs might be harvested using a variety of protocols and techniques and then provided via a common portal. X3 ML mappings [12] from standards such as ISO 19139 [10] and DCAT to CERIF [13] were used to automatically ingest metadata published by different RIs to produce a single resource catalogue.

The VRE4EIC Metadata Service was developed in accordance with the e-VRE Reference Architecture [14], providing the necessary components to implement the functionality of a metadata manager as prescribed by the architecture [17]. The purpose of the resulting portal was to provide faceted search over a single CERIF-based VRE catalogue containing metadata harvested from a selection of environmental science data sources. The search was therefore based on the composition of queries based on the context of the research data, filtered by organisations, projects, sites, instruments, and people as shown in Fig. 8.

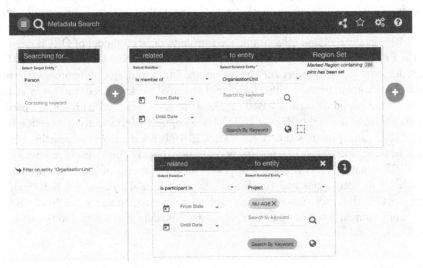

Fig. 8. The VRE4EIC metadata portal in action: searching for people that are members of an organisation which participated in the 'NU-AGE' project.

The portal (maintained at CNR-ISTI, Italy) supports geospatial search, export and storage of specific queries, and the export of results in various formats such as Turtle RDF and JSON. The CERIF catalogue itself was implemented in RDF (based on an OWL 2 ontology [15] using a Virtuoso data store[54], and was structured according to CERIF version 1.6[55]. Metadata harvested from external sources were converted to CERIF RDF

---

[54] https://virtuoso.openlinksw.com/.
[55] https://www.eurocris.org/cerif/main-features-cerif.

using the X3 ML mapping framework[56]; the mapping process itself was as illustrated in Fig. 9:

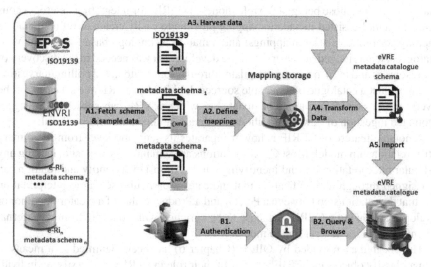

Fig. 9. e-VRE metadata acquisition and retrieval workflow: metadata records are acquired from multiple sources, mapped to CERIF RDF and stored in the VRE catalogue; authenticated VRE users then query data via the e-VRE.

1. Sample metadata, along with their corresponding metadata schemes, were retrieved for analysis. In addition to metadata from ENVRI and EPOS also records from CRIS (Current Research Information Systems which describe projects, persons, outputs, and funding) were harvested.
2. Mappings were defined that dictate the transformation of selected RDF and XML based schemas into CERIF RDF.
3. Metadata is retrieved from different data sources in their native formats, e.g. as ISO 19139 or CKAN[57] metadata (specifically as used in B2FIND within EUDAT in the context of ENVRI).
4. These mappings could then be used to transform the source metadata into CERIF format.
5. The transformed metadata was then ingested into the CERIF metadata catalogue.

Once ingested, these metadata became available to users of the portal, who could query and browse the metadata catalogue upon authentication via a front-end authentication/authorisation service. X3ML mappings were constructed using the 3M Mapping Memory Manager[58]. Among other functions, 3M supported the specification of generators to produce unique identifiers for new concepts constructed during translation of

---

[56] https://www.ics.forth.gr/isl/index_main.php?l=e&c=721.

[57] https://ckan.org/.

[58] https://github.com/isl/Mapping-Memory-Manager.

terms. Mappings into CERIF RDF were produced for Dublin Core, CKAN, DCAT-AP, and ISO 19139 metadata, as well as RI architecture descriptions in OIL-E.

The VRE4EIC Metadata Service demonstrated many desirable characteristics for a catalogue service, those being: a flexible model in CERIF for integrating heterogeneous metadata; a tool-assisted metadata mapping pipeline to easily create or refine metadata mappings or refine existing mappings; and a mature technology base for unified VRE catalogues. It was judged however that more development was needed in the discovery of new resources and the acquisition of updates through some automated polling/harvesting system against a catalogue of amenable sources. In this respect, RI-side services for the advertisement of new resources or updates to which a VRE can subscribe to trigger automated ingestion of new or modified metadata would be particularly useful.

A notable feature of CERIF is how it separates its semantic layer from its primary entity-relationship model. Most CERIF relations are semantically agnostic, lacking any particular interpretation beyond identifying a link. Almost every entity and relation can be assigned through a classification that indicates a particular semantic interpretation (e.g. that the relationship between a Person and a Product is that of a creator or author or developer), allowing a CERIF database to be enriched with concepts from an external semantic model (or several linked models).

The vocabulary provided by OIL-E (Chapter 6) has been identified as a means to further classify objects in CERIF in terms of their role in an RI, e.g. classifying individuals and facilities by the roles they play in research activities, datasets in terms of the research data lifecycle, or computational services by the functions they enable. This provides additional operational context for faceted search (e.g. identifying which processes generated a given data product) but providing additional context into the scientific context for data products (e.g. categorising the experimental method applied or the branch of science to which it belongs) is also necessary. Environmental science RIs such as AnaEE and LTER-Europe are actively developing better vocabularies for describing ecosystem and biodiversity research data, building upon existing SKOS vocabularies.

6   Future Directions and Challenges for Cataloguing

To demonstrate cataloguing capabilities a two-pronged approach was adopted.

Some records describing 'data products' were created from several RIs and ingested by B2FIND. This exposed the effort of metadata mapping but also the capability of a catalogue with metadata from different domains with unified syntax (but not necessarily unified semantics). This catalogue certainly demonstrated the potential for a homogeneous view over heterogeneous assets described by their metadata converted to a common format. However, the relatively limited schema used in EUDAT B2FIND means that some richness from the original ENVRI RI metadata records was lost.

Separately the EPOS metadata catalogue of services was used as an exemplar of the use of CERIF for integrated cataloguing, curation and provenance and via the associated VRE4EIC project the harvesting, mapping and conversion to CERIF of heterogeneous assets from multiple sources was demonstrated. Furthermore, CERIF provided a richer metadata syntax and semantics although - of course - if the source ENVRI RI catalogue had only limited metadata the full richness could not be achieved. There was some investigation in VRE4EIC of enhancing metadata by inferential methods since the formal syntax, referential and functional integrity and declared semantics of CERIF lend themselves to logic processing.

The objective of these two parallel exercises was to allow RIs to see what can be achieved – and what effort is necessary - in the integration of heterogeneous metadata describing assets to permit homogeneous cross-domain (re-)use of assets.

Further enhancements and improvements of the mapping (from various metadata formats used by the RIs to a canonical format) are necessary before the ENVRIplus records could be published and be searchable in the production B2FIND portal. Within EPOS 17 different metadata formats had to be mapped and converted to be ingested into the CERIF catalogue and made available for (re-)use and in VRE4EIC further heterogeneous assets were added. The effort of correct matching and mapping between metadata standards should not be underestimated but – once achieved – can provide homogeneous access over heterogeneous asset descriptions and hence support a portal functionality allowing the end-user to gain interoperability.

As indicated by K. Jeffery (see Chapter 7: the choice of the metadata elements in the catalogue (including their syntax and semantics) is crucial for the processes not only of curation but also of provenance and catalogue management and utilisation for dataset discovery and download. The RIs have different metadata formats and each has its own roadmap or evolution path improving metadata as required by their community. Unfortunately, there are many metadata standards, some general (and usually too abstract for scientific use) and some detailed and domain-specific (but not easily mapped against other formats). The need for rich metadata is becoming generally accepted. As mentioned by authors from the EOSC Pilot project [16] "Minimum and common metadata is useful for data discovery and data access. Rich metadata formats can be complex to adopt, but have the advantage of making data more "usable" by both humans and machines".

It is planned to continue – in the ENVRI community - with the EUDAT B2FIND catalogue (maintained by EUDAT) and also to continue the work with CERIF (maintained by EPOS), anticipating the need for richer metadata than the B2FIND schema for at least some of the ENVRI RIs. CERIF already can handle the functionality associated with services – and other RI assets - as required in the EOSC (European Open Science Cloud). In particular, EUDAT/B2FIND is concentrated on datasets whereas the EPOS CERIF catalogue - while also handling datasets, workflows, software, equipment and other assets - initially concentrated on services to ensure alignment with the emerging EOSC. A mapping between CERIF and the draft metadata standard for EOSC services has been done.

The overall strategy is to make cataloguing technology available to the ENVRI RIs for them to choose how they wish to proceed, considering also other International obligations for interoperability which may determine particular metadata standards. This means that it is likely for the foreseeable future that ENVRI will need to support a range of metadata standards - among the RIs, internationally and also to align with general efforts such as schema.org from Google and associated dataset search - but that to interoperate them a canonical rich metadata schema will be required. The work is open to be shared among any in the ENVRI community who wish to avail themselves of the software, techniques and know-how.

Acknowledgements. This work was supported by the European Union's Horizon 2020 research and innovation programme via the ENVRIplus project under grant agreement No. 654182.

References

1. DIRECTIVE 2003/4/EC OF THE EUROPEAN PARLIAMENT AND OF THE COUNCIL of 28 January 2003 on public access to environmental information and repealing Council Directive 90/313/EEC. https://eur-lex.europa.eu/LexUriServ/LexUriServ.do?uri=OJ:L:2003:041:0026:0032:EN:PDF. Accessed 04 Dec 2019
2. Riley, J.: NISO: Understanding Metadata (2017). https://groups.niso.org/apps/group_public/download.php/17443/understanding-metadata
3. Alemu, G., Stevens, B.: An Emergent Theory of Digital Library Metadata - Enrich then Filter. Chandos Information Professional Series. Elsevier, Amsterdam (2015)
4. Haslhofer, B., Klas, W.: A survey of techniques for achieving metadata interoperability. ACM Comput. Surv. (CSUR) **42**(2) (2010). http://eprints.cs.univie.ac.at/79/1/haslhofer08_acmSur_final.pdf
5. ISO 19115-1:2014: Geographic information—Metadata—Part 1: Fundamentals. ISO standard, International Organization for Standardization (2014)
6. DataCite Metadata Working Group. DataCite Metadata Schema for the Publication and Citation of Research Data. Version 4.1 (2017). http://doi.org/10.5438/0015
7. Parnell, P., et al.: Dublin Core: An Annotated Bibliography (2011). https://pdfs.semanticscholar.org/a614/cfb06d53ed8f0829370eab47bef02639f191.pdf
8. Erickson, J., Maali, F.: Data catalogue vocabulary (DCAT). W3C recommendation, W3C (2014). http://www.w3.org/TR/2014/REC-vocab-dcat-20140116/
9. Wilkinson, M., Dumontier, M., Aalbersberg, I., et al.: The FAIR Guiding Principles for scientific data management and stewardship. Sci. Data **3**, 160018 (2016). https://doi.org/10.1038/sdata.2016.18
10. ISO 19139:2007: Geographic information—Metadata—XML schema implementation. ISO/TS standard, International Organization for Standardization (2007)
11. Trani, L., Atkinson, M., Bailo, D., Paciello, R., Filgueira, R.: Establishing core concepts for information-powered collaborations. FGCS **89**, 421–437 (2018)
12. Marketakis, Y., et al.: X3ML mapping framework for information integration in cultural heritage and beyond. Int. J. Digit. Libr. **18**(4), 301–319 (2016). https://doi.org/10.1007/s00799-016-0179-1
13. Jörg, B.: CERIF: the common European research information format model. Data Sci. J. **9**, CRIS24–CRIS31 (2010). https://doi.org/10.2481/dsj.CRIS4
14. Remy, L., et al.: Building an integrated enhanced virtual research environment metadata catalogue. J. Electron. Libr. (2019). https://zenodo.org/record/3497056
15. W3C OWL Working Group: OWL 2 web ontology language. W3C recommendation, W3C (2012). https://www.w3.org/TR/2012/REC-owl2-overview-20121211/
16. Asmi, A., et al.: 1st Report on Data Interoperability - Findability and Interoperability. EOSCpilot deliverable report D6.3. Submitted on 31 December (2017). https://eoscpilot.eu/sites/default/files/eoscpilot-d6.3.pdf
17. Martin, P., Remy, L., Theodoridou, M., Jeffery, K., Zhao, Z.: Mapping heterogeneous research infrastructure metadata into a unified catalogue for use in a generic virtual research environment. Future Gener. Comput. Syst. **101**, 1–13 (2019). https://doi.org/10.1016/j.future.2019.05.076

18. Zhao, Z., et al.: Reference model guided system design and implementation for interoperable environmental research infrastructures. In: 2015 IEEE 11th International Conference on e-Science, pp. 551–556. IEEE, Munich (2015). https://doi.org/10.1109/eScience.2015.41

19. Chen, Y., et al.: A common reference model for environmental science research infrastructures. In: Proceedings of EnviroInfo 2013 (2013)

20. Martin, P., et al.: Open information linking for environmental research infrastructures. In: 2015 IEEE 11th International Conference on e-Science, pp. 513–520. IEEE, Munich (2015). https://doi.org/10.1109/eScience.2015.66

Identification and Citation of Digital Research Resources

Margareta Hellström[1] ⓘ, Maria Johnsson[2] ⓘ, and Alex Vermeulen[3](✉) ⓘ

[1] Department of Physical Geography and Ecosystem Science, Lund University, Sölvegatan 12, 22362 Lund, Sweden
margareta.hellstrom@nateko.lu.se
[2] Department of Scholarly Communication, Lund University Library, Box 3, 221 00 Lund, Sweden
maria.johnsson@ub.lu.se
[3] ICOS ERIC - Carbon Portal, Sölvegatan 12, 22362 Lund, Sweden
alex.vermeulen@icos-ri.eu

Abstract. Environmental research infrastructures are often built on a large number of distributed observational or experimental sites, run by hundreds of scientists and technicians, financially supported and administrated by a large number of institutions. It becomes very important to acknowledge the data sources and their providers. There is also a strong need for common data citation tracking systems that allow data providers to identify downstream usage of their data so as to demonstrate their importance and show the impact to stakeholders and the public. This chapter highlights identification and citation in environmental RIs, reviews available technologies and develops common services for these operations. This chapter presents a suggested common system design for Identification and Citation, as well as an outline for negotiations and discussions with publishers and other actors in the scholarly data management and curation world.

Keywords: Identification · Citation · Persistent identifiers

1 Introduction

To perform data intensive sciences one often requires data that are managed by different institutions. Observations and measurements from infrastructures in environmental and earth sciences are a particularly strong example of this multitude of sources. The result of the scientific analysis based on this data depends heavily on the access to high quality data and the proper citing of those data sources or data sets when publishing the final results becomes an important practice for acknowledging the original data providers and for keeping the study reproducible.

Environmental research infrastructures are often construed from a large number of distributed observational or experimental sites, run by hundreds of scientists and technicians, financially supported and managed by a large number of institutions. If these data are shared under an open access policy this becomes another important reason to

Z. Zhao and M. Hellström (Eds.): Towards Interoperable Research
Infrastructures for Environmental and Earth Sciences, LNCS 12003, pp. 162–175, 2020.
https://doi.org/10.1007/978-3-030-52829-4_9

acknowledge the data sources and their providers. A data citation tracking system that allows the data providers to identify downstream usage of their data and assess the impact of their data to stakeholders and the public is then a strong requirement.

Furthermore, a common policy model is needed for persistent identifiers for publishing and citing data. Moreover, the services for assigning and handling identifiers and for retrieving data content based on identifiers will also have to be provided.

In this chapter we will discuss the building blocks to fulfil theses needs, building on existing approaches and current activities undertaken by ENVRI partners, and—if needed—synchronise with developments that arise from up-coming studies and projects from both service providers (e.g. ePIC, DataCite and EUDAT) and initiatives based in research organizations (e.g. THOR and OpenAIRE). The work described has been operated in close cooperation with existing initiatives (e.g. Research Data Alliance and ICSU WDS) and will elaborate a common data citation solution for the involved RIs.

This chapter presents a strategy developed during the ENVRIplus project to negotiate with external organisations. The content is mainly based on the public deliverable D6.1, D6.2 and D6.3 of the ENVRIplus project[1].

2 Background

2.1 Identification

A number of approaches have been applied to solve the question of how to unambiguously identify digital research data objects [1]. Traditionally, researchers have relied on their own internal identifier systems, such as encoding identification information into filenames and file catalogue structures, but this is neither comprehensible to others, nor sustainable over time and space [2]. Instead, data object identifiers should be unique "labels", registered in a central registry database that contains relevant basic metadata about the object, including a pointer to the location where the object can be found as well as basic information about the object itself. Exactly which metadata should be stored in the identifier registry, and in which format, is a topic under discussion, see e.g. [3]. Many environmental observational datasets pose a special challenge in that they are not reproducible, which means that also fixity information (checksums or even "content fingerprints") should be tied to the identifier [4].

As a complement to the registry database, a lookup, or resolver, service is essential. When supplied with a valid identifier, the service should either return the associated metadata, or – as is more common – redirect to the supplied resource location. This can either be a direct link to the persistently identified object itself (e.g. a path to a file stored on a disk), or to a so-called landing page. The latter typically contains some basic metadata about the object, as well as information about how to access it.

In [1], the authors provide a comprehensive summary of the pros and cons of different identifier schemes, and also assess nine persistent identifier technologies and systems. Based on a combination of technical value, user value and archive value, DOIs (Digital Object Identifiers provided by DataCite) scored highest for overall functionality, followed by general handles (as provided by e.g. CNRI and DONA) and ARKs (Archive

---

[1] EU H2020 ENVRIplus www.envriplus.eu.

Resource Keys). DOIs have the advantage of being well-known to the scientific community via their use for scholarly publications, and this has contributed to their successful application to e.g. geoscience datasets over the last decade [5]. General Handle PIDs have up to now mostly been used to enable referencing of data objects in the pre-publication steps [6] of the research data life cycle (illustrated in Fig. 1). They could however in principle equally well be applied to finalised "publishable" data.

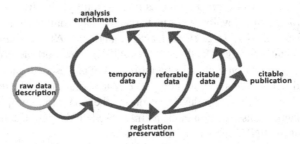

Fig. 1. The research data life cycle. Data intensive research is highly collaborative. Allocating persistent identifiers to data objects supports (re-)use and sharing of data also in early stages of the research life cycle [6].

Persistent identifiers systems are also available for research-related resources other than digital data & metadata, articles and reports—it is now possible to register many other objects, including physical samples (IGSN), software, workflow processing methods—and of course also people and organisations (ORCID, ISNI). In the expanding "open data world", PIDs are an essential tool for establishing clear links between all entities involved in or connected with any given research project [7].

2.2 Citation

The FORCE11 Joint Declaration of Data Citation Principles (JDDCP) [8] states that in analogy to articles, reports and other written scholarly work, also data should be considered as legitimate, citable products of research. (There is however currently an ongoing discussion as to whether datasets are truly "published" if they haven't undergone a standardised quality control or peer-review, see e.g. [9].) Thus, any claims in scholarly literature that rely on data must include a corresponding citation, giving credit and legal attribution to the data producers, as well as facilitating the identification of, access to and verification of the used data (subsets). A generic workflow for data citation is presented in Fig. 2. The workflow consists of a citation from a document to a dataset, a landing page in the repository where the dataset is stored, and the dataset itself.

Data citation methods must be flexible, which implies some variability in standards and practices across different scientific communities [8]. However, to support interoperability and facilitate interpretation, the citation should preferably contain a number of metadata elements that make the dataset discoverable, including author, title, publisher, publication date, resource type, edition, version, feature name and location. Especially important, the data citation should include a persistent method of identification that is

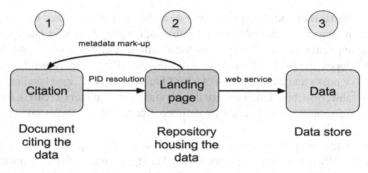

Fig. 2. A generic data citation workflow. (http://force11.github.io/data-citation-primer/)

globally unique and contains the resource location as well as (links to) all other pertinent information that makes it human and machine actionable. In some (sensitive) cases, it may also be desirable to add fixity information such as a checksum or even a "content fingerprint" in the actual citation text [4].

Finding standards for citing subsets of potentially very large and complex datasets poses a special problem, as outlined by [10], as e.g. granularity, formats and parameter names can differ widely across disciplines. Another very important issue concerns how to unambiguously refer to the state and contents of a dynamic dataset that may be variable with time, e.g. because new data are being added (open-ended time series) or corrections introduced (applying new calibrations or evaluation algorithms) [11].

Both these topics are of special importance for environmental research today.

A number of surveys have indicated that the perceived lack of proper attribution of data is a major reason for the hesitancy felt by many researchers to share their data openly [4, 12, 13]. This attitude also extends to allowing their data to be incorporated into larger data collections, as it is often not possible to perform micro-attribution – i.e., to trace back the provenance of an extracted subset (that was actually used in an analysis) to the individual provider – through the currently used data citation practices.

3   Components of PID Systems

3.1   Common PID Types: The Persistent Identifier Zoo

In this section, we present an overview of seven of the most commonly used persistent identifier types. The underlying study was performed in the summer of 2016 by Huber and co-workers, and the numbers and statistics represent the status of the re3data.org registry[2] at that time.

The Handle System (HS)
Arguably the biggest impact in the field of persistent identification of digital research resources was achieved by the Handle System [14]. The Handle System (HS) describes a minimal set of requirements for an infrastructure for the identification of objects in

---

[2] http://www.re3data.org/.

a digital infrastructure and how the identity of an object can be related to its location. The system is agnostic to the contents of the objects, keeping it open for interoperability with future applications. The HS separates the identifier from the resolving mechanism, making it independent of HTTP and DNS but in practice, the system is mostly leveraged using a HTTP proxy that allows the use of a RESTful API and URLified handles. The HS supplies a stable, distributed platform for the resolution of identifiers to URLs, including methods more sophisticated than HTTP redirects like template handles and embedded metadata.

In the sample of 1381 repositories listed in the re3data repository at the time of the study, the HS is used by 102 repositories. Handle is mainly used by institutional repositories, which might be linked to the role of Handle as an identifier in repository software like DSpace[3].

Besides the governance of top-level namespaces the HS does not provide more than the technical platform and comes with no obligations with respect to policies, for instance towards the persistence of the resolution of identifiers towards their targets.

Digital Object Identifier (DOI)
Looking at the 475 repositories using any kind of PID system, the most commonly implemented identifier type was the digital object identifier (DOI). DOIs, which were introduced in 1998 by the International DOI Foundation (see http://www.doi.org/), were used by 275 out of those 475 repositories, meaning that the use of the DOI eclipsed all other persistent identifiers. The use of DOI persistent identification of data initiated by a project funded by the German research foundation in 2003 [5]. DOI were chosen because of their already established part in the scholarly publication infrastructure.

The Digital Object Identifier (DOI) makes use of the Handle system and uses its namespace "10.[subnamespace]/". DOI is distinguished from other uses of the HS by the underlying social contract. In this social contract, participating parties pledge to maintain the resolution of identifiers to web endpoints indefinitely. This means that identifiers will theoretically always resolve to somewhere even though the referenced object might no longer exist. (See Sect. 4.3 and Sect. 4.4 for a discussion of "tombstones".)

"Cool" Uniform Resource Identifiers (CoolURIs)
Compared to the strict criteria of Nestor [15] and other related efforts, "cool" (meaning unchanging or static) Uniform Resource Identifiers (CoolURIs) somehow represent an anarchic view on identifiers. Similar to URN, the idea of CoolURIs goes back to early ideas about identification and location of objects on the web. The idea of CoolURI [16] is fundamental for the Semantic Web. It is based on Uniform Resource Identifiers (URIs) which, by proclamation, will not change. They make use of standard HTTP functionalities, in particular content negotiation[4], to enable the URI to be resolved to different representations (RDF, HTML) of the same object. CoolURIs allow webmasters to maintain the persistence of their resource identifiers, the URIs, with a minimum of effort and without a centralised PID system.

---

[3] http://www.dspace.org/.
[4] https://en.wikipedia.org/wiki/Content_negotiation.

Advocates of the CoolURI system reasoned that the use of HTTP functionalities is a bonus, suggesting that URI should be actionable. However, over the years this has proven to be unstable, the main reason for this being the fragility of base URL. The result of unstable base URLs will be "link-rot on steroids". There is already anecdotal evidence of base URL failures from the validation of xml schemas in long-term archiving of XML documents by the national libraries.

The CoolURI concept relies on HTTP as resolving mechanism and assumes that the HTTP protocol will be around for a long time.

3.2 Identifiers for Non-data Entities

Persistent identifiers are useful for many other entities than data objects and scientific articles. In the following, we list a selection of such entities which have a special interest to ENVRIplus partner RIs.

Identifiers for People
During the last five years, more and more researchers have become used to register-ing with ORCID and then using their ORCID IDs for communications with journal publishers, their funding agencies and in other research contexts. However, also other individuals associated with research projects (and active in producing research outputs) – such as research engineers, data curators, programmers and many others – should be encouraged to sign up for ORCID or similar persistent identifiers schemes for individu-als such as ISNI. The personal IDs can then be stored in RI catalogues, and be included in metadata objects and DataCite records.

Identifiers for Organizations
The organisational entities involved in research projects should in principle obtain per-sistent identifiers, for example via ISNI and ROR. However, this may not be as simple and clear-cut as for persons, since reorganisations and restructuring may occur at any time. For more information about ISNI, see Sect. 6.1.4.

Identifiers for Instrumentation and Sensors
By assigning unique and persistent identifiers to sensors and other instrumentation, and using these PIDs consistently in both cataloguing and curation, researchers can simplify the management and collection of observation metadata records, and facilitate property lookup and provenance tracing throughout all steps of the research data processing cycle.

Identifiers for Physical Samples
In order to simplify the referencing of physical samples, they can be registered and assigned a unique identifier. One initiative that provides this possibility in Earth sciences is System for Earth Sample Registration (SESAR), which allocates IGSNs (International Geological Sample Numbers) to environmental samples. (See http://www.geosamples.org/igsnabout for more details.)

Identifiers for Data Content Types

In order to facilitate (re-)use of datasets, especially in the context of machine-actionable workflows, it is useful to make use of persistently identified Data Type definitions. These should include a basic description of the characteristics of a given data or variable, but can also contain information on which software should be used to process it further. See e.g. the recommendations of the RDA Data Type Registries working group (https://www.rd-alliance.org/group/data-type-registries-wg/outcomes/data-type-registries).

Identifiers for Software

GitHub[5] and similar software repositories support versioning, and as such allow the code author to link directly by URL to a specific code package or file. In GitHub, objects can themselves be linked to dataset DOIs, so there are possibilities of cross-referencing. However, at the moment it is not yet possible to provide a DOI or any other PID to software codes or packages in GitHub. Notably, the German Climate Computing Centre DKRZ (see Sect. 7.1.2) is about to apply for national project funding to offer sustainable production and long term storage of scientific software. This will account for versioning and include the use of persistent IDENTIFIERS. See also Sect. 8.3.1.

Identifiers for Workflows

Workflows and workflow engines are being increasingly used also in environmental and Earth sciences as a means of organising and sharing scientific computations and analysis procedures. Referring to specific workflows simplifies the collection of provenance records associated with datasets. Registering workflows and assigning them PIDs promotes efficient documentation of workflows, allows making unambiguous references to them in e.g. provenance descriptions, and supports their reuse by both humans and machines. See Sect. 8.3.1 for more information.

4   Identification and Citation in Practice—Recommendations to RIs

Specifically, which type of persistent identifier is used by any RI should be dictated by the needs of both the RI and its typical end user communities. There are many different options (see Sect. 4.1). In general, those based on the Handle System (for example, DOIs from DataCite and PIDs from e.g. ePIC), as well as ORCIDs for people are at present the most commonly used by ENVRIplus partners (based on the questionnaire). The amount of metadata that is mandatory to provide at the time of identifier registration ("minting") varies.

4.1   Identification Best Practices for RIs

Research Infrastructures should strive to implement the use of PIDs for all of the following categories. (In some cases such as organisational entities, it may not yet be practical to assign PIDs, as the currently relevant registration schemes are poorly equipped to handle entities that frequently change names, stewardship etc.)

---

[5] https://github.com/.

- data objects (files, databases etc.)
- metadata objects
- articles, reports and other documents related to the data
- people, including everyone involved in the data production chain
- organizations (agencies, institutes, and RIs themselves) involved in the data production chain
- sensors and sensor platforms, measurement stations, cruises, measurement campaigns
- physical samples.

In addition, comprehensive use of PIDs should be considered for

- queries used for accessing and retrieving (subsets of) datasets
- data content types
- software releases used in the data processing
- workflows used in the data processing.

4.2 Citation Best Practices for RIs

RIs should strive to follw the following recommendations for data citation, based on the review of data citation best practices and recommendations from relevant organisations including [4, 8, 17]:

Technical aspects:

- All datasets intended for citation have a globally unique PID that can be expressed as an unambiguous URL
- A PID expressed as a URL resolves to a landing page for a dataset
- The landing page of a dataset is both human-readable and machine-readable (and preferably machine-actionable) and contains the dataset's PID
- PIDs for datasets support multiple levels of granularity (including fine-grained subsets as well as collections)
- Datasets are described with rich metadata (to track provenance information and to create meaningful citations and (including the identifier of the dataset))
- Metadata are accessible even if a dataset is no longer accessible
- RIs provide a robust resolver and registry for resolving PIDs and for data discovery
- Metadata protocols and standards are used, that ensure interoperability with related stakeholders, e.g. cataloguing and indexing services
- Data are published with a clearly defined data usage license.

Citation practices:

- RIs actively promote data citation (to users, publishers and other stakeholders in their research community (e.g. by providing documentation and how-tos) and by providing common citation formats to users)
- Citation methods are flexible to support each community while still ensuring interoperability across communities.

5 Cases in ENVRI

5.1 Development of a Citation and Usage Tracking System for Greenhouse Gas

ICOS, Integrated Carbon Observation System, is a pan-European research infrastructure with a mission to provide standardised, long term, high precision and high-quality observations on the carbon cycle and Green House Gas (GHG) budgets and their perturbations. The ICOS observing network consists of over 140 observation stations, each related to one or more of the three domains Atmosphere, Ecosystem and Ocean, and operated by its (currently 14) member countries. The collected data is processed and quality controlled at Thematic Centres (one for each domain), before being openly distributed via the ICOS data centre named Carbon Portal (CP).

The ICOS Carbon Portal is designed to manage and/or distribute data objects of a number of different categories. All data objects are assigned a Handle System-based persistent identifier at the time of ingestion into the Carbon Portal repository. The PID of these individual data objects can be resolved via e.g. the handle.net resolver service, which redirects to the object's landing page hosted by the Carbon Portal. The landing page lists the most relevant metadata of the object, including a direct link to access to the data object.

Objects can be registered with DataCite, either as single objects or as collections of data objects. This means that they are also assigned a DOI and that associated metadata are stored in the DataCite catalogue following the DataCite Metadata Schema[6]. This allows the data objects to be found also through searches on the DataCite portal, and also provides full integration with the Citation Formatter service from Crossref and DataCite.

Resolving the DOI, e.g. via the resolving services of handle.net and the DOI Foundation, results in a redirect to the landing page of the object or object collection, hosted by the Carbon Portal. The DataCite DOI identifiers have the form "10.18160/<suffix>", where "10.18160" is the ICOS-specific DataCite prefix, and <suffix> is a globally unique string. (The strings used for DOIs are also computed starting from the data object hash sum, but are designed to be shorter and easier to read for humans. As in the case of the Handle PIDs mentioned above, additional tests for uniqueness are of course performed before submitting the DOI registration request.)

Any updated versions of a given object, for example dynamic, growing time series or corrected data sets, are assigned a completely new PID. In order to provide unbroken provenance chains, the metadata record of the old version is updated with a link to the superseding object, and vice versa. This strategy is applied to data objects of all types described above.

The Carbon Portal provides a landing page (as shown in Fig. 3) for any digital object described in the metadata store – including data sets, but also stations, data type specifications and concepts. All landing pages are created dynamically, i.e. at the moment that their URL is accessed. This means that the displayed information always reflects the current, most up-to-date information. The format, and what information is shown, will vary between the type of objects, with the richest content provided for data objects.

---

[6] https://schema.datacite.org/.

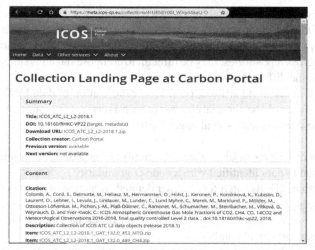

Fig. 3. Screenshot of the ICOS landing page for the data object ICOS_ATC_L2_L2-2018.1 - Collection of ICOS ATC L2 data objects (release 2018.1). The page exposes basic metadata about the data set itself, its contents, and a recommended citation string – the latter includes the data set's DataCite DOI number, https://doi.org/10.18160/RHKC-VP22.

ICOS data is distributed using a Creative Commons Attribution 4.0 International (CC BY 4.0) license[7], which users have to accept before they can access the data. The CC BY license requires end users to give appropriate credit (i.e. citing data when it is used), provide a link to the license, and indicate if changes were made when re-distributing the data. When citing ICOS data, at a bare minimum the data object's persistent digital identifier should be given in a machine-actionable form (as a HTTP URL, for example http://hdl.handle.net/11676/6PrNhZelwXKHLqO41QRsbheu or https://doi.org/10.18160/RHKC-VP22); this minimal form is sufficient for inclusion in provenance records, but for use in scientific literature much more information (contributors, data set name etc.) is of course required.

5.2   Facilitating Quantitatively Correct Data Usage Accounting

In a world of open and free data sharing, it is often necessary to document the use of data products and give this as a quantitative merit to data producers and providers. Since all entities involved in the data production chain face the challenge of having to find sources of continued funding for their efforts while "selling" their data is not an option. They need to justify to funding agencies and users the relevance of their observations and contributions to data production.

An existing analogy to such a use-based merit system are scientific journal publications, where authors receive merit based on the number of "uses" of the article, i.e. based

---

[7] https://creativecommons.org/licenses/by/4.0/.

on the number of citations. Journals are selected by authors and institutions based on the aggregation of those citation scores in recent years, i.e. how visible the result becomes by using a given publication channel. However, aggregating scores, i.e., citation numbers accumulated across repositories, may be difficult to compose if data is stored at different granularities in the different archives.

By analogy to scientific articles, persistent identification of data by Digital Object Identifiers (DOIs) would be a crucial element of such a service for quantitative accounting of data use. However, at least 4 challenges exist:

- **DOI granularity:** This would make usage numbers based on a fine granularity biased in comparison to data identified with a coarser granularity.
- **Data collections:** DOIs can refer to a user defined collection of other datasets, which themselves may be identified by DOIs. The data collection approach makes data very convenient to cite. However, the contribution of different data producers to such a collection can vary significantly.
- **Accounting mechanism:** Indexing agencies will or have been setting up services for counting of use events involving (DOI) identified data. From here, services need to be implemented that break down data use events into the contribution of single data producers, and with a fixed granularity allowing comparisons between data producers.
- **Nature of data use events:** Scientific data can be used in many different ways, e.g. illustration for outreach purposes, trend analysis, constraining models of environmental processes, event analysis, just to mention a few, and data can be accessed once or multiple times for the same use case event. A list of data use types counted towards use accounting, including weighting factors, would typically be agreed on and continuously updated by a cross-domain working group consisting of experts on data production, data management, and data indexing.

In order to meet the challenges, and to work towards implementing accounting services for data use, the project team defined the following tasks in the early project phase:

Data Identification with Homogeneous Granularity in Primary Archives

These DOIs, in this context called primary DOIs, would be used as reference for setting up data use accounting. The ambition in the early project phase was to achieve homogeneous granularity, and thus comparable data use metrics, across repositories and RIs. During the implementation, it turned out that the goal of achieving homogeneous granularity of primary data identifiers across atmospheric RIs was too ambitious. Data products are simply too different in nature among repositories, sometimes even within a single RI. However, primary data identifier granularity should be homogeneous at least within one repository, preferably comparable also among repositories of a single RI.

Transparent Data Accounting When Using Data Collections

When identifying data in larger studies, e.g. global climatology of atmospheric parameters, using primary DOIs, requires quoting hundreds of DOIs, which would be rather inconvenient. The DOI specification provides for coining DOIs for user specified data collections, which are ideally suited to identify data used in larger studies. Ideally, the references would also include further provenance information in order to identify and acknowledge contributors to the data product used.

By interacting with the relevant RDA working group on research data collections[8], this work resulted in a fully finished recommendation for issuing and handling persistent identifiers for data collections that meet the requirement of referencing back to the primary identifiers of the data contained in the collection[9].

Performing Correct Accounting of Data Use

For scientific publications, accounting of use is performed by the indexing agencies. If they offer a similar service for data, it needs to be assured that references to collection identifiers are resolved to the primary identifiers to ensure correct accounting of data use. The task involves a dialogue with the indexing agencies to implement this policy. A dialogue with DataCite as an indexing agency collecting use events involving data DOIs revealed that indexing agencies show little willingness to resolve references to primary identifiers contained in collection DOIs when accounting for data use. From the indexing agencies perspective, this approach makes sense due to the issue of heterogeneous granularity of primary data identifiers across or even within domains. As a result, the task needed to be modified. The service of calculating metrics for data use is moved from the indexing agency to the primary data repository. Based on its own primary DOIs with homogeneous granularity, the primary repository can access data use events stored at the indexing agency, resolve references in collection DOIs, and thus calculate data use metrics comparable across the repository. A prerequisite for this approach would be machine-to-machine access to the indexing agencies data holdings by the primary data archives. A dialogue about this is ongoing and needs to be continued in the near future.

6 Conclusion

This chapter discussed the basic concepts of the data identification and citation, and related standards and best practices. The chapter presented two cases studied in the ENVRIplus project.

Acknowledgements. This work was supported by the European Union's Horizon 2020 research and innovation programme via the ENVRIplus project under grant agreement No 654182.

---

[8] https://www.rd-alliance.org/groups/research-data-collections-wg.html.

[9] https://github.com/RDACollectionsWG/specification/blob/master/Recommendation%20pack age/rda-collections-recommendation.pdf.

References

1. Duerr, R.E., et al.: On the utility of identification schemes for digital Earth science data: an assessment and recommendations. Earth Sci. Inform. **4**, 139–160 (2011). https://doi.org/10.1007/s12145-011-0083-6
2. Stehouwer, H., Wittenburg, P.: RDA Europe Analysis Programme: Survey of EU Data Architectures, Deliverable D2.5 from the RDA Europe project (FP7-INFRASTRUCTURES-2012-1) (2015). https://www.rd-alliance.org/data-architecture-survey-report.html. Accessed 16 May 2020
3. Weigel, T., DiLauro, T., Zastrow, T.: PID Information Types WG final deliverable (2015). http://dx.doi.org/10.15497/FDAA09D5-5ED0-403D-B97A-2675E1EBE786
4. Socha, Y.M. (ed.): Out of cite, out of mind: the current state of practice, policy, and technology for the citation of data. Data Sci. J. **12** (2013). Available at https://www.jstage.jst.go.jp/article/dsj/12/0/12_OSOM13-043/_pdf. Accessed 30 Jan 2017
5. Klump, J, Huber, R., Diepenbroek, M.: DOI for geoscience data - how early practices shape present perceptions. Earth Sci. Inform. **9** (2015). https://doi.org/10.1007/s12145-015-0231-5
6. Schwardmann, U.: Epic - Persistent Identifiers For Eresearch (2015). https://doi.org/10.5281/ZENODO.31785
7. Dodds, L., Phillips, G., Hapuarachchi, T., Bailey, B., Fletcher, A.: Creating Value with Identifiers in an Open Data World. Report from Open Data Institute and Thomson Reuters (2014). http://innovation.thomsonreuters.com/content/dam/opeweb/documents/pdf/corporate/Reports/creating-value-with-identifiers-in-an-open-data-world.pdf
8. Data citation synthesis group: joint declaration of data citation principles. In: Martone M. (ed.) FORCE11, San Diego, CA (2014). https://www.force11.org/group/joint-declaration-datacitation-principles-final. Accessed 30 Dec 2016
9. Parsons, M.A., Duerr, R., Minster, J.-B.: Data citation and peer review. EOS Trans. AGU **91**, 297 (2010). https://doi.org/10.1029/2010EO340001
10. Huber, R., Asmi, A., Buck, J., Lucas, J.M.D., Diepenbroek, M., Michelini, A.: Participants Of The Joint COOPEUS/ENVRI/EUDAT PID Workshop: Data citation and digital identifiers for time series data/environmental research infrastructures (2015). http://figshare.com/articles/Data_citation_and_digital_identifiers_for_time_series_data_environmental_research_infrastructures/1285728, https://doi.org/10.6084/M9.FIGSHARE.1285728
11. Rauber, A., Asmi, A., van Uytvanck, D., Proell, S.: Data Citation of Evolving Data: Recommendations of the Working Group on Data Citation (WGDC) (2015). https://zenodo.org/record/1406002
12. Uhlir, P.F.: Rapporteur, For Attribution—Developing Data Attribution and Citation Practices and Standards: Summary of an International Workshop (August 2011), National Research Council (2012). http://www.nap.edu/openbook.php?record_id=13564. Accessed 16 May 2020
13. Gallagher, J., Orcutt, J., Simpson, P., Wright, D., Pearlman, J., Raymond, L.: Facilitating open exchange of data and information. Earth Sci. Inform. **8**, 721–739 (2015). https://doi.org/10.1007/s12145-014-0202-2
14. Kahn, R., Wilensky, R.: A framework for distributed digital object services. Int. J. Digit. Libr. **6**, 115–123 (2006). https://doi.org/10.1007/s00799-005-0128-x
15. Bütikofer, N.: Catalogue of criteria for assessing the trustworthiness of PI systems, nestorMaterialien, Niedersächsische Staats und Universitätsbibliothek Göttingen, Göttingen, Germany. http://nbn-resolving.de/urn:nbn:de:0008-20080710227. Accessed 16 May 2020

16. Berners-Lee, T.: Cool URIs don't change, World Wide Web Consortium (W3C), Cambridge, MA. https://www.w3.org/Provider/Style/URI.html. Accessed 30 Jan 2017
17. Fenner, M.: A data citation roadmap for scholarly data repositories. Sci. Commun. Educ. (2016). https://doi.org/10.1101/097196

Data Processing and Analytics for Data-Centric Sciences

Leonardo Candela(✉) , Gianpaolo Coro , Lucio Lelii , Giancarlo Panichi ,
and Pasquale Pagano

National Research Council of Italy, Istituto di Scienza e
Tecnologie dell'Informazione "A. Faedo", Via G. Moruzzi, 1, 56124 Pisa, Italy
{leonardo.candela,gianpaolo.coro}@isti.cnr.it

Abstract. The development of data processing and analytics tools is heavily driven by applications, which results in a great variety of software solutions, which often address specific needs. It is difficult to imagine a single solution that is universally suitable for all (or even most) application scenarios and contexts. This chapter describes the data analytics framework that has been designed and developed in the ENVRIplus project to be *(a)* suitable for serving the needs of researchers in several domains including environmental sciences, *(b)* open and extensible both with respect to the algorithms and methods it enables and the computing platforms it relies on to execute those algorithms and methods, and *(c) open-science-friendly*, i.e. it is capable of incorporating every algorithm and method integrated into the data processing framework as well as any computation resulting from the exploitation of integrated algorithms into a "research object" catering for citation, reproducibility, repeatability and provenance.

Keywords: Data analytics · Open science · Virtual Research Environment

1 Introduction

Data processing and analytics are playing a key role in modern science [8]. The paradigms of data-intensive science [20] and open science [5] as well as the opportunities offered by cloud computing and the *as-a-Service* delivery model [1, 21] continuously drive the development of ever new and diverse solutions to data analytics needs. In fact, a plethora of solutions has been designed and developed ranging from programming models to analytics frameworks and systems, notebooks, workflow management systems and science gateways [6, 9, 22, 25]. Such heterogeneity provides a rich possibility for the scientific community to select suitable solutions, case by case; however, it also results in high costs to support the sharing of analytics methods and algorithms across domains.

In order to provide scientists with an analytics platform aiming at offering both (a) the freedom to develop analytics methods by using a rich array of programming approaches (e.g. R scripts, Python programs and Unix compiled code) and (b) a simple use and re-use of analytics methods developed by others according to open science practices, advanced

© The Author(s) 2020
Z. Zhao and M. Hellström (Eds.): Towards Interoperable Research
Infrastructures for Environmental and Earth Sciences, LNCS 12003, pp. 176–191, 2020.
https://doi.org/10.1007/978-3-030-52829-4_10

infrastructures like D4Science have been equipped with a set of services realising a solution for data analytics promoting the above principles [3, 4]. This chapter presents the data analytics solutions developed as part of D4Science infrastructure in the context of ENVRIplus [37], and discusses the benefits resulting from its uptake and exploitation in several use cases.

The rest of the chapter is organised as follows. Section 2 overviews existing solutions for data analytics. Section 3 discusses the data analytics solution developed in the project. Section 4 examines the proposed solution and discusses some use cases. Finally, Sect. 5 concludes the chapter by highlighting future works.

2 State of the Art

Khalifa et al. [22] surveyed existing solutions for data analytics and observed that (i) "existing solutions cover bits-and-pieces of the analytics process" and (ii) when devising an analytics solution there are six pillars representing issues to deal with. The six pillars identified by Khalifa et al. include (i) *Storage*, i.e., how data to be analysed are going to be handled; (ii) *Processing*, i.e., how the pure processing task is going to happen; (iii) *Orchestration*, i.e., how computing resources are going to be managed to reduce processing time and cost; (iv) Assistance, i.e., how scientists and practitioners are provided with facilities helping them to perform their task; (v) *User Interface*, i.e., how scientists and practitioners are provided with the data analytics system to run their analytics, monitor the execution and get back the results; (vi) *Deployment Method*, i.e., how the analytics system is offered to the end-users.

A lot of technologies and approaches have been developed to support data processing and analytics tasks including (i) High-performance computing (HPC) solutions, i.e., aggregated computing resources to realise a "high-performance computer" (including processors, memory, disk and operating system); (ii) Distributed Computing Infrastructures, i.e., distributed systems characterised by heterogeneous networked computers that offer data processing facilities. This includes High-throughput computing (HTC) and cloud computing; (iii) Scientific workflow management systems (WMS), i.e., systems enacting the definition and execution of scientific workflows consisting of: a list of tasks and operations, the dependencies between the interconnected tasks, control-flow structures and the data resources to be processed; (iv) Data analytics frameworks and platforms, i.e., platforms and workbenches enabling scientists to execute analytic tasks. Such platforms tend to provide their users with implementations of algorithms and (statistical) methods for the analytics tasks. These classes of solutions and approaches are not isolated, rather they are expected to rely on each other to provide end-users with easy to use, efficient and effective data processing facilities, e.g. scientific WMS rely on distributed computing infrastructures to actually execute their constituent tasks.

Liew et al. [25] have recently analysed selected scientific WMSs that are widely used by the scientific community, namely: Airavata [30], Kepler [27], KNIME [7] Meandre [26], Pegasus [18], Taverna [31], and Swift [34]. Such systems have been analysed with respect to a framework aiming at capturing the major facets characterising WMSs: (a) processing elements, i.e., the building blocks of workflows envisaged to be either web services or executable programs; (b) coordination method, i.e., the mechanism controlling the execution of the workflow elements envisaged to be either orchestration or choreography; (c) workflow representation, i.e., the specification of a workflow that can meet the two goals of human representation and/or computer communication; (d) data processing model, i.e., the mechanism through which the processing elements process the data that can be bulk data or stream data; (e) optimisation stage, i.e., when optimization of the workflow (if any) is expected to take place that can either be build-time or runtime.

A series of *platforms and frameworks* have been developed to simplify the execution of (scientific) distributed computations. This need is not new; it is actually rooted in high-throughput computing which is a well-consolidated approach to provide large amounts of computational resources over long periods of time. The advent of Big Data and Google MapReduce in the first half of 2000 brings new interests and solutions. Besides taking care of the smart execution of user-defined and steered processes, platforms and environments start offering ready to use implementations of algorithms and processes that benefit from a distributed computing infrastructure.

There exist many data analytics frameworks and platforms, including:

- Apache Mahout[1] is a platform offering a set of machine-learning algorithms (including collaborative filtering, classification, clustering) designed to be scalable and robust. Some of these algorithms rely on Apache Hadoop, others are relying on Apache Spark.
- Apache Hadoop[2] is a basic platform for distributed processing of large datasets across clusters of computers by using a MapReduce strategy [24]. In reality, this is probably the most famous open-source implementation of MapReduce, a simplified data processing approach to execute data computing on a computer cluster. Worth to highlight that one of the major issues with MapReduce – resulting from the "flexibility" key feature, i.e., "users" are called to implement the code of map and reduce functions – is the amount of programming effort. In fact, other frameworks and platforms are building on it to provide users with data analytics facilities (e.g. Apache Mahout).
- Apache Spark [33] is an open-source, general-purpose cluster- computing engine which is very fast and reliable. It provides high-level APIs in Java, Scala, Python and R, and an optimised engine that supports general execution graphs. It also supports a rich set of higher-level tools including Spark SQL for SQL and structured data processing, MLlib for machine learning, GraphX for graph processing, and Spark Streaming.
- iPython/Jupyter [29] is a notebook-oriented interactive computing platform which enacts to create and share "notebooks", i.e., documents combining code, rich text, equations and graphs. Notebooks support a large array of programming languages (including R) and communicate with computational kernels by using a JSON-based computing protocol. Similar solutions include knitr [32] which works with the R programming language and Dexy[3], a notebook-like program that focuses on helping users to generate papers and presentations that incorporate prose, code, figures and other media.

---

[1] http://mahout.apache.org/.

[2] https://hadoop.apache.org/.

[3] http://www.dexy.it.

3 DataMiner: A Distributed Data Analysis Platform

DataMiner (DM) [3, 11, 14] aims to provide a cloud-based data analysis platform. It is designed to support the following key principles:

- *Extensibility*: the platform is "open" with respect to (i) the analytics techniques it offers and supports and (ii) the computing infrastructures and solutions it relies on to enact the processing tasks. It is based on a plug-in architecture to support adding new algorithms/methods, new computing platforms;
- *Distributed processing*: the platform is conceived to execute processing tasks by relying on "local engines"/"workers" that can be deployed in multiple instances and execute parallel tasks in a seamless way to the user. The platform is able to rely on computing resources offered by both well-known e-Infrastructures (e.g. EGI) as well as resources made available by the Research Infrastructure to deploy instances of the "local engines"/"workers". This is key to make it possible to "move" the computation close to the data;
- *Multiple interfaces*: the platform offers its hosted algorithms as-a-service and via a Web graphical user interface. This allows using the algorithms from software capable to execute processing tasks from well-known applications (e.g. R and KNIME);
- *Cater for scientific workflows*: the platform is exploitable by existing WorkFlow Management System (WFMS) [36] (e.g. a node of a workflow can be the execution of a task/method offered by the platform) and supports the execution of a workflow specification (e.g. by relying on one or more instances of WFMSs);
- *Easy to use*: the platform was designed in order to accommodate usability requirements for different types of users ranging from undergraduate students to expert scientists;
- *Open science-friendly*: the platform aims at supporting the open science paradigm in terms of repeatability and reproducibility of the experiments and the re-usability of the produced results. This goal is achieved via computational provenance tracking and the production of "research objects" that make it possible for users to repeat the experiment done by other users while respecting access policies to data and processes.

3.1 Development Context

The development of DataMiner is in the context of the D4Science infrastructure [3, 4] which is designed and developed to provide researchers and practitioners with working environments where resources of interests (datasets, computing, and services) for each designated community are easily made available by coherent and aggregated views and where open science practices are transparently promoted. This infrastructure is built and operated by relying on gCube technology [4], a software system specifically conceived to enable the construction and development of *Virtual Research Environments* (*VREs*) [10], i.e. web-based working environments dynamically build thus to be tailored to support the needs of their designated communities, each working on a research question. Every VRE offers domain-specific facilities, i.e. datasets and services suitable for the specific research question, as well as a rich array of basic services supporting collaboration and cooperation among its users, namely: (i) a *shared workspace* to store and organise any version of a research artefact; (ii) a *social networking area* to have discussions on any topic (including working version and released artefacts) and be informed on happenings; (iii) a *data analytics platform* to execute processing tasks either natively provided by VRE users or borrowed from other VREs to be applied to VRE users' cases and datasets, and (iv) a *catalogue-based publishing platform* to make the existence of a certain artefact public and disseminated. These facilities are at the fingerprint of VRE users.

The data analytics part mainly comprises two major components further discussed below, i.e. *DataMiner* realising the analytics engine supporting both providers and consumers of analytics methods to make use of a distributed computing infrastructure; and *Statistical Algorithms Importer (SAI)* realising the facilitator helping analytics methods owner to onboard their methods into the DataMiner engine. These components are nicely integrated with the services discussed above, thus the overall analytics platform result from the intertwining of them with the rest.

3.2 Architecture

DataMiner is based on the 52° North WPS implementation[4] but extends this implementation in order to meet Open Science oriented requirements. DM is a Java-based Web service running on an Apache Tomcat instance, enhanced with libraries to make it work in the D4Science e-Infrastructure. The DM architecture is made up of two sets of machines (clusters) that operate in a Virtual Research Environment: the Master and the Worker clusters, as shown in Fig. 1. The Master and the Worker clusters are dynamically provisioned by D4Science through an orchestration engine based on the OpenStack platform for Cloud computing. A load balancer distributes the computational requests uniformly to these machines. When a WPS computational request comes to the Master cluster balancer, this distributes the request to one of the services in the Master DM cluster. The DMs host processes provided by several developers. Each machine is endowed with a DM service that communicates with the D4Science Information System to notify its presence and capabilities. The load balancer is the main access point to interact with DM services. The machines of the Worker cluster have the same computational power as the other machines but can run only one (parallelised) process at a time and principally serve Map-Reduce computations.

Two types of algorithms are hosted by DataMiner: "Local" and "Cloud" algorithms. Local algorithms are directly executed on the Master machines and possibly use parallel processing on several cores and a large amount of memory. Instead, Cloud algorithms use distributed computing with a Map-Reduce approach and rely on the Worker cluster (Cloud nodes). With respect to the standard 52° North implementation, DM adds a number of features. First, it returns a different list of processes according to the VRE in which the service is invoked. When an algorithm is installed on a DM, it is also indexed in the D4Science Information System as an infrastructural resource. Thus, an e-Infrastructure manager can make it available to a number of VREs according to the algorithm's publication policies. When invoked in a VRE, DM returns only the subset of processes that have been assigned to that VRE.

---

[4] 52° North Web Processing Service software website https://52north.org/software/software-projects/wps/.

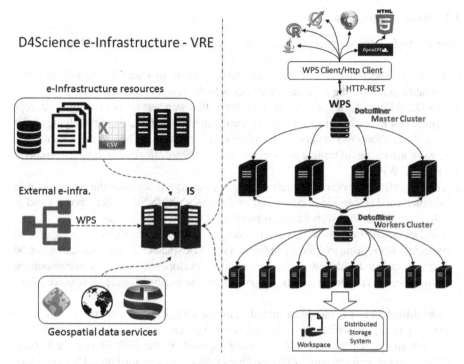

Fig. 1. The architecture of the D4Science DataMiner system.

The DMs are stateful services, i.e. when a computation starts, they communicate the status of the execution to asynchronous clients. The WPS standard embeds a direct URL to the DM machine that is carrying out the computation to check for the status. For each executed computation, DM releases an "equivalent HTTP-GET" request to repeat the experiment via a Web browser.

The DataMiner services use the security services of D4Science and require a user token to be specified via HTTPS-access authentication for each operation. The token identifies both a user and a Virtual Research Environment and is used to size the resources available to the user.

The DataMiner computations can use inputs loaded onto the D4Science Workspace, which may come also from Workspace folders shared among several users. This enables collaborative experimentation already at the input selection phase. Inputs can also come from external repositories because a file can be provided either as an HTTP link or embedded in a WPS execution request. The computational outputs are written onto the D4Science Workspace at the end of the computation. A Workspace folder is created that contains the input, the output, the parameters of the computation, and a provenance document in the Prov-O format summarizing this information. This folder can be shared with other people and can be used to execute the process again.

3.3 System Implementation

Overall, the following system components have been developed in DataMiner:

- The GUI: a web-based user interface enacting users to select an existing process, execute it, monitor the execution and access to the results (cf. Sect. 3.4);
- The DataMiner Master computational cluster: this web service is in charge to accept requests for executing processes and executing them, either locally or by relying on the DataMiner Worker(s) depending from the specific process. The service is conceived to work in a cluster of replica services and is offered by a standard web-based protocol, i.e. OGC WPS;
- The DataMiner Worker: this web service is in charge to execute the processes it is assigned to. The service is conceived to work in a cluster of replica services and is offered by a standard web-based protocol, i.e. OGC WPS;
- The DataMiner Processes: this a repository of processes the platform is capable to execute. This repository if equipped with a set of off-the-shelf processes and it can be further populated with new processes either (a) developed from scratch in compliance with a specific API or (b) resulting from annotating existing processes (cf. Sect. 3.5).

DataMiner [11, 14] has been included in the D4Science environment as a cloud computing platform, which currently makes ~400 processes available as-a-service.

Every analytics process is automatically exposed by the Web Processing Service (WPS) standard protocol and API of the Open Geospatial Consortium[5]. Thanks to this, every single process can be exploited by a number of clients embedded in third-party software that can interact with the DataMiner hosted processes through WPS.

DataMiner allows the hosted processes to be parallelised for execution both on multiple virtual cores and on multiple machines organised as a cluster. A typical DataMiner cluster to support big data processing is made up of 15 machines with Ubuntu 16.04.4 LTS x86 64 operating system, 16 virtual cores, 32 GB of RAM and 100 GB of disk space. A further cluster with a similar configuration is available to manage Map-Reduce computations. The DataMiner machines are hosted principally by the National Research Council of Italy[6], the Italian Network of the University and Research (GARR)[7], and the European Grid Infrastructure (EGI)[8]. A load balancer distributes computational requests uniformly to the machines of the computational cluster. Each machine hosts a processing request queue that allows a maximum of 4 concurrent executions running on one machine. With this combination of parallel and distributed processing, DataMiner allows processing big data while enabling provenance tracking and results sharing.

At the end of computation, the meta-information about the input and output data, and the parameters used (i.e. the computational provenance) are automatically saved on the D4Science Workspace and are described using the Prov-O XML ontological standard [23].

---

[5] Open Geospatial Consortium https://www.opengeospatial.org/standards/wps.
[6] National Research Council of Italy website www.cnr.it.
[7] GARR Consortium website www.garr.it.
[8] EGI Foundation website www.egi.eu.

A Web interface is available for each process, which is automatically generated based on the WPS interpretation. Through this interface, users can select the data to process from the Workspace and conduct experiments based on shared folders that allow automatic sharing of results and provenance with other users.

DataMiner also offers tools to integrate algorithms written in a multitude of programming languages [17].

3.4 Data Provenance During Data Processing

The research objects produced by the DataMiner computations are a set of files organised in folders and containing every input, output, an executable reference to the executed method as well as rich metadata including a PROV-O provenance record. The research objects are saved together with their metadata on the D4Science Workspace. Objects in the Workspace can be shared with co-workers and can be published by a catalogue with a license governing their usage.

DataMiner can operate within one or more Virtual Research Environments, i.e. it interoperates with additional VRE-specific services that manage multi-tenancy, communities, social networking communications, and collaboration among the VRE members.

From the end-user perspective, DataMiner offers a working environment oriented to collaborative experimentation where users:

- can easily execute and monitor data analytics tasks by relying on a rich and open set of available methods, either via WPS service endpoints or via Web GUI;
- can easily share & publish their analytics methods (e.g. implemented in R, Java and Python) within a Virtual Research Environment and make them usable by other processes supporting WPS;
- are provided with a "research object" describing every computational job executed by the workbench, which enables repeatability, reproducibility, re-use, citation, and provenance tracking for the experiments.

Overall DataMiner gives access to two types of resources:

- A distributed, open, and heterogeneous computing infrastructure for the execution of data analysis and mining tasks;
- A large pool of methods integrated with the platform, each made available as-a-service under the WPS standard according to VRE-specific access policies, i.e. public, private, shared with selected users, unavailable.

3.5 The Web-Based User Interface

DataMiner offers a web-based GUI to its users, as shown in Fig. 2. On the left panel (Fig. 2 a), the GUI presents the list of capabilities available in the specific "application context", which are semantically categorised (the category is indicated by the process provider). For each capability, the interface calls the WPS DescribeProcess operation to get the descriptions of the inputs and outputs. When a user selects a process in the right panel, the GUI on-the-fly generates a form with different fields corresponding to the inputs. Input data can be selected from the Workspace (the button associated with the input opens the Workspace selection interface). The "Start Computation" button sends the request to the DM Master cluster, which is managed as explained in the previous section. The usage and the complexity of the cloud computations are completely hidden from the user, but the type of computation is reported as metadata in the provenance file.

A view of the results produced by the computations is given in the "Check the Computations" area (Fig. 2 b), where a summary sheet of the provenance of the experiment can be obtained ("Show" button, Fig. 2 c). From the same panel, the computation can be also resubmitted. In this case, the Web interface reads the XML file containing the PROV-O information associated with computation and rebuilds a computation request with the same parameters. The computation folders may also include computations executed and shared by other users. Finally, the "Access to the Data Space" button allows obtaining a list of the overall input and output datasets involved in the executed computations (Fig. 2 d), with provenance information attached that refers to the computation.

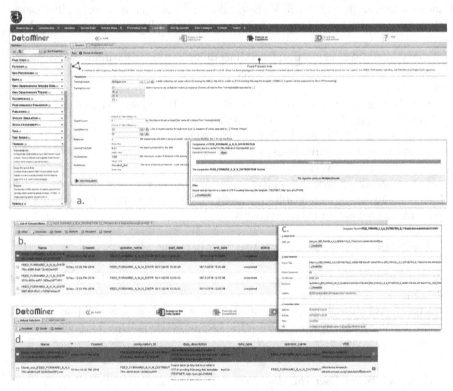

Fig. 2. The web interface of DataMiner in a D4Science VRE.

3.6   The Algorithms Importer

DataMiner allows importing prototype and production scripts written in a variety of languages (R, Java, Python, .Net, Fortran, Linux bash scripts, etc.). The tool was initially conceived to support scientists making prototype scripts, who needed to share results and provide their models and methods for use by other scientists. To this aim, DataMiner allows scientists to publish as-a-service scripts usually running on private desktop machines.

The Statistical Algorithms Importer (SAI) is the DataMiner interface that allows scientists to easily and quickly integrate their software with DataMiner, which in turn publishes this software-as-a-service under the WPS standard while managing multi-tenancy and concurrency. Additionally, it allows scientists to update their software without following long software re- deploying procedures each time.

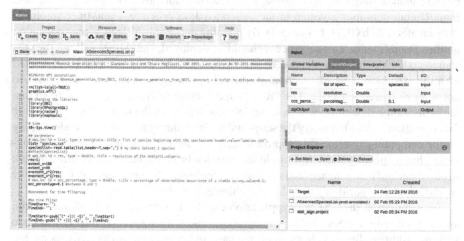

Fig. 3. The interface of the DataMiner Algorithms Importer tool.

The SAI interface (shown in Fig. 3 for R Script importing) is managed through a control panel in the top bar. The "Project" button allows creating, opening and saving a working session. A user uploads a set of files and data on the workspace area (lower-right panel). Upload can be done by dragging and dropping local desktop files. As a next step, the user indicates the "main script", i.e. the script/program that will be executed on DataMiner and that will use the other scripts and files. After selecting the main script, the left-side editor panel visualises it (when this is not compiled software) and allows modifying it. Afterwards, the user indicates the input and output of the script by highlighting variable definitions in the script and pressing the +Input (or +Output) button: behind the scenes the application parses the script strings and guesses the name, description, default value and type of the variable. This information is visualised in the top-right side Input/Output panel, where the user can modify the guessed information. Alternatively, for R scripts SAI can automatically compile the same information based on WPS4R annotations in the script. Other tabs in this interface area allow setting global variables and adding metadata to the process. In particular, the Interpreter tab allows indicating the programming language interpreter version and the packages required by the script and the Info tab allows indicating the name of the algorithm and its description. In the Info tab, the user can also specify the VRE in which the algorithm should be available.

Once the metadata and the variables information have been compiled, the user can create a DataMiner as-a-service version of the software by pressing the "Create" button in the Software panel. The term "software", in this case indicates a Java program that implements an as-a-service version of the user-provided scripts. The Java software contains instructions to automatically download the scripts and the other required resources on the server that will execute it, configure the environment, execute the main script and return the result to the user. The computations are orchestrated by the DataMiner computing platform that ensures the program has one instance for each request and user. The servers will manage concurrent requests from several users and execute code in a closed sandbox folder, to avoid damage caused by malicious code. Based on the SAI Input/Output definitions written in the generated Java program, DataMiner automatically creates a Web GUI (cf. Sect. 3.4). By pressing the "Publish" button, the application notifies DataMiner that a new process should be deployed. DataMiner will not own the source code, which is downloaded on-the-fly by the computing machines and deleted after the execution. This approach meets the policy requirements of those users who do not want to share their code. The "Repackage" button re-creates the software so that the computational platform will be using the new version of the script. The repackaging function allows a user to modify the script and to immediately have the new code running on the computing system. This approach separates the software updating and deployment phases, making the script producer completely independent on e-Infrastructure deployment and maintenance issues. However, deployment is necessary again whenever Input/Output or algorithm's metadata are changed [38, 39].

To summarise, the SAI Web application enables a user integrated software to run as-a-service features. SAI reduces integration time with respect to direct Java code writing. Additionally, it adds (i) multi-tenancy and concurrent access, (ii) scope and access management through Virtual Research Environments, (iii) output storage on a distributed, high-availability file system, (iv) graphical user interface, (v) WPS interface, (vi) data sharing and publication of results, (vii) provenance management, (viii) accounting facilities, and (ix) Open Science compliance.

4 Discussion

The heterogeneity characterising existing data analytics systems makes it evident that when discussing data processing "solutions" there are different angles, perspectives and goals to be considered. When analysing technologies from the scientist-perspective, the following trends should be considered:

- Technology should be "*ease of (re-)use*", i.e., it should not distract effort from the pure processing task. Scientists should be exposed to technologies that are flexible enough to enable them to quickly specify their processing algorithm/pipeline. It should not require them to invest effort in learning new programming languages or in deploying, configuring or running complex systems for their analytics tasks. Methods and algorithms are expected to be reused as much as possible, thus data processing should enable them to be "published" and shared.

- *"as-a-Service"* rather than *"do-it-yourself"*, i.e., scientists should be provided with an easy to use working environment where they can simply inject and execute their processing pipelines without spending effort in operating the enabling technology. This makes it possible to rely on economies of scale and keep the costs low.
- Solutions should be "hybrid", i.e., it is neither suitable nor possible to implement one single solution that can take care of any scientific data processing need. Certain tasks must be executed on specific infrastructures while other tasks are conceived to crunch data that cannot be moved on other machines from where they are stored.

These trends actually suggest that scientists are looking for things like "workbenches" or "virtual research environments" or "virtual laboratories" [10] providing them with easy to use tools for accessing and combining datasets processing workflows that behind the scene and almost transparently exploit a wealth of resources residing on multiple infrastructures and data providers (according to their policies). Such environments should not be pre-cooked or rigid, rather they should be flexible to enable scientists to enact their specific workflows. They should provide their users with appropriate and detailed information enacting to monitor the execution of such a workflow and be informed of any detail occurring during the execution. Finally, they should promote "open science" practices, e.g. they should record the entire execution chain leading to a given result, they should enact others to repeat/repurpose an existing process.

When analysing the overall solution to be developed and operated by RIs, the following aspects (going beyond the technology) are worth being considered:

- Provide support for research developers who produce and refine the code and workflows that underpin many established practices, scientific methods and services. Without their efforts in understanding issues, in explaining software behaviour, and improving quality, scientists would struggle to continue to handle existing methods and explore new opportunities. They need tools that inform them about the operational use of their products and technologies that protect their invested effort as platforms evolve. They are in danger of being undervalued, overwhelmed by the complexity and the pace of change, and of being attracted to the "big data" industry.
- Provide support for operations teams who need to keep the complex systems within and between RIs running efficiently as platforms change and communities' expectations rise while funders become more miserly. The tools and support they need are similar to those discussed in the previous bullet. They are not the same as the e-Infrastructure providers, they deploy and organise above those resources, but depend on them.
- Provide support for scientific innovators. They need to play with ideas, work on samples in their own favourite R&D environment, and then test their ideas at a moderate and growing scale. The provided facilities should allow them to move easily between developing ideas and proto-deployment, and eventually, when their ideas work out, to production deployment.
- The majority of researchers do not want to innovate, they just want to get on with their daily job. As much care as possible must be invested in protecting their working practices from change. However, if tools become available, e.g. driven from provenance

data, which help their work by removing chores, such as naming, saving, moving and archiving data, without them feeling they have lost responsibility for quality, then they will join in, and that eventually leads to fewer errors and better-curated data [28].

• There are some computational processes that require expert oversight while they are running, that can save substantial waste or steer to better results.

All in all, data processing is strongly characterised by the "one size does not fit all" philosophy. There is no, and there will arguably never be, a single solution that is powerful and flexible enough to satisfy the needs arising in diverse contexts and scenarios.

The tremendous velocity characterising technology evolution calls for implementing sustainable data processing solutions that are not going to require radical revision by specialists whenever the supporting technologies evolve. Whenever a new platform capable of achieving better performance compared to existing ones becomes available, users are enticed to move to the new platform. However, such a move does not come without pain and costs.

Data analytics tasks tend to be complex pipelines that can require combining multiple processing platforms and solutions. Exposing users to the interoperability challenges resulting from the need to integrate and combine such heterogeneous systems strongly reduces their productivity.

There is a need to develop data processing technologies that address the problem by abstracting from (and virtualising) the platform(s) that take care of executing the processing pipeline. Such technologies should go in tandem with optimisation technologies and should provide the data processing designer with fine-grained processing directives and facilitate detailed specification of processing algorithms [35].

The solution for data analytics we presented so far was designed and implemented by taking all of this into account thus resulting suitable for several application scenarios. Concrete use cases have been discussed in previous works, e.g. [2, 12, 13, 15, 16, 19, 35]. Coro et al. [11] discuss how the overall solution presented here has been exploited to implement a complex use case in computational biology reducing the time from months to a few hours.

ENVRIplus use cases developed by relying on this solution are discussed in Chapter 17.

5   Conclusion and Future Work

Data Processing is a wide concept embracing tasks ranging from (systematic) data collection, collation and validation to data analytics aiming at distilling and extracting new "knowledge" out of existing data by applying diverse methods and algorithms. When devising a solution suitable for the heterogeneity characterising science nowadays it is immediate to realise that it is almost impossible to envisage a solution that is powerful and flexible enough to satisfy the needs arising in diverse contexts and scenarios.

In this chapter, we presented a solution for data analytics that is open by design, i.e. conceived to (a) host and enact data analytics processes implemented by relying on several languages, and (b) transparently offer computing capacity from several and heterogeneous providers. Moreover, the envisaged solution has been intertwined with other services thus to facilitate the implementation of open science practices. Such a solution proved to be effective in several application contexts.

Future work includes the need to enhance the facilities aiming at exploiting integrated processes into notebooks and WFMS. In fact, although WPS facilitates this activity some development is needed to invoke every process. Moreover, mechanisms aiming at transforming the platform into a proactive component that by considering the user task can suggest suitable processes to play with.

Acknowledgements. This work was supported by the European Union's Horizon 2020 research and innovation programme via the ENVRIplus project under grant agreement No. 654182.

References

1. Allen, B., et al.: Software as a service for data scientists. Commun. ACM **55**(2), 81–88 (2012)
2. Assane, M., et al.: Realising a science gateway for the agri-food: the aginfra plus experience. In: 11th International Workshop on Science Gateway (IWSG) (2019)
3. Assante, M., et al.: Enacting open science by D4science. Future Gener. Comput. Syst. **10**(1016), 555–563 (2019). http://www.sciencedirect.com/science/article/pii/S0167739X183 1464X
4. Assante, M., et al.: The gcube system: delivering virtual research environments as-a-service. Future Gener. Comput. Syst. **95**, 445–453 (2019)
5. Bartling, S., Friesike, S. (eds.): Opening Science: The Evolving Guide on How the Internet is Changing Research, Collaboration and Scholarly Publishing. Springer, Cham (2014). https://doi.org/10.1007/978-3-319-00026-8
6. Belcastro, L., Marozzo, F., Talia, D.: Programming models and systems for big data analysis. Int. J. Parallel Emergent Distrib. Syst. **34**(6), 632–652 (2019)
7. Berthold, M.R., et al.: Knime-the konstanz information miner: version 20 and beyond. AcM SIGKDD Explor. Newsl. **11**(1), 26–31 (2009)
8. Bordawekar, R., Blainey, B., Apte, C.: Analyzing analytics. ACM SIGMOD Record **42**(4), 17–28 (2014)
9. Calegari, P., Levrier, M., BalczyÅski, P.: Web portals for high-performance computing: a survey. ACM Trans. Web **13**(1), 1–5 (2019). http://doi.acm.org/10.1145/3197385
10. Candela, L., Castelli, D., Pagano, P.: Virtual research environments: an overview and a research agenda. Data Sci. J. **12**, GRDI–013 (2013)
11. Coro, G., Candela, L., Pagano, P., Italiano, A., Liccardo, L.: Parallelizing the execution of native data mining algorithms for computational biology. Concurrency Comput. Pract. Exp. **27**(17), 4630–4644 (2015). https://onlinelibrary.wiley.com/doi/abs/10.1002/cpe.3435
12. Coro, G., Masetti, G., Bonhoeffer, P., Betcher, M.: Distinguishing violinists and pianists based on their brain signals. In: Tetko, I.V., Kůrková, V., Karpov, P., Theis, F. (eds.) ICANN 2019. LNCS, vol. 11727, pp. 123–137. Springer, Cham (2019). https://doi.org/10.1007/978-3-030-30487-4_11
13. Coro, G., Pagano, P., Ellenbroek, A.: Combining simulated expert knowledge with neural networks to produce ecological niche models for Latimeria chalumnae. Ecol. Model. **10**(1016), 55–63 (2013)
14. Coro, G., Panichi, G., Scarponi, P., Pagano, P.: Cloud computing in a distributed e-infrastructure using the web processing service standard. Concurrency Comput. Pract. Exp. **29**(18) (2017). https://onlinelibrary.wiley.com/doi/abs/10.1002/cpe.4219

15. Coro, G., Webb, T.J., Appeltans, W., Bailly, N., Cattrijsse, A., Pagano, P.: Classifying degrees of species commonness: North Sea fish as a case study. Ecol. Model. **10**(1016), 272–280 (2015)

16. Coro, G., Large, S., Magliozzi, C., Pagano, P.: Analysing and forecasting fisheries time series: purse seine in Indian Ocean as a case study. ICES J. Mar. Sci. **73**(10), 2552–2571 (2016)

17. Coro, G., Panichi, G., Pagano, P.: A web application to publish R scripts as-a-service on a cloud computing platform. Boll. di Geofis. Teorica ed Appl. **57**, 51–53 (2016)

18. Deelman, E., et al.: Pegasus, a workflow management system for science automation. Future Gener. Comput. Syst. **46**, 17–35 (2015)

19. Froese, R., Thorson, J.T., Reyes, J.R.: A bayesian approach for estimating length-weight relationships in fishes. J. Appl. Ichthyol. **30**(1), 78–85 (2014)

20. Hey, A.J., Tansley, S., Tolle, K.M., et al.: The Fourth Paradigm: Data-Intensive Scientific Discovery, vol. 1. Microsoft Research Redmond, Redmond (2009)

21. Josep, A.D., Katz, R., Konwinski, A., Gunho, L., Patterson, D., Rabkin, A.: A view of cloud computing. Commun. ACM **53**(4), 50–58 (2010)

22. Khalifa, S., et al.: The six pillars for building big data analytics ecosystems. ACM Comput. Surv. (CSUR) **49**(2), 33 (2016)

23. Lebo, T., et al.: Prov-o: the PROV ontology. W3C Recommendation **30** (2013)

24. Li, F., Ooi, B.C., Özsu, M.T., Wu, S.: Distributed data management using MapReduce. ACM Comput. Surv. (CSUR) **46**(3), 31 (2014)

25. Liew, C.S., Atkinson, M.P., Galea, M., Ang, T.F., Martin, P., Hemert, J.I.V.: Scientific workflows: moving across paradigms. ACM comput. Surv. **49**(4), 1–66 (2016). http://doi.acm.org/10.1145/3012429

26. Llorà, X., Ács, B., Auvil, L.S., Capitanu, B., Welge, M.E., Goldberg, D.E.: Meandre: semantic-driven data-intensive flows in the clouds. In: 2008 IEEE Fourth International Conference on eScience, pp. 238–245. IEEE (2008)

27. Ludäscher, B., et al.: Scientific workflow management and the Kepler system. Concurrency Comput. Pract. Exp. **18**(10), 1039–1065 (2006)

28. Myers, J., et al.: Towards sustainable curation and preservation: the SEAD project's data services approach. In: 2015 IEEE 11th International Conference on e-Science, pp. 485–494. IEEE, Munich, Germany (2015). https://doi.org/10.1109/eScience.2015.56

29. Pérez, F., Granger, B.E.: Ipython: a system for interactive scientific computing. Comput. Sci. Eng. **9**(3), 21–29 (2007)

30. Pierce, M.E., et al.: Apache airavata: design and directions of a science gateway framework. Concurrency Comput. Pract. Exp. **27**(16), 4282–4291 (2015)

31. Wolstencroft, K., et al.: The taverna workflow suite: designing and executing workflows of web services on the desktop, web or in the cloud. Nucleic Acids Res. **41**(W1), W557–W561 (2013)

32. Xie, Y.: Dynamic Documents with R and Knitr. Chapman Hall/CRC, Boca Raton, Florida (2015)

33. Zaharia, M., et al.: Apache spark: a unified engine for big data processing. Commun. ACM **59**(11), 56–65 (2016)

34. Zhao, Y., et al.: Swift: fast, reliable, loosely coupled parallel computation. In: 2007 IEEE Congress on Services (Services 2007), pp. 199–206. IEEE (2007)

35. Zhou, H., et al.: CloudsStorm a framework for seamlessly programming and controlling virtual infrastructure functions during the DevOps lifecycle of cloud applications. Softw. Pract. Exper. **49**, 1421–1447 (2019). https://doi.org/10.1002/spe.2741

36. Evans, K., et al.: Dynamically reconfigurable workflows for time-critical applications. In: Proceedings of the 10th Workshop on Workflows in Support of Large-Scale Science (WORKS 2015), pp. 1–10. ACM Press, Austin, Texas (2015). https://doi.org/10.1145/2822332.2822339

37. Zhao, Z., et al.: Reference model guided system design and implementation for interoperable environmental research infrastructures. In: 2015 IEEE 11th International Conference on e-Science, pp. 551–556. IEEE, Munich, Germany (2015). https://doi.org/10.1109/eScience.2015.41
38. Hu, Y., et al.: Deadline-aware deployment for time critical applications in clouds. In: Rivera, F.F., Pena, T.F., Cabaleiro, J.C. (eds.) Euro-Par 2017. LNCS, vol. 10417, pp. 345–357. Springer, Cham (2017). https://doi.org/10.1007/978-3-319-64203-1_25
39. Hu, Y., Zhou, H., de Laat, C., Zhao, Z.: Concurrent container scheduling on heterogeneous clusters with multi-resource constraints. Future Gener. Comput. Syst. **102**, 562–573 (2020). https://doi.org/10.1016/j.future.2019.08.025

Virtual Infrastructure Optimisation

Spiros Koulouzis(iD), Paul Martin(iD), and Zhiming Zhao$^{(\boxtimes)}$(iD)

Multiscale Networked Systems, University of Amsterdam, 1098XH Amsterdam,
The Netherlands
{s.koulouzis,z.zhao}@uva.nl, pwmartin.research@gmail.com

Abstract. The increasing volumes of data being produced, curated and made
available by research infrastructures in the environmental science domain require
services able to optimise the delivery staging and process of data on behalf of
researchers. Specialised data services for managing the data lifecycle, for creating
and delivering data products, and for customised data processing and analysis,
all play a crucial role in how these research infrastructures serve their communities, and many of these activities are time-critical needing to be carried out
frequently within specific time windows. We describe our experiences identifying
the time-critical requirements of environmental scientists making use of computational research support environments. We also present a microservice-based
infrastructure optimisation suite, the Dynamic Real-time Infrastructure Planner,
used for constructing virtual infrastructures for research applications on demand.
This chapter is partially based on a recent paper presented in [1].

Keywords: Infrastructure optimization · Cloud computing

1 Introduction

The ENVRI community works together to provide shared technological and governance
solutions for data-driven science, in particular defining common operations for environmental research infrastructures and identifying and adopting technologies that implement
those operations. Addressing the need for interoperable services for such diverse topics
as identification and citation, curation, provenance and cataloguing, the Data for Science
theme of ENVRIplus brought together a cluster of environmental research infrastructures (RIs) and (Information and Communication Technologies) institutions to come up
with practical solutions to long-standing problems in such diverse areas as identification
and citation, curation, cataloguing, processing and provenance. One particular area of
interest, however, was optimisation; particularly the optimisation of virtual infrastructure
used to support scalable data workflows needed both by RIs as part of their own internal
data pipelines, and by RI users as part of their data science applications. Therefore, it
is necessary to provide sufficiently advanced computational networked infrastructure
to manage both the transportation of large (distributed) datasets and the data-intensive
processing of such datasets.

© The Author(s) 2020
Z. Zhao and M. Hellström (Eds.): Towards Interoperable Research
Infrastructures for Environmental and Earth Sciences, LNCS 12003, pp. 192–207, 2020.
https://doi.org/10.1007/978-3-030-52829-4_11

Performance is a crucial factor for many scenarios involving research support environments, influencing the quality of experience factors such as responsiveness to requests, to more system-level concerns such as efficient load distribution across distributed nodes in a confederation of data services. An example of a performance-critical system involving environmental data would be an early warning system where real-time sensor data have to be analysed quickly enough to identify events and provide adequate time for response. Even in non-emergency contexts, there are many cases where RIs collect real-time data continuously from sensors for swift processing to provide "nearly real-time" services to researchers. The specific example used in this paper is that of a data subscription service whereby updates to tailored subsets of a dataset are pushed to subscribers within a requested deadline. Notably, these services often cut across research support environments; RIs provide the service but delegate the hosting and management of the data processing pipeline to an e-infrastructure, generally to take advantage of elastic infrastructure resources rather than provide dedicated infrastructure within their data centres (which often operate as loose confederations with limited budgets for services beyond data curation and publication). Virtual Research Environments (VREs) may also be involved as part of the interface with researchers: for example, to subscribe to RI services or retrieve (and process) the results from such services.

To deliver acceptable performance, time-critical applications thus rely not only on the infrastructure for parallel computing or fast communication between components but also on optimisation of system-level application behaviour [2, 3]. The customisation of the infrastructure must consider performance constraints on applications at run-time as well as the utilisation and cost of the underlying resources across applications [4, 5]. In this chapter, we present a smart infrastructure optimisation engine, called Dynamic Real-time Infrastructure Planner (DRIP), that has been developed to bridge the gap between application requirements and service delivery on the part of research support environments, to optimise the quality of service at all levels. DRIP can be used to deploy, control and manage the kinds of distributed data pipelines needed for advanced RI services on the Cloud-based infrastructures now being provided by e-infrastructures.

2 Requirements and State of the Art

In this section, we analyse the basic requirements for service performance optimisation for time-critical data services in research support environments, review the state-of-the-art in real-time systems that may bear an impact on the development or operation of such data services, and summarise the essential challenges for time-critical data services on modern e-infrastructures.

2.1 Requirements

When we refer to time-critical applications, we do not usually mean speed-critical applications in the sense of applications that simply need to minimise the completion time (i.e. must be run fast). True "real-time" applications are characterised by bounded response time constraints on inputs, with certain consequences upon failure to meet deadlines [6]. Based on those consequences, real-time applications are referred to as hard real-time if any missed deadline leads to immediate failure of the application, soft real-time if missing deadlines merely leads to degradation of user experience, and firm real-time if failure is brought about by too many missed deadlines in succession. Nearly real-time applications meanwhile are those with an intrinsic yet bounded delay introduced by data processing or transmission. Note that this does not make all nearly real-time applications "soft"-such applications can still impose a hard requirement for processing to fall within the permitted bounds. We might consider most processes in research support environments to be soft insomuch as failure to meet deadlines is usually not immediately disastrous. However, processes that are continuously run in tandem with real-time data acquisition can be seen to be "firm" due to the cascading impact of repeated failure to process their inputs on time; similarly, any highly parallelised workflow with bottlenecks in the data pipeline can suffer a precipitous drop in general quality of service if delays in one parallel element impact a non-parallelised bottleneck downstream. When we refer to time-critical applications, we therefore generally mean real-time or nearly real-time applications that are "firm" (or harder) in terms of the consequences of failure to meet the quality of service requirements. True hard real-time constraints in research infrastructure are rare but may emerge in particular for safety-critical applications that depend on real-time observational data.

In practical terms, the "firmness" of a response time constraint dictates the degree of a limited resource that should be allocated to ensuring the constraint. Isolated failures do not have the same impact as failures that beget further failures. It may be possible (and desirable) in specific research support environments to be able to assign a metric to constraints based on firmness that can be translated into concrete resource level requirements or adaptation strategies so that this information can be passed to optimisation services that must prioritise particular services or metrics. The requirements for optimising performance in research support environments are mainly dominated by the requirements of the data-centric research activities (the simplest but most important being the retrieval of specific datasets on request) that demand high performance or responsiveness. Within RIs, services are often developed with time constraints imposed on the acquisition, processing and publishing of real-time observations, in scenarios such as disaster early warning [7]. For VREs and RIs, performance factors are strongly influenced by the time needed to customise the runtime environment and to schedule the workflow applications [8]. Steering of applications during complex experiments is also temporally bounded [9]. Computing tasks and services provided by e-infrastructures are managed and offered to clients based on service-level agreements (SLAs). Time constraints are also imposed on the scheduling and execution of tasks that require high performance or high throughput computing (HPC/HTC). The overhead introduced by the customisation, reservation and provisioning of suitable infrastructure, the monitoring of runtime behaviour for infrastructure, and the support for runtime control also needs to be reduced and maintained within minimum levels. Failure recovery for deployed services and applications in real-time is also important when supporting time-critical applications; time constraints are not only imposed on failure detection, but also on decision-making and recovery.

2.2 Related Work

Within the Cloud context, many approaches have been proposed to address the scheduling, scaling and execution of tasks with deadlines. The majority of these proposals, however, adopt the viewpoint of the Cloud provider, which is often concerned with optimal VM placement on physical machines [10, 11]. In other cases, complex scheduling algorithms consider only the planning phase and do not react at runtime to changes in performance or failures [12]. Moreover, the majority of these approaches consider either synthetic tasks and workloads, simulated Cloud environments or both [13, 14].

2.3 State of the Art

The fulfilment of most time-critical requirements for research support environments relies on optimal execution of tasks on e-infrastructures, as well as efficient movement of data across networks. We identify several categories of time-critical application.

Time-Critical Information Search and Query. Typical technologies for real-time data query and search model the search activities of users, and their projected needs, predicting future queries [15], optimising catalogues [16] or prioritising urgent tasks [17], as well as optimising the presentation of contextual information [18]. Information retrieval is a core part of many data services and may require the retrieval of multiple datasets to answer a given query or considerable internal processing of data files for document-oriented data.

Time-Critical Workflow Execution. Time-critical constraints on workflows are typically expressed as deadlines for completing (part of) the workflow, or for responding to invocations or events within a certain time window. Scheduling the execution of such workflows requires consideration of not only individual task deadlines but also cost and occupation of resources [19]. Algorithms based on partial critical paths can be used to solve such problems [20, 21], applying meta-heuristic approaches, e.g. particle swarm optimisation [22]. When customising virtual infrastructures, a common approach will 1) select suitable virtual machines (VMs) based on specific task-VM performance metrics, 2) minimise communication costs between tasks by grouping tasks needing frequent communication in the same VM, and 3) refine the selection based on the calculation of new critical paths. Most current work focuses on guaranteeing a single deadline encompassing the entire application, e.g. the Critical Path-based Iterative (CPI) [23] and Complete Critical paths (CPIS) [24] algorithms. All these technologies have been widely investigated for applications modelled as directed acyclic graphs (DAGs) as DAG-based methods are popular for building data-flows for data-intensive applications.

Real-Time Modelling and Simulation. In data science, coupling different simulation models of individual systems can be performed to study the behaviours of complex systems, e.g. combining species distribution models with weather models to study how diseases are distributed via insects and species migration at different times. Simulating physical systems does not always require the simulation to run at wall clock rates [25], but executing such simulations on distributed infrastructure does impose requirements on managing the simulation times of different sub-components, e.g. to control the relationships among events and time [26].

Real-Time Computational Steering. Real-time steering of a computing system requires monitoring of the runtime status of both application and infrastructure. Infrastructure-level monitoring takes place at the network level and on computing and storage nodes [27]. Monitoring service quality of Cloud environments allows providers or users to evaluate compliance with SLAs [28]. At application level monitoring often requires embedded probes within application components [29]. Logging and provenance subsystems often capture the runtime status of the overall system as well [30]. To visualise the runtime status and to allow a user to make correct decisions regarding system control, different kinds of monitoring information together with the context of the system execution have to be harmonised based on the timestamp. Semantic technologies are often used to integrate such information and to offer query interfaces to link them [29]. Runtime steering of computing systems can take the form of adaptations of application logic at specific control points where the system actively provides time windows for users to intercede, or else the system can be interrupted by the user during execution [31]. The controllability of infrastructures e.g. dynamically configuring or scaling nodes [32], or controlling network flows [33], offer applications opportunities to refine the system performance.

Real-Time Data Acquisition. Acquiring real-time observations is important for many RIs. The quality of communication between sensors and data processing units is crucial for timely acquisition. Software-defined sensor networks can be used to optimise communication between sensors [34], as can applying edge computing solutions to tightly coupled sensors with data processing [35]. To make sensor data available to users in near real-time, partially automating data quality control and annotation is important [36], but currently, most data quality control is performed manually. Standardising this process and exploiting scalable virtualised infrastructure are recurring requirements for environmental RI [37].

Real-Time Data Transfer. Real-time data transfer between components occurs frequently within e-infrastructures. At the network level, real-time data protocols [38], multi-path TCP and other protocols are used to optimise data streaming throughput. SDN [50] technologies are used to adapt network flows between data nodes dynamically, and traffic programming models such as co-flow [39] are used to reschedule runtime data transfer. At the transfer service level, dynamic schedule data transfer workers are used in the LOBCDER service to handle the balance of downloads [40].

Infrastructure Provisioning for Time-Critical Applications. Fast provisioning of virtualised infrastructure opens the possibility of runtime adaptation to meet time-critical requirements. Optimising VM image size [41], directly forking runtime images from memory [42], or parallelising the provision procedure [43]. Using P2P or SND technology to optimise image transfer among data centres is also possible. Zhou et al. describe a transparent networked virtual infrastructure graph partitioning and parallel provisioning approach to map infrastructure across data centres [44].

Real-Time Service Level Agreement. Real-time support of virtualised infrastructure has attracted increasing interest [45]. SLAs for real-time applications and their negotiation at runtime will be crucial for supporting real-time applications in Cloud. Most approaches are based on graph mapping using key quality parameters such as execution time; improving the mapping procedure can be done by parallelising the search procedure for matching resources and applications [46], pre-processing the resource information by clustering the resource information based on the SLA request, and multi-objective optimisation for searching out alternative solutions [47]. Rich contextual information and semantic annotation is another key issue influence the search quality [48].

3 Challenges for Time-Critical Applications on e-Infrastructure

In data science, the research data lifecycle is considered to be of primary importance, but at each stage of that lifecycle, we must also consider the lifecycle of the data pipelines or data processing workflows that are needed to support each stage. Given the increasing availability and adoption of virtualised e-infrastructure and Cloud services targeted towards RIs and the general research community, we are particularly interested in the life cycle for applications on virtual infrastructure (i.e. configurations of networked VMs upon which data processing workflows are deployed on behalf of researchers either for specialised tasks or as part of the general data lifecycle managed by RIs).

For static infrastructures, the development and configuration of a particular application (e.g. a data processing pipeline or workflow) can be adapted with respect to the hardware and host architecture known to the developers. This may still require considerable technical expertise of course, but can nonetheless be considered to at least represent a single initial investment to providing an efficient, performant technical solution.

In contrast, deploying application workflows on virtual infrastructures allows RIs to make use of commodity e-infrastructure resources as and when needed, rather than requiring an investment in dedicated hardware, and in principle offers the additional advantages of scalability and seamless migration which can to some extent be managed almost entirely by the e-infrastructure provider. It is difficult, however, to optimise generic virtual infrastructure for specific applications, and so difficult to guarantee a certain level of performance, which is particularly of concern for time-critical applications.

Figure 1 illustrates the lifecycle of application workflows on virtual infrastructure. Five main phases are identified:

1. Virtual infrastructure planning. Regarding the scheduling of application workflows onto a topology of virtual machines that ensure the availability and suitability of virtual resources at all stages of the workflow.
2. Virtual infrastructure optimisation. Regarding the iterative refinement of an infrastructure plan to meet all (or a maximal subset of) requirements for performance, reliability, quality of service, etc.
3. Virtual infrastructure provisioning. Regarding the actual provisioning of planned infrastructure across one or more data centres or Clouds in such a way as to create a network of resources that meet the control and data-flow requirements of a distributed application.
4. Software platform deployment. Regarding the actual deployment of application elements onto the provisioned infrastructure, as well as the initialisation and control of such elements at runtime.

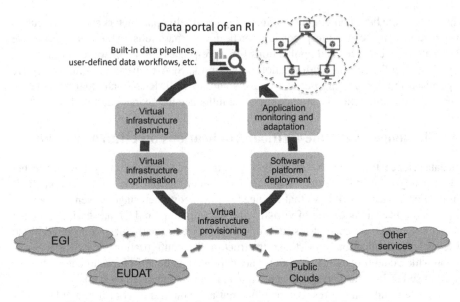

Fig. 1. The lifecycle of application workflows on virtual infrastructure.

5. Application monitoring and adaptation. Regarding the monitoring of a running application with respect to selected metrics necessary for evaluating the performance and liveness of the application, as well as the ability to intelligently take measures to improve and regain a desired quality of service, e.g. by automatically scaling or migrating application elements in the virtual infrastructure, or re-configuring components where practical.

While there exist a number of general solutions for managing each of these phases for the most common technologies for providing virtual infrastructure, or even subsets thereof, there is no single integrated solution for managing the entire lifecycle just described. Moreover, if we want to apply such a solution to time-critical applications, then it is necessary to address additional challenges:

- To meet time requirements for discovering and retrieving data from distributed access/storage services and virtual infrastructures provided by different RIs it is necessary to be able to define deadlines throughout the application deployment lifecycle both individually and collectively
- To develop a time-critical application, either the developer needs to be able to describe how constraints at application level propagate down to the level of restrictions on infrastructure and quality of service, or else the optimisation services developed for the infrastructure must be able to do that for the developer.
- During the execution of time-critical applications, data sources, software components, and the execution engines of some parts of the application may have to be handled by different underlying infrastructures, making it difficult to calculate and enforce quality of service constraints across the entire application/infrastructure stack.

To help address these concerns, we have developed the Dynamic Real-time Infrastructure Planner, which provides a set of services to optimise the automation from infrastructure customisation and provisioning to application deployment and runtime control.

4 Dynamic Real-Time Infrastructure Planner

The Dynamic Real-time Infrastructure Planner (DRIP) is a service suite for planning and provisioning networks of virtual machines and then deploying distributed applications across those networks, managing the virtual infrastructure during runtime based on certain time-critical constraints defined with the application workflow. The DRIP service provides an engine for automating all these procedures by making use of pluggable microservices for providing specific functionalities orchestrated via a single manager component behind a RESTful Web API for easy use and retrieval of results. This approach enables a more holistic approach to the optimisation of resources and meeting application-level constraints such as deadlines or SLAs. It also allows application developers to seamlessly plan a customised virtual infrastructure based on constraints on QoS, time-critical constraints or constraints on budget. Based on such a plan DRIP can provision a virtual infrastructure across several Cloud providers, and then be used for deploying application components, starting execution on-demand, and managing the runtime application deployment state. Therefore, DRIP is not bound to any particular application. Instead, it is flexible and can deploy a wide range of applications on top of a customised and heterogeneous virtual infrastructure composed of multiple Clouds providers to meet the application's constraints.

4.1 Architecture and Functional Components

The DRIP services include a number of components, interacting via an internal message brokering service orchestrated by a single manager. These components and their interaction are shown in Fig. 2.

All components of DRIP under the control of the DRIP manager are designed to be independently replaceable, to allow for improved or alternative implementations of e.g. the planner or the provisioner. Indeed, multiple versions of a component could coexist, allowing for greater flexibility or even simply to better balance the load of requests to DRIP. The types of service that can be included in DRIP, and their current implementations, are now detailed:

- **The DRIP manager** is a Web service that allows DRIP functions to be invoked by external clients. Each request is directed to the appropriate component by the manager, which coordinates the individual components and scales them up if necessary. Resource information, credentials, performance profiles and application workflows used by the manager and other components are all internally managed via the knowledge base as described below.
- **The planner** uses a partial critical path algorithm [1] optimised for workflows with multiple internal deadlines in order to produce efficient infrastructure topologies, selecting the most cost-effective virtual machines [20]. Multiple planner components

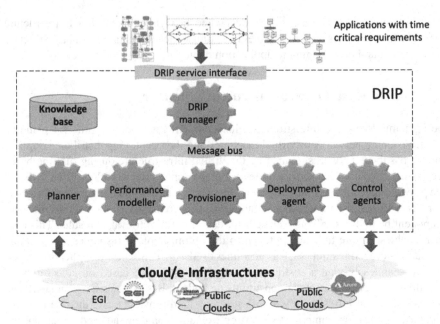

Fig. 2. How services provided by DRIP are invoked by RI services to provide downstream services to users.

can be attached to DRIP in order to manage different kinds of application workflow or infrastructure topology, taking of different technologies such as software-defined networking [49] to customise the network topology among VMs and optimally place network controllers for the networked VMs [50].

- **The performance modeler** is a tool which automates the execution of a given application on a virtual machine and collects the performance information of the application. In this way, it can profile the performance characteristics of specific applications. The output will be used by the planner to select virtual machines during its planning procedure.
- **The provisioner** is responsible for automating the provisioning of infrastructure plans produced by the planner(s) onto the underlying Cloud or e-infrastructure. The current provisioner can decompose the infrastructure description and provision it across multiple data centres (possibly from different providers) with transparent network configuration [44].
- **The deployment agent** installs application components onto provisioned infrastructure. The current deployment agent is able to schedule the deployment sequence based on network bottlenecks, and maximise the fulfilment of deployment deadlines for all the Cloud providers currently supported by the default DRIP provisioner [51].
- **Infrastructure control agents** provide sets of APIs that DRIP can then provide to applications to control the scaling of containers or VMs and for adapting network flows or to use itself in conjunction with a monitoring framework to automatically maintain the quality of service of the deployed application.

- **A DRIP knowledge base** is employed by DRIP for storing information about user credentials, the types of the resource offered by Clouds or e-infrastructure, and other useful data that the DRIP manager or any other component can retrieve or contribute to.
- **The message bus** connects all the components in the DRIP to enable the communication among them.
- **The service interface** provides a standardised API to the application developers, or software clients (e.g. data portal or workflow system) to invoke the function.

4.2 Implementation Details

DRIP was developed in the context of EU H2020 projects SWITCH[1] (as part of a workbench for time-critical, self-adaptive applications on Cloud) and ENVRIPLUS[2] (to provide e-infrastructure optimisation services for scientific workflows).

The prototype of DRIP adopts industrial and community standards. The infrastructure planner uses the TOSCA specification[3] to get descriptions of applications and their constraints. The infrastructure provisioner uses OCCI as its default provisioning interface, and currently supports the Amazon EC2, EGI FedCloud and ExoGen[4] Clouds. Since DRIP relies on multiple Cloud providers it offers a best-effort approach for the provision, stability and performance of the underlying virtual infrastructure. However, using performance and reliability models for each provider and each region, DRIP is able to provide an optimal, stable and responsive vitalised infrastructure for time-critical applications [52]. The deployment agent can deploy overlay Docker clusters such as Docker Swarm or Kubernetes. It may also deploy any type of customised application based on Ansible playbooks [53]. The infrastructure control agents are a set of APIs that DRIP provides to applications to allow for scaling of containers or VMs and for adapting network flows. The manager provides a RESTful interface to allow integrated interaction with all components and uses RabbitMQ as its internal message broker to direct requests appropriately. All DRIP software is available via open source repository[5] under the Apache-2.0 license.

4.3 How DRIP Works

Figure 3 demonstrates how DRIP works. We choose an example of disaster early warning, in which a legacy application needs to be migrated in Cloud environments using DRIP.

1. The application developer needs to identify the application components to be deployed in Cloud, describe the dependencies between components, and specify the time-critical constraints (the deadlines between specific application tasks), as shown in the step in Fig. 3.

---

[1] www.switchproject.eu.

[2] www.envriplus.eu.

[3] https://www.oasis-open.org/committees/tosca/.

[4] http://www.exogeni.net/.

[5] https://github.com/QCAPI-DRIP/.

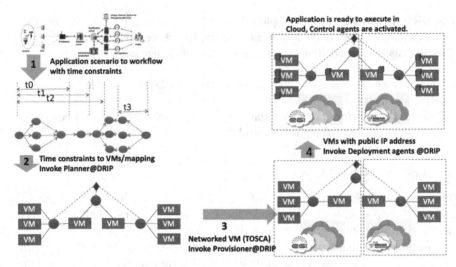

Fig. 3. A conceptual demonstrator of DRIP.

2. The DRIP planner will plan the virtual infrastructure for the application based on the description provided by step 1. Currently, the description structure is based on the template derived from TOSCA standard. The planned virtual infrastructure including a) a set of virtual machines (VM), with specific size of CPU, memory and storage, b) the network topology among VMs, and c) the controller for the network. The output of this step will use the same TOSCA format, but with concrete information of the virtual infrastructure.

3. The DRIP provisioner will continue with the step 2; it will select specific data centres or Cloud providers to provision the planned VM. The provisioner is able to parallelise the provisioning procedure and automate the network configuration among VMs. The step 3 modifies the TOSCA description from step 2 with concrete public IP address. After step 3, all the VM will be remotely accessible.

4. The DRIP deployment agent will use the provisioned virtual infrastructure to deploy the application components identified in step 1. After the fourth step, the application is ready to execute.

The application topology is currently described using TOSCA and must be part of the request made to DRIP. When a planning request comes, the manager will direct the request to the *infrastructure planner* to generate a plan, which can be sent back to the user for further confirmation. If the constraints cannot be satisfied the planner informs the user that a plan cannot be generated. The DRIP manager stores the necessary Cloud credentials on behalf of the user. The *provisioning agent* can provision the virtual infrastructure via interfaces offered by the Cloud providers. Once this has finished, the deployment agent will deploy all necessary components onto the provisioned infrastructure from designated repositories and set up the control interfaces needed for runtime control of application and infrastructure. Figure 4 shows a detailed sequence diagram.

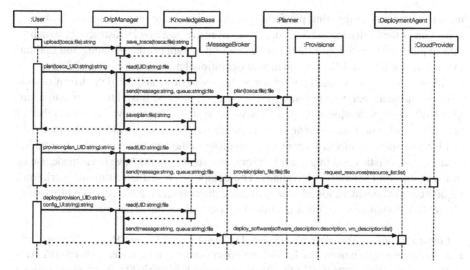

Fig. 4. A detailed sequence diagram of DRIP.

4.4 Future Work: Workflow Reproducibility

One of the key objectives in environmental research facilitated by the contextualisation of processes and actors in research infrastructure is that of workflow reproducibility, whereby the provenance of experiments conducted by scientists using the infrastructure is recorded so as to allow the experiments to be re-executed with minimal difficulty. Recording provenance in various workflow systems has been explored before, but more effort is needed to adopt standard provenance frameworks with standard vocabulary (e.g. based on PROV) that can be implemented independently by different workflow systems and data processing platforms. The use of ontologies for verification and validation of workflows has already been explored. For example, Miksa and Rauber [54] provide such ontologies and accompanying tool support for just this kind of activity.

Virtualisation, of both computing resources and the interstitial network, provide a good basis for fluid intelligent infrastructure, providing flexible logical networks of computing and data nodes on demand that can be optimised based on some co-extant knowledge fabric.

5 Summary

The ability to comprehend, facilitate and augment how researchers use research infrastructure to support data-driven science is crucial, but also extremely challenging to acquire given the proliferation of competing systems and standards in the world computational landscape. This problem is not limited to the logical aggregation of research data products for the purposes of cataloguing and access, but also extends to the use of information to promote the efficient operation of research infrastructure and the effective use of 'e-infrastructure' including the compute, data and network resources provided via initiatives such as EGI and PRACE. The seamless, easy access to distributed data and the use of community computing platforms requires significant automation via brokering agents and other software services operating over the baseline infrastructure, which in turn requires substantial knowledge infrastructure to support the planning and optimal execution of a host of different concurrent operational and data-driven workflows.

Most scientific investigations follow a clear workflow, and for data-driven or otherwise computational workflows, different processes can be linked together into a single distributed application managed by a single system. There have been a number of scientific workflow management systems developed in order to address the manifold challenges raised by modern scientific computing in the last two decades, all with different characteristics and target applications and such systems have been made use of in many different scientific disciplines. The composition and execution of workflows require careful consideration of how to manage the communication between processing elements and maintain sufficient quality of service across the entire workflow.

Acknowledgements. This work was supported by the European Union's Horizon 2020 research and innovation programme via the ENVRIplus project under grant agreement No 654182. This work was also partially supported by the European Union's Horizon 2020 research and innovation programme via the SWITCH project under grant agreement No 643963, and ARTICONF project under grant agreement No 825134.

References

1. Koulouzis, S., et al.: Time-critical data management in clouds: challenges and a Dynamic Real Time Infrastructure Planner (DRIP) solution. Concurr. Comput. Pract. Exp. e5269 (2019). https://doi.org/10.1002/cpe.5269
2. Foster, I., Kesselman, C.: Scaling system-level science: scientific exploration and it implications. Computer **39**(11), 31–39 (2006)
3. Štefanič, P., et al.: SWITCH workbench: a novel approach for the development and deployment of time-critical microservice-based cloud-native applications. Future Gener. Comput. Syst. **99**, 197–212 (2019). https://doi.org/10.1016/j.future.2019.04.008
4. Koulouzis, S., et al.: Distributed data management service for 0 applications. IEEE Internet Comput. **20**(2), 34–41 (2015)
5. Zhao, Z., Belloum, A., De Laat, C., Adriaans, P., Hertzberger, B.: Using Jade agent framework to prototype an e-Science workflow bus. In: Seventh IEEE International Symposium on Cluster Computing and the Grid (CCGrid 2007), Rio de Janeiro, Brazil, pp. 655–660. IEEE (2007). https://doi.org/10.1109/CCGRID.2007.120
6. Laplante, P.A., Ovaska, S.J.: Real-Time Systems Design and Analysis: Tools for the Practitioner. Wiley, Hoboken (2011)
7. Poslad, S., Middleton, S.E., Chaves, F., Tao, R., Necmioglu, O., Bügel, U.: A semantic IoT early warning system for natural environment crisis management. IEEE Trans. Emerg. Top. Comput. **3**(2), 246–257 (2015)
8. Hu, Y., Zhou, H., de Laat, C., Zhao, Z.: Concurrent container scheduling on heterogeneous clusters with multi-resource constraints. Future Gener. Comput. Syst. **102**, 562–573 (2020). https://doi.org/10.1016/j.future.2019.08.025
9. Evans, K., et al.: Dynamically reconfigurable workflows for time-critical applications. In: Proceedings of the 10th Workshop on Workflows in Support of Large-Scale Science - WORKS 2015, Austin, Texas, pp. 1–10. ACM Press (2015). https://doi.org/10.1145/2822332.2822339

10. Dashti, S.E., Rahmani, A.M.: Dynamic VMS placement for energy efficiency by PSO in Cloud computing. J. Exp. Theor. Artif. Intell. **28**(1–2), 97–112 (2016)
11. Usmani, Z., Singh, S.: A survey of virtual machine placement techniques in a cloud data center. Procedia Comput. Sci. **78**, 491–498 (2016). https://doi.org/10.1016/j.procs.2016.02.093
12. Gao, Y., Wang, Y., Gupta, S.K., Pedram, M.: An energy and deadline aware resource provisioning, scheduling and optimization framework for Cloud systems. In: Proceedings of the Ninth IEEE/ACM/IFIP International Conference on Hardware/Software Codesign and System Synthesis, p. 31 (2013)
13. Li, D., Chen, C., Guan, J., Zhang, Y., Zhu, J., Yu, R.: DCloud: deadline-aware resource allocation for Cloud computing jobs. IEEE Trans. Parallel Distrib. Syst. **27**(8), 2248–2260 (2016)
14. Shi, J., Luo, J., Dong, F., Zhang, J.: A budget and deadline aware scientific workflow resource provisioning and scheduling mechanism for Cloud. In: Proceedings of the 2014 IEEE 18th International Conference on Computer Supported Cooperative Work in Design (CSCWD), pp. 672–677 (2014)
15. Downey, D., Dumais, S.T., Horvitz, E.: Models of searching and browsing: languages, studies, and application. IJCAI **7**, 2740–2747 (2007)
16. Agichtein, E., Brill, E., Dumais, S.: Improving web search ranking by incorporating user behavior information. In: Proceedings of the 29th Annual International ACM SIGIR Conference on Research and Development in Information Retrieval, pp. 19–26 (2006)
17. Mishra, N., White, R.W., Ieong, S., Horvitz, E.: Time-critical search. In: Proceedings of the 37th International ACM SIGIR Conference on Research & Development in Information Retrieval, pp. 747–756 (2014)
18. Ingwersen, P., Järvelin, K.: The Turn: Integration of Information Seeking and Retrieval in Context, vol. 18. Springer, Heidelberg (2006). https://doi.org/10.1007/1-4020-3851-8
19. Yu, J., Buyya, R., Tham, C.K.: Cost-based scheduling of scientific workflow applications on utility grids. In: e-Science and Grid Computing (2005)
20. Wang, J., et al.: Planning virtual infrastructures for time critical applications with multiple deadline constraints. Future Gener. Comput. Syst. **75**, 365–375 (2017)
21. Abrishami, S., Naghibzadeh, M., Epema, D.H.: Deadline-constrained workflow scheduling algorithms for infrastructure as a service Clouds. Future Gener. Comput. Syst. **29**(1), 158–169 (2013)
22. Rodriguez, M.A., Buyya, R.: Deadline based resource provisioning and scheduling algorithm for scientific workflows on Clouds. IEEE Trans. Cloud Comput. **2**(2), 222–235 (2014)
23. Cai, Z., Li, X., Gupta, J.: Heuristics for provisioning services to workflows in XaaS Clouds. IEEE Trans. Serv. Comput. **9**(2), 250–263 (2016)
24. Taal, A., Wang, J., de Laat, C., Zhao, Z.: Profiling the scheduling decisions for handling critical paths in deadline-constrained cloud workflows. Future Gener. Comput. Syst. **100**, 237–249 (2019). https://doi.org/10.1016/j.future.2019.05.002
25. Alawneh, S., Dragt, R., Peters, D., Daley, C., Bruneau, S.: Hyper-real-time ice simulation and modeling using GPGPU. IEEE Trans. Comput. **64**(12), 3475–3487 (2015)
26. Zhao, Z., Albada, D.V., Sloot, P.: Agent-based flow control for HLA components. Simulation **81**(7), 487–501 (2005)
27. Bulut, A., Koudas, N., Meka, A., Singh, A.K., Srivastava, D.: Optimization techniques for reactive network monitoring. IEEE Trans. Knowl. Data Eng. **21**(9), 1343–1357 (2009)
28. Zhou, H., Ouyang, X., Su, J., Laat, C., Zhao, Z.: Enforcing trustworthy cloud SLA with witnesses: a game theory–based model using smart contracts. Concurr. Comput. Pract. Exp. (2019). https://doi.org/10.1002/cpe.5511
29. Taherizadeh, S., Jones, A.C., Taylor, I., Zhao, Z., Stankovski, V.: Monitoring self-adaptive applications within edge computing frameworks: a state-of-the-art review. J. Syst. Softw. **136**, 19–38 (2018). https://doi.org/10.1016/j.jss.2017.10.033

30. Chebotko, A., Lu, S., Chang, S., Fotouhi, F., Yang, P.: Secure abstraction views for scientific workflow provenance querying. IEEE Trans. Serv. Comput. **3**(4), 322–337 (2010)
31. Kritikakou, A., Pagetti, C., Baldellon, O., Roy, M., Rochange, C.: Run-time control to increase task parallelism in mixed-critical systems. Real-Time Syst. (ECRTS) **2014**(26), 119–128 (2014)
32. Serrano, N., Gallardo, G., Hernantes, J.: Infrastructure as a service and Cloud technol-ogies. IEEE Softw. **32**(2), 30–36 (2015)
33. Nussbaum, A., Choodamani, S.M.C., Schwan, K.: ObsCon: Integrated monitoring and control for parallel, real-time applications. Clust. Comput. (CLUSTER) **2015**, 474–477 (2015)
34. Anadiotis, A.C.G., Galluccio, L., Milardo, S., Morabito, G., Palazzo, S.: Towards a software-defined network operating system for the IoT. In: 2015 Internet of Things (WF-IoT), pp. 579–584 (2015)
35. Papageorgiou, A., Cheng, B., Kovacs, E.: Real-time data reduction at the network edge of Internet-of-Things systems. Netw. Serv. Manag. (CNSM) **2015**(11), 284–291 (2015)
36. Hu, S., et al.: Data acquisition for real-time decision-making under freshness constraints. In: 2015 IEEE Real-Time Systems Symposium, pp. 185–194. IEEE (2015)
37. Laranjeiro, N., Soydemir, S.N., Bernardino, J.: A survey on data quality: classifying poor data. Dependable Comput. (PRDC) **2015**, 179–188 (2015)
38. Shamani, M.J., Zhu, W., Naghshin, V.: TMPTCP: tailless multi-path TCP. In: Broadband and Wireless Computing, Communication and Applications (BWCCA), vol. 2015, no. 10, pp. 325–332 (2015)
39. Fu, Z., Song, T., Wang, S., Wang, F., Qi, Z.: Seagull–a real-time coflow scheduling system. Cyber Secur. Cloud Comput. (CSCloud) **2015**, 540–545 (2015)
40. Koulouzis, S., Belloum, A.S.Z., Bubak, M.T., Zhao, Z., Živković, M., de Laat, C.T.A.M.: SDN-aware federation of distributed data. Future Gener. Comput. Syst. **56**, 64–76 (2016). https://doi.org/10.1016/j.future.2015.09.032
41. Tang, C.: FVD: a high-performance virtual machine image format for Cloud. In: USENIX Annual Technical Conference, vol. 2 (2011)
42. Lagar-Cavilla, H.A., et al.: SnowFlock: rapid virtual machine cloning for Cloud computing. In: Proceedings of the 4th ACM European Conference on Computer Systems, pp. 1–12. ACM (2009)
43. Zhou, H., et al.: CloudsStorm: a framework for seamlessly programming and controlling virtual infrastructure functions during the DevOps lifecycle of cloud applications. Softw.: Pract. Exp. **49**, 1421–1447 (2019). https://doi.org/10.1002/spe.2741
44. Zhou, H., Hu, Y., Wang, J., Martin, P., Laat, C.D., Zhao, Z.: Fast and dynamic resource provi-sioning for quality critical Cloud applications. In: 2016 IEEE 19th International Symposium on Real-Time Distributed Computing (ISORC), pp. 92–99 (2016)
45. Wartel, R., et al.: Image distribution mechanisms in large scale Cloud providers. In: 2010 IEEE Second International Conference on Cloud Computing Technology and Science, pp. 112–117. IEEE (2010)
46. Müller, C., et al.: Comprehensive explanation of SLA violations at runtime. IEEE Trans. Serv. Comput. **7**(2), 168–183 (2013)
47. Casale, G., et al.: Current and future challenges of software engineering for services and applications. Procedia Comput. Sci. **97**, 34–42 (2016). https://doi.org/10.1016/j.procs.2016.08.278
48. Liao, X., Zhao, Z.: Unsupervised approaches for textual semantic annotation, a survey. ACM Comput. Surv. **52**, 1–45 (2019). https://doi.org/10.1145/3324473
49. Kreutz, D., Ramos, F.M., Verissimo, P.E., Rothenberg, C.E., Azodolmolky, S., Uhlig, S.: Software-defined networking: a comprehensive survey. Proc. IEEE **103**(1), 14–76 (2015)

50. Wang, J., de Laat, C., Zhao, Z.: QoS-aware virtual SDN network planning. In: 2017 IFIP/IEEE Symposium on Integrated Network and Service Management (IM), Lisbon, Portugal, pp. 644–647. IEEE (2017). https://doi.org/10.23919/INM.2017.7987350
51. Hu, Y., et al.: Deadline-aware deployment for time critical applications in clouds. In: Rivera, F.F., Pena, T.F., Cabaleiro, J.C. (eds.) Euro-Par 2017. LNCS, vol. 10417, pp. 345–357. Springer, Cham (2017). https://doi.org/10.1007/978-3-319-64203-1_25
52. Martin, P., et al.: Information modelling and semantic linking for a software workbench for interactive, time critical and self-adaptive Cloud applications. In: 2016 30th International Conference on Advanced Information Networking and Applications Workshops (WAINA), pp. 127–132. IEEE (2016)
53. Miller, M.A., Pfeiffer, W., Schwartz, T.: The CIPRES science gateway: enabling high-impact science for phylogenetics researchers with limited resources. In: Proceedings of the 1st Conference of the Extreme Science and Engineering Discovery Environment: Bridging from the eXtreme to the Campus and Beyond, p. 39 (2012)
54. Mayer, R., Miksa, T., Rauber, A.: Ontologies for describing the context of scientific experiment processes. In: 2014 IEEE 10th International Conference on e-Science, vol. 1, pp. 153–160. IEEE (2014)

Data Provenance

Barbara Magagna[1]([✉]) [iD], Doron Goldfarb[1] [iD], Paul Martin[2] [iD], Malcolm Atkinson[3] [iD],
Spiros Koulouzis[2] [iD], and Zhiming Zhao[2] [iD]

[1] Environment Agency Austria, Vienna, Austria
{barbara.bagagna,doron.goldfarb}@umweltbundesamt.at
[2] Multiscale Networked Systems, University of Amsterdam,
1098XH Amsterdam, The Netherlands
paulmartin.research@gmail.com, {s.koulouzis,z.zhao}@uva.nl
[3] University of Edinburgh, Edinburgh, UK
malcolm.atkinson@ed.ac.uk

Abstract. The provenance of research data is of critical importance to the reproducibility of and trust in scientific results. As research infrastructures provide more
amalgamated datasets for researchers and more integrated facilities for processing
and publishing data, the capture of provenance in a standard, machine-actionable
form becomes especially important. Significant progress has already been made
in providing standards and tools for provenance tracking, but the integration of
these technologies into research infrastructure remains limited in many scientific
domains. Further development and collaboration are required to provide frameworks for provenance capture that can be adopted by as widely as possible, facilitating interoperability as well as dataset reuse. In this chapter, we examine the
current state of the art for provenance, and the current state of provenance capture
in environmental and Earth science research infrastructures in Europe, as surveyed
in the course of the ENVRIplus project. We describe a service developed for the
upload, dissemination and application of provenance templates that can be used
to generate standardised provenance traces from input data in accordance with
current best practice and standards. The use of such a service by research infrastructure architects and researchers can expedite both the understanding and use
of provenance technologies, and so drive the standard use of provenance capture
technologies in future research infrastructure developments.

Keywords: Provenance · Scientific workflow management · Research data

1 Provenance in the Environmental Domain

One particularly sensitive issue in the context of environmental research data lifecycles
is the provenance of offered data products. In order to allow scientific reuse, published
research datasets need clear annotations detailing their genesis and any additional processing applied afterwards. This includes information about the methodology, instrumentation and software used in data acquisition, subsequent processing and preservation, covering all steps of the typical research data lifecycle. The collected information

© The Author(s) 2020
Z. Zhao and M. Hellström (Eds.): Towards Interoperable Research
Infrastructures for Environmental and Earth Sciences, LNCS 12003, pp. 208–225, 2020.
https://doi.org/10.1007/978-3-030-52829-4_12

should not only be targeted at a human audience but should also be machine-processable in order to support various forms of analysis for a variety of purposes such as the choice of suitable data sources or the assessment of patterns of re-use. The increasing availability of reusable, provenance-enabled datasets moreover requires researchers and engineers to consider their "second hand" provenance in addition to "first hand" locally-generated traces and the subsequent combination of these different streams for further reuse. This, even more, underscores the importance of consistent usage of dedicated and interoperable standards for representing provenance, especially in light of recent developments regarding requirements for to-be-published research data, such as the FAIR data principles.

While issues of reproducibility and scientific integrity of research results have traditionally been a central concern for any scientific domain, current environmental developments on global scales often trigger controversies about the underlying cause-effect scenarios. This sometimes even leads to mutual accusations of politically or ideologically driven manipulations of data and resulting scientific evidence. Such a strong political relevance of contemporary environmental research data thus underscores the importance of adequate protocols allowing to trace the respective results back to their origin, acting as evidence for their soundness.

Given the scenarios sketched above, there is a clear and increased need for environmental research infrastructures to develop and maintain well-established provenance generation, provision and tracking infrastructure which is interoperable across the overall landscape of involved domains. Unfortunately, this requirement is hampered by the great heterogeneity of approaches to environmental research and the resulting spectrum of environmental research infrastructures, characterised by a wide variety of objects of interest, applied acquisition and overall research methodologies. Services aiming to cater the needs of the individual research workflows would thus either have to be very specific, or as generic as possible.

In this chapter, we survey the state of the art of provenance gathering and visualisation technologies and standards and describe how we addressed the heterogeneity of research infrastructure in the context of the ENVRIplus project[1], which was charged with the development of generic common services to assist the development and interoperability for environmental and Earth science research infrastructures (RIs) in Europe. We review some of the requirements of RIs regarding research data and data process provenance, and we describe a system for producing, sharing and instantiating provenance templates online, which we believe can help RI architects and engineers, as well as general researchers, to produce better-standardised provenance traces that can be interpreted in a broad range of different contexts.

2 State of the Art

Although there exist several provenance models used in specific settings promoted by different international initiatives, the main basic standard widely-used and referred to is the W3C's PROV recommendation[2], which evolved from the Open Provenance Model

---

[1] https://www.envriplus.eu/.

[2] https://www.w3.org/TR/prov-overview/.

(OPM). After three international workshops [1] on provenance standardisation, OPM was developed in 2010 and subsequently adopted by many workflow systems. PROV is very much influenced by OPM and was released as a standard by the W3C Provenance Working Group in April 2013. The W3C PROV Recommendation [2] consists of some constituent standards including for PROV XML[3] and PROV as an ontology for RDF-based data (PROV-O)[4].

The essential elements (see Fig. 1) of the PROV ontology are called Starting Point Terms and consist of three primary classes with unique and mandatory identifiers and nine properties to describe the relations between the classes. The three classes are prov:Entity which is the central concept and represents resources, prov:Activity representing actions performed upon entities, and prov:Agent representing persons or machines who bears some form of responsibility for an activity. The most important relationships are: *used* (an activity used some artefact), *wasAssociatedWith* (an agent participated in some activity), *wasGeneratedBy* (an activity generated an entity), *wasDerivedFrom* (an entity was derived from another entity), *wasAttributedTo* (an entity was attributed to an agent), *actedOnBehalfOf* (an agent acted on behalf of another agent) and *wasInformedBy* (an activity used an entity produced by another activity). Expanded terms are used to specialise agents and entities and to introduce time validity descriptions for activities. *Qualified terms* are used to provide additional attributes of the binary relations introducing so-called qualified patterns. In this way it is possible to add for example the concept 'plan', an association class, to describe more in detail how an activity was carried out.

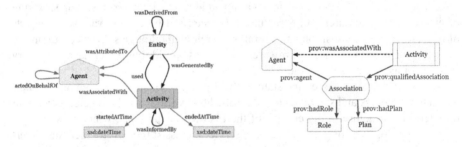

Fig. 1. PROV-O, starting point terms and qualified patterns.

W3C PROV has primarily been designed to describe retrospective provenance (r-prov) which refers to a-posteriori descriptions of provenance traces of data resources, i.e. provenance as an extended log of all the steps executed to generate the data entity. The concept of provenance can, however, also refer to tracing the genesis of workflows used for generating data, and moreover even to the a-priori description of such workflows, in which case it is called prospective provenance (p-prov) which can be considered to be a form of workflow description language.

In order to be able to represent workflow templates and workflow instances, Garijo and Gil extended PROV [3] to P-plan. OPMW [4], an extension of P-plan, PROV

---

[3] https://www.w3.org/TR/2013/NOTE-prov-xml-20130430/.

[4] https://www.w3.org/TR/2013/REC-prov-o-20130430/.

and OPM, is designed to represent prospective provenance of scientific workflows at a fine granularity. D-PROV [5] extended PROV with workflow structure, later being replaced by ProvONE[5], which can track all different types of provenances including the graph structure of the dataflow itself. S-PROV [6] is built upon the PROV and ProvONE models, helping the scientist to analyse the workflow at different levels of granularity and capturing runtime change. It is the underlying model for S-ProvFlow [7], a provenance framework for storage, access and discovery of data-intensive streaming lineage, used in the VERCE Earthquakes Simulation Portal[6] used by the EPOS[7] community. PROV-Wf [8], another specialisation of the W3C PROV-Data Model, allows the capture of both prospective and retrospective provenance but also supports domain-specific data provenance increasing the potential of provenance data analysis. Not all contemporary approaches to provenance are based on W3C PROV. One different approach is the WF4Ever Research Object description[8], conceived as self-contained units of knowledge, aggregating information about the generation workflow at a general level, not directly aligned but still mappable to W3C PROV. CERIF[9] (the Common European Research Information Format) is an entity-relationship model with temporal additions of the research domain used in the EPOS community. It supports the management of Research Information, including details on people, projects, organisations, publications and products. Instances of this representation provide some provenance information because of the time-stamped linking entities used to assets when certain relationships were formed. Nevertheless, CERIF needs further development of some provenance aspects related to the integration of causal-effect relationships among entities [9].

A variety of online tools are made available on dedicated websites[10] in order to support the use of the PROV standard in data management. Examples are the public provenance data storage based on ProvStore [10], the validator against the PROV standard as well as conversion services for various standard output formats. Dedicated libraries are in turn provided for including provenance functionality in local applications. ProvToolbox[11] is an example for a Java library providing different means for manipulating provenance descriptions and converting them between RDF, PROV-XML, PROV-N, and PROV-JSON encoding; comparable functionality for Python-based environments is provided by the PROV[12] Python package.

Starting an overview about technologies and approaches in use from the point of data acquisition, the first phase of the ENVRI RM research data life cycle (see figure page 14 of [11]), it is important to distinguish between manual scenarios and automated settings. Manual measurements and observations made using pen and paper would need to be transferred to spreadsheets or databases before existing provenance recording tools such as InSituTrac [12] could be applied. Shifting from pen-and-paper data collection

---

[5] http://tinyurl.com/ProvOne.

[6] http://portal.verce.eu.

[7] https://www.epos-eu.org/.

[8] http://wf4ever.github.io/ro/2016-01-28/.

[9] https://www.eurocris.org/cerif/main-features-cerif.

[10] https://openprovenance.org/ and https://provenance.ecs.soton.ac.uk/.

[11] https://github.com/lucmoreau/ProvToolbox.

[12] https://github.com/trungdong/prov.

to acquisition via handheld devices could therefore also improve provenance related aspects, such as demonstrated by the EcoProv [13] approach. Handheld applications such as developed in the Urbanopoly project [14] are moreover a potential platform for collecting provenance data in citizen science settings. As an example for tracing the genesis of manually curated scientific databases, the "copy-paste model" approach from [15] captures chains of insertions, deletions and imports from sources to a target database.

It is clear therefore that in many cases, data acquisition includes both manual and automatic aspects. In sample-based data collection, provenance capture should ideally start with the human sampling process and continue with the subsequent analysis taking place in laboratories. In this regard, the alignment of the ISO 19156 Sampling Features Schema with the W3C PROV has resulted in the sam-lite ontology[13] allowing the recording of specimen preparation chains via PROV [16]. As far as automatic data collection is concerned, internal processes are not always accessible for provenance recording, which is often the case with proprietary measurement devices hiding internal data transformations. In such cases, the specific information about the method and technology applied in measurement devices could be made available as contextual information via device type registries such as ESONET Yellow Pages[14]. Reliability of transmission is another aspect relevant for example in wireless sensor networks where collecting provenance becomes essential to ensure the integrity of the data packages transmitted [17].

Provenance is a crucial aspect in heterogeneous sensor infrastructures on the Web, also referred to as the Sensor Web, requiring the adaptation of existing data models. Integrating lineage information with observation descriptions may be done in a number of ways: Cox [16] for example aligned the Observations and Measurements model (O&M) with W3C PROV, while Jiang et al. [18] suggested directly extending PROV-O to cover O&M related concepts.

Increasingly, data are not acquired from one source but are derived from chains of services, resembling so-called "Virtual Data Products" (VDP). As an example for such cases, Yue et al.in [19] proposed the description of provenance via process models with a fixed structure which should be instantiated whenever a VDP is retrieved. This would enable prospective provenance based on the individual service descriptions and the process model or retrospective provenance derived from individual instantiations.

The tracking of provenance information during process execution can often be relatively straightforward, as many scientific workflow management platforms have already integrated this functionality based on provenance standards in their system. Example include Kepler [20], Pegasus [21], Taverna [22] (used by the LifeWatch[15] community) and dispel4py [23] (used in the seismology community). But if researchers run their processes on their private machines outside any particular provenance framework, then that provenance can only be tracked manually, which may be cumbersome and error-prone. One option is to use tools to extract provenance data from specially annotated scripts. Examples are the NoWorkflow system [24] for retrospective provenance and the

[13] http://def.seegrid.csiro.au/static/ontology/om/sam-lite.html.
[14] https://www.esonetyellowpages.com.
[15] https://www.lifewatch.eu/.

YesWorkflow system [25] for prospective provenance which can easily be integrated into interactive notebooks like Jupyter/IPython [26] in use by many researchers today.

The use of the PROV standard by workflow management systems allows provenance information from multiple workflow management systems to be stitched together, but it is still challenging to produce a single cohesive provenance trace without some kind of overarching processing framework in place to orchestrate the provenance generation and storage. A promising approach is described in [27] where the Swift framework for parallel processing is augmented with provenance query frameworks such as MTCProv. Another possibility is to embed provenance recording at the operating system level as demonstrated by CamFlow, a Linux Security Module. Other approaches aim to wrap the entire process within a sandbox operating environment that enables the replication of the process via Docker virtual containers [28].

The PROV-AQ specification[16] provides recommendations related to the annotation of data objects with information on how to retrieve their provenance and to the discovery and query of PROV data. It expects that provenance is served via URIs provided via HTTP response headers, which either directly resolve to the provenance content or point to a dedicated query service. Another technology called Prov-pingback is a mechanism to track client derivations delivering URLs alongside each dataset which should be used to upload the provenance about the data transformations back to the provider [29].

The full visualization of provenance data as graphs of PROV-O triples may often not be satisfactory because of the potential complexity of data lineage. The provision of aggregated representations and thus large-scale overviews of the provenance information can instead substantially support users in the analysis of data generation.

Provenance Map Orbiter [30] is a technology that uses techniques for graph summarization, exploiting intrinsic hierarchies in the graphs, and semantic zoom. Other approaches are direct visualizations of subsets of the full provenance graph focusing on the temporal representation of chains of PROV activities linked together by the entities via Sankey diagrams [31]. Focusing on filesystem provenance, InProv [32] is a technology which transforms provenance graph data into temporally related aggregations visualizing them via a dedicated radial layout diagram. This approach allows navigation on the succession of different temporal visualizations including the storage of a visual protocol. It is being used in the seismology context in conjunction with the Bulk Dependency Visualiser [33] which provides large scale views on data dependencies in distributed stream processing environments such as data reuse between different users.

It is clear that there already exist a variety of tools for provenance gathering and visualisation, mostly based around a core group of standards that have been broadly accepted by the scientific community. It becomes necessary then to ask whether these tools and standards are seeing sufficient adoption in practice by the environmental and Earth science RI community, and if not, what the major barriers are to their adoption. In the following sections, we address how, in the context of ENVRIplus, we analysed the use cases and requirements of current European RI projects, and then drew upon the relevant standards and software libraries to provide a common provenance templating service for RIs and associated users.

---

[16] https://www.w3.org/TR/prov-aq/.

3 ENVRI RI Use Cases and Requirements

The provenance-related section of the questionnaire for the requirements elicitation process in the early phase of the ENVRIplus project (carried out in autumn 2015) intended to collect whether provenance was already considered in the data management plans of the RIs in the ENVRI cluster and if any related implementations were already in use [34]. Among the nine RIs that gave feedback, only two already had a data provenance tracking system integrated in their data processing workflows (EPOS and IS-ENES[17]). For those RIs where the latter was not the case, the next set of questions were focused on their potential interest in provenance: which type of information should be recorded, which standards to rely on and finally what sort of support was expected from ENVRIplus ICT staff. Most RIs considered provenance information as essential and some of them already stored provenance related information for certain aspects like data lineage following metadata standards such as ISO19139 or O&M. This information, however, was not considered sufficient to reproduce data since individual processing steps were not documented in enough detail. The outcomes suggested that it was highly relevant to learn more about what kind of information data provenance should provide, especially in contrast to what was already present in existing metadata about datasets. Another identified need was to get more insight into existing provenance recording systems.

As the outputs from the first requirement collection were rather moderate and unspecific, a second-round was undertaken in spring 2018 in order to retrieve more concrete information from the RIs (see Table 1). The objective was to understand the individual RI needs related to the potential implementation of provenance management systems. Regular teleconferences with live demos of implemented provenance services were thus offered to raise the awareness of the benefits and potential of data provenance techniques. Nine of 20 RIs sent their feedback, five of them have already participated in the first round of requirements collection, but this time giving a deeper insight into their needs, while the remaining four addressed this topic for the first time. As already anticipated, EPOS and IS-ENES, both quite advanced regarding this topic in comparison to other RIs, were able to provide more specific information about their requirements, but also about their existing implementations.

Table 1. Requirements collection (R1: 2015, R2: 2018).

	ACTRIS	AnaEE	EISCAT-3D	EMBRC	EMSO	EPOS	EURO-ARGO	EURO GOOS	IAGOS	ICOS	IS-ENES2	LTER	SeaDataNet
R 1	x					x	x	x	x	x	x	x	x
R 2	x	x	x	x	x	x				x	x	x	

The RI representatives were asked to provide use cases with specific requirements considering that provenance information may be relevant in all phases of the research data life-cycle (DLC), from acquisition and curation to processing and use. Seven RIs provided specific use cases and requirements. Use cases were defined in this case as

---

[17] https://is.enes.org/.

descriptions of a set of interactions between a system and one or more actors, representing a user-perspective specification of functions in a system. For each use case, more than one requirement could be identified. The latter is understood as a functional perspective to approach the problem from a solution angle, providing a formal description of what users expect from the system [35].

The most evident differences in the provenance collection use cases and requirements between ENVRIplus RIs were found to be their varying focus in specific data life-cycle phases but also their varying level of automation. Some RIs included observation networks of scientists and/or instruments producing data (e.g. ACTRIS, EISCAT-3D and LTER-Europe), while others provided advanced processing services (e.g. AnaEE and IS-ENES2). Some RIs had fully automated sensor networks in place whereas human intervention was limited to monitoring, interpretation and/or maintenance tasks. Other RIs, in turn, encompassed considerably more manual steps occurring in the data acquisition but also during the processing phase. This diversity was clearly reflected in the use cases provided by the different communities.

In less automated settings, different aspects of provenance collection itself were reported and less on subsequent analysis and visualisation of such data. Respective use cases included scenarios for tracking lineage for script-based workflows, provenance for automated and non-automated data acquisition such as human observation and physical sample-based data collection, as well as provenance for data publishing and reuse.

In more automated settings (like in EPOS), the reported use cases were often addressing user needs and system features to address them, such as "discovery of experiments" or "navigation through data and dependencies" which were more relevant in the processing phase of the DLC.

Use cases mentioned by more than two RIs (highlighted in bold in Table 2) aimed at automated data collection via sensors, QC measures on instruments, data curation steps including QA/QC flagging procedures, data lineage of data products or aggregations as well as at model runs and their parameter settings.

As far as regards requirements, the different RIs converged more. Recurring requirements were various types of registries since recording provenance for processes with different agents and entities usually requires unique identifiers for each involved instance. Registries for any type of entity including persons, measurement sensors, software, etc. can thus be considered a prerequisite for any meaningful provenance approach. Other commonly expressed requirements were provenance tracking techniques, including domain-specific metadata from controlled vocabularies in the provenance tracks and recording of errata and of data use/citations [35, 42].

Based on the requirements of the various RIs in the ENVRI community and the resources available for development and innovation in the context of the ENVRIplus project, it was considered how best to support better provenance recording at a community level. With regard to this, a generic provenance service was developed that allowed for the generation of provenance traces based on pre-defined templates, which we will now describe in more detail.

Table 2. Use cases and requirements of RIs.

DLC phase	USE CASES	ACTRIS	AnaEE	EISCAT-3D	EMBRC	EPOS	ICOS	IS-ENES2	LTER
acquisition	method		x				x		x
	non-automated data collection						x		x
	non-automated physical samples						x		x
	automated data collection via sensors			x			x		x
	QC measures at instruments	x					x		x
	Eddy Covariance data algorithms						x		
	measurement station changes						x		
curation	**curation**		x	x		x		x	
	annotation		x						
	metadata		x						
	QA/QC	x		x			x	x	
	transfer to data centers							x	
	versioning							x	
publishing	**data products**				x	x	x	x	
	data lineage in scientific publications								x
	discovery of experiments					x			x
	interact. exploration of data dependencies					x			
processing	**model runs and configuration**		x	x		x	x	x	x
	data lineage in scripts		x						x
	track provenance in excel								x
	monitor workflow runs						x		
use	**data usage**				x	x	x		
	collaborative interactions					x			

REQUIREMENTS		ACTRIS	AnaEE	EISCAT-3D	EMBRC	EPOS	ICOS	IS-ENES2	LTER
provenance tracking			x	x	x	x	x	x	x
selective generation of traces						x			
ingestion of provenance							x		
errata tracking						x		x	
registries									
	datasets		x					x	x
	instruments/sensors	x				x	x		x
	physical samples						x		x
	persons					x	x		x
	sites/facilities		x						x
	lab equipment								x
	software/tools	x		x		x			
	publications						x		x
	vocabularies		x						
archives	long term data archival incl. provenance					x		x	

4   A Generic Provenance Service for the ENVRI Community (and Beyond)

As shown by the results of the ENVRIplus provenance use case gathering and requirements analysis, one main provenance-related distinction between research infrastructures can be drawn along the level of automation: highly automated research infrastructures, such as those operating on large-scale sensor networks, often feature dedicated
software environments for executing clearly defined workflows, while less automated and
smaller-scale infrastructures, such as those relying on human observation and sampling
procedures, are often characterised by heterogeneous workflows consisting of alternat-
ing human and machine activities. The former is, therefore, better suited to becoming
adapted for large scale provenance collection, while the latter represents a challenge in
this regard.

Moreover, infrastructures are often still lacking important functionality required for meaningful provenance collection, an example being registries for relevant entities such as physical samples, sensors, instrumentation or personnel, important elements for the creation of provenance traces incorporating well defined and resolvable identifiers. Given the present scenario, the consideration of dedicated provenance services in the context of heterogeneous RI landscapes leaves the choice between concentrating on individual prototypes limited to a few selected research infrastructures only or on focusing on a more generic service concept suitable to a wide variety of RIs, ideally applicable to various levels of maturity. For ENVRIplus, the latter approach has been followed for being more in line with the overall project goals.

Generic approaches can, amongst other aspects, include means to create, store, query or visualise provenance collections. The latter three already require a collection of provenance data to be in place, suggesting that the creative aspect should be considered first when starting from scratch. Correspondingly, this also applied to the ENVRIplus context, motivating related activities accordingly. From an application-level perspective, there are three approaches to generating provenance [36]. "Passive Monitoring" refers to tracing a specific process solely based on the existing information it exchanges with its environment, not requiring any modifications of the original setup. "Overriding" is in turn about adding explicit provenance output to parts of the underlying execution environment (e.g. used software libraries) but not to the process itself, while "Instrumentation" refers to its direct provenance related modification. As far as the latter two are concerned, the heterogeneous landscape of environmental research infrastructures would thus require the direct modification of a wide variety of individual processes or underlying libraries, present either as compiled source code or via a scripting or workflow description language, for enabling the output of provenance. Although generic tools such as YesWorkflow [25] exist for annotating script-based code sequences, they do not cover the full range of possible workflow configurations and require deep knowledge of the code to be augmented. In turn, the notion of Passive Monitoring requires the identification of existing process output and its retrospective translation into a standardised form, potentially allowing the generation of provenance information without modifying underlying processes and their environments. Although having the disadvantage of being limited to the available existing output, this approach suggests itself as a low-threshold starting point for generating initial provenance traces for existing processes.

4.1 Using PROV-Template to Support the Generation of Provenance

One existing approach to turn existing process output into standardised provenance traces is called PROV-Template[18] [37]. As its name implies, it is based on the idea of creating templates which predefine the structure of the intended provenance information using variables which are later instantiated with appropriate data extracted from existing process output. As stated in [36], PROV-Template refers to prior descriptions of how retrospective provenance is to be collected and it is thus not related to the concept of prospective provenance outlined above. Closely related to the W3C PROV data model introduced in Sect. 2, the approach uses the model constructs specified there to define

---

[18] https://provenance.ecs.soton.ac.uk/prov-template/, retrieved March 6th 2019.

the templates, making them valid W3C PROV documents themselves. This, on the one hand, has the advantage that the outcomes of modelling activities to represent provenance traces for specific processes in PROV can be used as both templates and as a blueprint for implementing provenance output directly. On the other hand, existing libraries/services for storing, translating, manipulating, visualizing or validating PROV documents, such as the ProvToolbox[19] or the Python PROV library[20], can be applied to template documents as well.

PROV-Templates are instantiated via bindings which substitute template variables with actual values. The instantiation process is referred to as expansion, illustrated in Fig. 2 with an example template shown in the centre of the figure, representing an activity transforming one or more source datasets into a target dataset, featuring a responsible agent acting on behalf of an organization. As the "var:" prefixes of the element IDs suggest, they all serve as variables to be substituted with values extracted from the runtime log of a process corresponding to the template. This includes the mentioned PROV elements and their attributes, for which both keys and values can be specified as variables as well.

Example values for an appropriate process output are shown in the table at the top of Fig. 2 with the orange and green arrows indicating their bindings to the respective variables, resulting from a mapping effort which has to be done by suitable experts. The two different arrow colours emphasise that in this example, variables for attribute keys are substituted with column names and the remaining variables with column content, respectively. As visible in the table, the sets of substitute values for each variable can have different cardinality, such as for example in the columns "Source" and "Match Column" featuring the IDs of three source data files and the names of their data columns to be used for merging them, in which case the expansion results in multiple instantiations of the respective variables. While n to 1 mappings such as the one presented in this example are expanded in a straightforward manner, the reader is referred to [37] for a formal description of the underlying expansion rules which also apply to more complex n to m mappings.

The result of the example instantiation is shown at the bottom of Fig. 2, illustrating how the expansion of the n to 1 case yields three source file entities connected to the single transformation activity. Where no identifier is explicitly provided by the source data for a given element (such as for the activity variable itself), an expansion mechanism to provide an automatically generated unique ID can be used instead; this is indicated by the use of the "vargen" namespace. Although a useful feature for situations where the input data does not provide unique identifiers for certain elements, it has to be handled with care when considered for entities that potentially appear in multiple process instantiations, such as persons for example.

By design, PROV-Template enables the separation of concerns between the actual process and the generation of its provenance trace: as long as the process output contains sufficiently granular information in at least semi-structured form, it can be externally converted to W3C PROV via appropriate templates, relieving the process developers

---

[19] https://lucmoreau.github.io/ProvToolbox/, retrieved March 5th 2019.

[20] https://github.com/trungdong/prov, retrieved March 5th 2019.

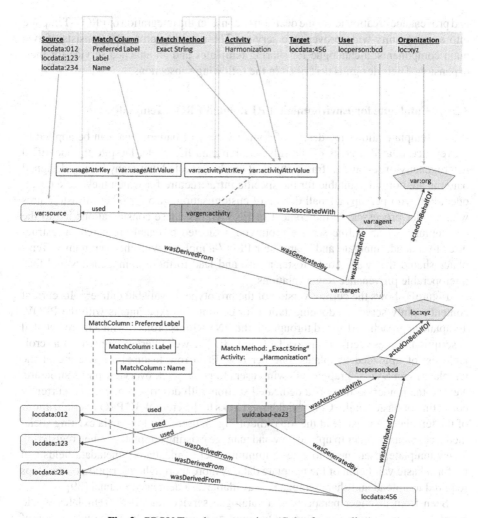

Fig. 2. PROV-Template expansion. (Color figure online)

themselves from the necessity to adhere to a specific data standard in this regard, potentially outsourcing the mapping effort to others. A resulting advantage is that intended changes in the provenance output at a later point in time can in many cases be achieved by modifying only the templates instead of having to touch the process implementation itself.

In the context of ENVRIplus, using PROV-Template appeared as a good starting point for provenance related activities. Assuming a general adherence to the W3C PROV standard, any modelling effort put into experimenting with templates wouldn't be lost even if other approaches than PROV-Template would be adopted in the end, since the created templates could then serve as a data model for any other endeavour to create PROV output. The suggested flexibility stemming from the separation of process output and provenance creation led to the decision to consider the integration of PROV-Template into a community-wide provenance service. The resulting prototype consists of two main components: a catalogue for sharing templates and an attached service for their expansion. Their design is described in the following subsections.

4.2   A Catalogue for Environmental RI Related PROV-Templates

PROV-Template allows the design of a wide variety of patterns that can be applied to provenance related aspects of the full research data life cycle. Despite the identified heterogeneity of research infrastructures, it can be expected that many locally designed patterns are not only suitable for the specific infrastructure for which they were developed, but, with perhaps a small degree of customisation, also for other infrastructures with similar processes. In the context of ENVRIPlus, these considerations led to the development of an online service prototype dedicated to enable research infrastructures to upload, annotate and share their PROV-templates with the community. Templates shared that way should foster re-use and lead to more homogeneous and thus interoperable provenance representations.

Figure 3 shows the current version of the prototype[21] available online[22]. Its current content mainly serves as documentation for community experiments with the PROV-Template approach performed throughout the ENVRIplus project, with a more detailed description of these activities is available in [38]. The web interface consists of a scrollable list of uploaded templates, one per row. Each row features a rendering of the template as an SVG[23] graphic, allowing users to get a quick overview on its structure. Next to the rendering there is a dedicated section with descriptive metadata, currently consisting of basic Dublin Core[24] fields, and links to different W3C PROV serializations of the template. As visible at the top right of Fig. 3, users can log-in via existing social media accounts in order to upload new and manage existing templates. When registering a new template, users need to enter a minimum set of mandatory metadata fields and perform basic validation of the template data. A more thorough description of the steps required to upload and share templates is available as a dedicated manual [39].

Seen from a longer perspective, a catalogue service for PROV-Templates would benefit from various improvements. One important aspect would be the integration of vocabulary suitable for a more thorough description of templates in the context of their specific purpose within an RI's data life-cycle. It is expected that the use of more dedicated authoritative terminology would enable a more consistent annotation of templates, leading to better findability and the retrieval of more adequate templates for specific use cases. The ENVRI Reference Model described in [11] could serve as an important foundation in this regard and the integration of its fine-grained views on research infrastructures with the notion of W3C PROV-Templates potentially mutually beneficial. The availability of templates with fine-grained annotations would subsequently, however, require the adaptation of the search interface to efficiently make use of the increased

---

[21] https://github.com/EnviPlus-PROV/ProvTemplateCatalog, retrieved March 7th 2019.

[22] https://www.envri.eu/provenancetemplates, retrieved March 7th 2019.

[23] https://www.w3.org/TR/SVG11/, retrieved March 7th 2019.

[24] http://dublincore.org/documents/dces/, retrieved March 7th 2019.

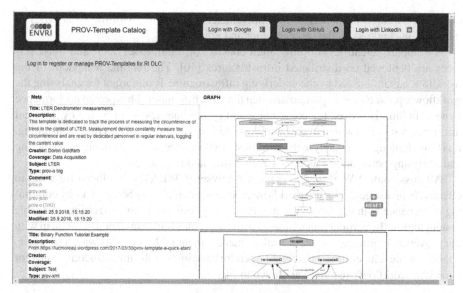

Fig. 3. ENVRIplus PROV-Template catalogue.

expressiveness. Another aspect for improvement would be community features such as rating, commenting and collaborative editing, potentially enabling users to go beyond mere re-use of each other's results.

4.3 Custom Expansion Service for PROV-Template

The second part of the ENVRIplus provenance service is dedicated to the expansion of PROV-Templates. Its basic component is a Python library[25] providing dedicated functions for translating a provided PROV-template and compatible bindings into an instantiated PROV document. Built on top of an existing Python library[26] for basic PROV handling, this implementation of the PROV-Template expansion mechanism is the first one of its kind and thus complements the Java-based proof-of-concept implementation available as part of the ProvToolbox.

The library follows the demand expressed by members of the ENVRI community for a way to directly integrate PROV-Template functionality into Python-based workflows without having to call an external service for that purpose. Besides being usable in standalone form, the library is nevertheless also integrated with the Template catalogue where it is encapsulated behind a dedicated web API, described in [39], for expanding the templates registered there.

5 Provenance and System Logs

A functional provenance service also requires other operations: provenance information capturing, storage, and query. Besides the provenance service presented in Sect. 4, we

---

[25] https://github.com/EnviPlus-PROV/EnviProvTemplates, retrieved March 7th 2019.
[26] https://github.com/trungdong/prov, retrieved March 8th 2019.

have also explored feasibility to link provenance information with the other types of information captured by the infrastructure and platform.

A complex scientific workflow often consists of multiple services, and those services are deployed on distributed infrastructures [40]. The runtime behaviour of the workflow, e.g. monitored by the underlying infrastructure, is important for analysing the workflow's provenance, in particular when the workflow has an unexpected performance issue or failure. However, the provenance and system metrics are provided by different information sources, which makes the integrated analysis difficult and time-consuming. It is thus challenging to analyse the workflow performance, due to difficulty in gathering and analysing performance metrics across distributed infrastructures.

A Cross-context Workflow Execution Analyser (CWEA) is developed for users to effectively investigate possible workflow execution anomalies or bottlenecks by combining provenance with available system metrics [41]. The tool is able to retrieve available system logs of the particular machines (virtual machines if in Cloud) and align them with the provenance provided by the workflow management system. In this way, a user (e.g. application developer or infrastructure operator) can inspect the infrastructure status for particular workflow execution, as shown in Fig. 4.

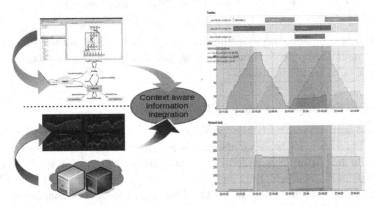

Fig. 4. The basic idea of the cross-context workflow execution analyser, and its output. In the right side of Fig. 4, user can interactively check the workflow processes (from the provenance), and check the system resource information (e.g. CPU and network).

6 Conclusion

In this chapter, we reviewed the state of the art of provenance tracking, focusing on provenance for research data and processes as needed for data-driven environmental science. The challenges of providing FAIR open data, particularly with regard to reproducibility, demonstrate a clear need for better and more extensive provenance gathering throughout the research data life-cycle. Much of the necessary research has already been accomplished, with the various methods, technology and standards ready to use in many contexts and ready to roll out and adopt in others. There is still however a need for development to establish consistent implementations for every system, tool and context into which provenance must be situated. Some technical research into how to handle scale and security issues may be needed as this wider adoption occurs, as will the development of better governance frameworks and best practices for new researchers to adopt as part of their day-to-day activities.

In the context of the ENVRIplus project, a survey of provenance gathering capabilities and needs across the cluster of European environmental and Earth science research infrastructures was carried out. This provided the basis for the development of a shared provenance template service, via which RI developers and researchers can share executable specifications of the provenance patterns used within their infrastructures and workflows. This service also provided the ability to directly instantiate templates with uploaded datasets in order to automatically generate provenance traces in accordance with the W3C PROV standard. It is hoped that this kind of service can assist RI developers in formalising their provenance gathering procedures, share their work, and synchronise how provenance traces for a similar type of dataset and process are constructed across RIs, improving interoperability and reusability of the resources they provide to their respective scientific communities.

Acknowledgements. This work was supported by the European Union's Horizon 2020 research and innovation programme via the ENVRIplus project under grant agreement No 654182.

References

1. Moreau, L., Freire, J., Futrelle, J., McGrath, R.E., Myers, J., Paulson, P.: The open provenance model: an overview. In: Freire, J., Koop, D., Moreau, L. (eds.) IPAW 2008. LNCS, vol. 5272, pp. 323–326. Springer, Heidelberg (2008). https://doi.org/10.1007/978-3-540-89965-5_31
2. Groth, P., Moreau, L.: PROV-overview. W3C. W3C Note, April 2013. http://www.w3.org/TR/2013/NOTE-prov-overview-20130430/
3. Garijo, D., Gil, Y.: Augmenting PROV with plans in p-plan: scientific processes as linked data. In: CEUR Workshop Proceedings (2012)
4. Garijo, Y., Gil, G., Corcho, O.: Towards workflow ecosystems through semantic and standard representations. In: Proceedings of the 9th Workshop on Workflows in Support of Large-Scale Science, pp. 94–104. IEEE Press (2014)
5. Missier, P., Dey, S., Belhajjame, K., Cuevas-Vicenttin, V., Ludäscher, B.: D-PROV: extending the PROV provenance model with workflow structure. In: 5th USENIX Workshop on the Theory and Practice of Provenance (TaPP 13), Lombard, IL (2013)
6. Spinuso, A.: S-ProvFlow and DARE management for data-intensive platforms. In: RDA-Europe Meeting on Data Provenance Approaches, Barcelona, 15–16th January (2018)
7. Spinuso, A.: Active provenance for data-intensive research, Ph.D. thesis, School of Informatics, University of Edinburgh (2018)
8. Costa, F., et al.: Capturing and querying workflow runtime provenance with PROV: a practical approach. In: Proceedings of the Joint EDBT/ICDT 2013 Workshops, pp. 282–289. ACM (2013)

9. Bailo, D., Ulbricht, D., Nayembil, L., Trani, L., Spinuso, A., Jeffery, K.: Mapping solid earth data and research infrastructures to CERIF. Procedia Comput. Sci. **106**, 112–121 (2017)
10. Huynh, T.D., Moreau, L.: ProvStore: a public provenance repository. In: Ludäscher, B., Plale, B. (eds.) IPAW 2014. LNCS, vol. 8628, pp. 275–277. Springer, Cham (2015). https://doi.org/10.1007/978-3-319-16462-5_32
11. de la Hidalga, A.N., et al.: The ENVRI Reference Model (ENVRI RM) version 2.2 (2017). http://doi.org/10.5281/zenodo.1050349
12. Asuncion, H.U.: Automated data provenance capture in spreadsheets, with case studies. Future Gener. Comput. Syst. **29**(8), 2169–2181 (2013)
13. Zhang, Q., et al.: WIP: provenance support for interdisciplinary research on the North Creek Wetlands. In: IEEE 11th International Conference on e-Science (e-Science), pp. 521–528 (2015)
14. Buneman, P., Chapman, A., Cheney, J., Vansummeren, S.: A provenance model for manually curated data. In: Moreau, L., Foster, I. (eds.) IPAW 2006. LNCS, vol. 4145, pp. 162–170. Springer, Heidelberg (2006). https://doi.org/10.1007/11890850_17
15. Celino, I.: Human computation VGI provenance: semantic web-based representation and publishing. IEEE Trans. Geosci. Remote Sens. **51**(11), 5137–5144 (2013)
16. Cox, S.: Ontology for observations and sampling features, with alignments to existing models. Semant. Web **8**(3), 453–470 (2017)
17. Wang, C., Zheng, W., Bertino, E.: Provenance for wireless sensor networks: a survey. Data Sci. Eng. **1**(3), 189–200 (2016)
18. Jiang, J., Kuhn, W., Yue, P.: An interoperable approach for Sensor Web provenance. In: 2017 6th International Conference on Agro-Geoinformatics, pp. 1–6 (2017)
19. Yue, P., Gong, J., Di, L.: Augmenting geospatial data provenance through metadata tracking in geospatial service chaining. Comput. Geosci. **36**(3), 270–281 (2010)
20. Altintas, I., Barney, O., Jaeger-Frank, E.: Provenance collection support in the Kepler scientific workflow system. In: Moreau, L., Foster, I. (eds.) IPAW 2006. LNCS, vol. 4145, pp. 118–132. Springer, Heidelberg (2006). https://doi.org/10.1007/11890850_14
21. Kim, J., Deelman, E., Gil, Y., Mehta, G., Ratnakar, V.: Provenance trails in the wings/pegasus system. Concurr. Comput.: Pract. Exp. **20**(5), 587–597 (2008)
22. Zhao, J., Goble, C., Stevens, R., Turi, D.: Mining Taverna's semantic web of provenance. Concurr. Comput.: Pract. Exp. **20**(5), 463–472 (2008)
23. Filgueira, R., Krause, A., Atkinson, M., Klampanos, I., Spinuso, A., Sanchez-Exposito, S.: dispel4py: an agile framework for data-intensive escience. In: 2015 IEEE 11th International Conference on e-Science (e-Science). IEEE, pp. 454–464 (2015)
24. Murta, L., Braganholo, V., Chirigati, F., Koop, D., Freire, J.: noWorkflow: capturing and analyzing provenance of scripts. In: Ludäscher, B., Plale, B. (eds.) IPAW 2014. LNCS, vol. 8628, pp. 71–83. Springer, Cham (2015). https://doi.org/10.1007/978-3-319-16462-5_6
25. McPhillips, T., et al.: YesWorkflow: a user-oriented, language-independent tool for recovering workflow information from scripts. arXiv preprint arXiv:1502.02403 (2015)
26. Pimentel, J., Braganholo, V., Murta, L., Freire, J.: Collecting and analyzing provenance on interactive notebooks: when IPython meets noworkflow. In: Workshop on the Theory and Practice of Provenance (TaPP), Edinburgh, Scotland, pp. 155–167 (2015)
27. Gadelha, L., Wilde, M., Mattoso, M., Foster, I.: MTCProv: a practical provenance query framework for many-task scientific computing. Distrib. Parallel Databases **30**(5–6), 351–370 (2012)
28. Pasquier, T., et al.: Practical whole-system provenance capture. In: Proceedings of the Symposium on Cloud Computing, pp. 405–418. ACM (2017)
29. Lebo, T., West, P., McGuinness, D.L.: Walking into the future with PROV pingback: an application to OPeNDAP using prizms. In: Ludäscher, B., Plale, B. (eds.) IPAW 2014. LNCS, vol. 8628, pp. 31–43. Springer, Cham (2015). https://doi.org/10.1007/978-3-319-16462-5_3

30. Macko, P., Seltzer, M.: Provenance map orbiter: interactive exploration of large provenance graphs. TaPP **2011**, 1–6 (2011)
31. Hoekstra, R., Groth, P.: PROV-O-Viz-understanding the role of activities in provenance, in International Provenance and Annotation Workshop, pp. 215–220 (2014)
32. Borkin, M.A., et al.: Evaluation of filesystem provenance visualization tools. IEEE Trans. Visual Comput. Graph. **19**(12), 2476–2485 (2013)
33. Spinuso, A., Fligueira, R., Atkinson, M., Gemuend, A.: Visualisation methods for large provenance collections in data-intensive collaborative platforms. In: EGU General Assembly Conference Abstracts, vol. 18, pp. 14793 (2016)
34. Zhao, Z., et al.: Reference model guided system design and implementation for interoperable environmental research infrastructures. In: 2015 IEEE 11th International Conference on e-Science, Munich, Germany, pp. 551–556. IEEE (2015). https://doi.org/10.1109/eScience.2015.41
35. Magagna, B., et al.: Deliverable 8.5: data provenance and tracing for environmental sciences: system design, a document of ENVRIplus project (2018)
36. Frew, J., Metzger, D., Slaughter, P.: Automatic capture and reconstruction of computational provenance. Concurr. Comput.: Pract. Exp. **20**(5), 485–496 (2008)
37. Moreau, L.: A templating system to generate provenance. IEEE Trans. Softw. Eng. **44**(2), 103–121 (2017)
38. Goldfarb, D., et al.: Deliverable 8.6 Data provenance and tracing for environmental sciences: prototype and deployment, a document of ENVRIplus project (2018)
39. Goldfarb, D., Martin, P.: PROV-template registry and expansion service manual (2018). https://envriplus-provenance.test.fedcloud.eu/static/EnvriProvTemplateCatalog_Manual_v2.pdf
40. Zhao, Z., Belloum, A., Bubak, M.: Special section on workflow systems and applications in e-Science. Future Gener. Comput. Syst. **25**, 525–527 (2009). https://doi.org/10.1016/j.future.2008.10.011
41. el Khaldi Ahanach, E., Koulouzis, S., Zhao, Z.: Contextual linking between workflow provenance and system performance logs. In: 2019 15th International Conference on eScience (eScience), San Diego, CA, USA, pp. 634–635. IEEE (2019). https://doi.org/10.1109/eScience.2019.00093
42. Tanhua, T., et al.: Ocean FAIR data services. Front. Mar. Sci. **6**, 440 (2019). https://doi.org/10.3389/fmars.2019.00440

Semantic Linking of Research Infrastructure Metadata

Paul Martin[1] ⓘ, Barbara Magagna[2] ⓘ, Xiaofeng Liao[1] ⓘ, and Zhiming Zhao[1(✉)] ⓘ

[1] Multiscale Networked Systems, University of Amsterdam,
1098XH Amsterdam, The Netherlands
`paulmartin.research@gmail.com`, `{x.liao,z.zhao}@uva.nl`
[2] Environment Agency Austria, Vienna, Austria
`barbara.magagna@umweltbundesamt.at`

Abstract. The use of metadata to characterise scientific datasets, making data easier to discover and use directly by researchers and via various online data services, is one of the primary concerns of research infrastructures (RIs); also, of concern is the use of metadata to describe equipment, facilities, services and other research assets. Metadata models and terminology differ greatly between different communities and infrastructures however, and so make synthesising complex interdisciplinary scientific workflows involving assets from multiple RIs very challenging.

'Semantic linking' addresses the need to enhance the interoperability of RI services and data by bridging metadata schemes, ontologies and vocabularies used by different research communities, whether by standardising the terminologies and schemes used by those communities, or by dynamically transforming metadata from one standard to another when retrieved by services on behalf of researchers executing their scientific workflows.

Multiple techniques for and modes of semantic linking have been investigated in the context of the ENVRI community cluster of environmental and Earth science RIs, including top-down modelling of entities and activities within a standard reference model, enrichment of existing metadata records with shared terminology, full transformation of metadata records from one standard to another, and the generation of additional links to existing online data. We review some of these activities and their application to the promotion of semantic interoperability between RIs, and discuss other possibilities and recent developments that may also be useful for enhancing interdisciplinary data science.

Keywords: Metadata · Semantics · Linking

1 Introduction

The adoption and use of metadata for characterising and cataloguing scientific data and other research assets is one of the primary concerns of modern scientific research infrastructures (RIs). The production and maintenance of good metadata has bearing on the

Z. Zhao and M. Hellström (Eds.): Towards Interoperable Research
Infrastructures for Environmental and Earth Sciences, LNCS 12003, pp. 226–246, 2020.
https://doi.org/10.1007/978-3-030-52829-4_13

entire research lifecycle, from acquisition and curation through to publishing, processing and use. The adoption of standard protocols, metadata schemes and controlled vocabularies for use in scientific data and their associated metadata by a given research community is supposed to expedite data sharing and the development of interoperable data services within a scientific discipline. The increasing need for interdisciplinary research makes such standardisation more challenging however, as the range and diversity of scientific products that should be normalised grows ever greater. Even mature standards do not always meet all community requirements, or else have ambiguous semantics that lead to variation in how they are applied. In addition, many communities have already adopted and adapted to their own preferred standards independently, and have their own established best practices and legacy systems. It therefore seems unavoidable that there will always be variation in metadata schemes, vocabularies and protocols, and thus a need to be able to translate information between different semantic contexts, as represented by specific data models and terminology, whether on request or performed dynamically out of sight of researchers. Regardless of how it is carried out, we refer to this kind of translation as *semantic linking*; techniques for bridging the gap between two or more semantic domains to permit cross-domain data science.

Semantic linking is of great importance in the development of an interdisciplinary 'data science commons' for researchers—a common environment for getting access to and contributing scientific data. The ideal scenario is that researchers can retrieve data, tools, models and other services from different RIs based on scientific requirements without having to know which specific infrastructure serves which specific data, and can use them in complex workflows without having to manually rework data inputs at each step [41]. Specifically, the use of semantic linking is necessary in the development of joint catalogues or indexes of research assets (needed for cross-RI search and discovery), to export data and metadata into different operational contexts, and to glue together services with different input and output formats.

Semantic linking was thus identified as one of the three main cross-cutting activities of the 'Data for Science' theme of the ENVRIplus project[1], alongside the development and exploitation of the ENVRI Reference Model (ENVRI RM) [1, 2] and the specification of common abstract architecture for the construction of interoperable services. One of the results of this activity was the development of Open Information Linking for Environmental Research Infrastructures (OIL-E) [3] as a kind of architectural hub ontology for RI descriptions. Using OIL-E as our baseline semantic model, we surveyed four different kinds of semantic linking during the project; in this chapter we review these four kinds in turn and consider how they reflect on the challenge of achieving semantic interoperability in data science research in general and within the environmental and earth sciences in particular.

In the next section (Sect. 2), we examine more closely the background and motivation for the investigation of semantics in environmental and Earth science RIs. We describe the methodology applied in ENVRIplus for surveying and rationalising the semantic landscape of RIs involved in the project (Sect. 3), before then moving on to discussing the four semantic linking scenarios we proceeded to investigate (Sect. 4). We discuss some of the technological developments that might have bearing on RI semantics and

---

[1] https://www.envriplus.eu/.

metadata and on semantic linking activities in general (Sect. 4) before finally drawing our conclusions (Sect. 6).

2 Background

Modern day environmental research depends on the collection and analysis of large volumes of data gathered via sensors, field observations, controlled experiments, simulation and modelling. In this context, the role of research infrastructures (RIs) is to support researchers with datasets, platforms and tools that allow them to engage effectively with the available data, but no single research infrastructure can hope to encompass fully the whole research ecosystem [4]. Consequently, today there is a host of different research infrastructures, each with their own intersecting speciality areas, but more broadly sharing many common scientific, technical, political and governance-oriented interests. Meanwhile, researchers are being called upon to address societal challenges that are inextricably tied to the stability of our native ecosystems. These challenges are intrinsically interdisciplinary in nature, requiring collaboration across traditional disciplinary boundaries. The challenge, therefore, is to help researchers to freely and effectively interact with the full range of research assets potentially available to them across many different research infrastructures, with the intention that they are allowing them to collaborate and conduct their research more effectively than ever was possible before. This is the challenge that initiatives such as the Research Data Alliance[2] and proposals for FAIR (Findable, Accessibility, Interoperable and Reusable) data [5] seek to address, and it is one that fundamentally relies on the proper elicitation and application of semantics in research data in general.

Data semantics are provided by the various schemas produced for datasets and metadata and are embedded in the choice of vocabulary used to describe different data elements. For metadata in particular, having well-defined and rigorous descriptions in a machine-actionable format confers a number of advantages to both the provider and user of the data or other resources being described. Publishing metadata about the resources (not only data, but also services, tools and facilities) that RIs offer online (indicating such information as the type of resource and their provenance) allows them to advertise their offerings and allows researchers to browse and discover resources (including data, models, tools, services and other kinds of resources both digital and physical) that could be useful to their research. It also permits comparison and the integration of resources into larger workflows or toolchains. More fundamentally however, it also ensures that the resource (and this is especially vital for scientific datasets) is and continues to be correctly understood, and not subject to confusion regarding the exact thing being measured or observed, the units used, or the time and location when/where a measurement or observation was made. Semantic rigour is thus vital for well-grounded, reproducible and accountable research.

In this space there are many metadata standards, old and new; some of which are *de facto* standards long adopted by particular communities, while others have achieved *de jure* status as recommendations by certain community institutions such as the International Organization for Standardization (ISO) and the Open Geospatial Consortium

---

[2] https://www.rd-alliance.org/.

(OGC). For example, in the geospatial area, which concerns many environmental and Earth science RIs, there exist established standards such as ISOs 19115 [6] (for geospatial data) and 19139 [7] (the accompanying XML profile), which form the basis for the INSPIRE[3] recommendation for spatial metadata in Europe. In practice, however, the implementation of these and other standards can sometimes be partial or idiosyncratic across communities, with resulting variations in how metadata elements are realised or terms applied. There are also standard protocols for accessing catalogues of metadata records used to describe data collections via the Web; standards such as DCAT [8] describe how data catalogues should be structured, and protocols such as CSW [9] and OAI-PMH [10] describe how they ought to be accessed. Many RIs use these established protocols, but some RIs also use Semantic Web [11] technologies such as OWL [12] and SKOS [13] to describe their resources and use SPARQL [14] to access them. These RIs adapt ontologies such as OBOE [15] (for observations) and vocabularies such as EnvThes [16] (for ecology) to meet their own community's needs while building upon the semantic harmonisation work of other neighbouring communities. Continuing harmonisation of vocabulary and metadata between research infrastructures thus remains an on-going concern; for example, the European Open Science Cloud initiative (EOSC) [17] considers it a major priority to integrate existing terminological resources with the services provided by European RIs to realise its goals for better cross-disciplinary open science, and a similar urgency can be seen in other open science initiatives around the world.

The integration of resources requires alignment of data formats and content. One of the roles of an RI within the context of its target community is to facilitate standardisation, and as such RIs are very useful vehicles for aligning the use of semantics within a community. Nevertheless, such standardisation activity becomes very difficult once boundaries between communities (even within the same scientific discipline) are crossed. This is because intrinsically, the requirements and usage of data products can be very different between communities. This means that the metadata models used, and indeed how the very datasets being described are even structured for use by researchers, likewise differ considerably between communities. A simple example would be how some communities gather all data related to a given location into a single dataset that might then partitioned by time period, while other communities may gather all of a single kind of observation into one dataset with the locality of each observation reduced to a single field within each row of data. Thus, it remains necessary, even in the presence of initiatives such as RDA (which provides a forum for discussion of best practices for addressing various data science challenges) and initiatives such as Copernicus[4] and GEOSS[5] (which act as aggregators for specific classes of data and thus promote certain standards for such data), to consider how to transform metadata between models in order to allow different data services and tools to work together as part of a cohesive operational workflow.

The semantic linking work of ENVRIplus was intended to guide the harmonisation of semantics across environmental science research infrastructures by providing both

---

[3] https://inspire.ec.europa.eu/.

[4] https://www.copernicus.eu/.

[5] https://www.earthobservations.org/geoss.php.

contextualisation and a standard 'connective' upper ontology for the different kinds of entities and activities commonly found in those infrastructures. Notably, there is no catch-all solution to the problem of mapping between different metadata schemes used by RIs, for which there has been considerable effort already expended and for which considerable effort will be expended in future. Instead, there exist many tools and frameworks for handling such mappings and a great body of research. Our concern then is rather with providing some baseline support for analysing the diversity of such schemes and mappings where they exist, and so help research infrastructure developers to focus their efforts on specific problem areas.

3 Semantic Linking in ENVRIplus

To even approach the topic of semantic linking, there is a need to understand the *semantic landscape* of research infrastructure at large. By 'landscape', we essentially mean information about not just which metadata schemes, ontologies and vocabularies are in use by different RIs, but also how they are used and for what purpose. Without an understanding of the landscape of the use of semantic instruments and standards, it is impossible to identify where to target semantic linking activities—to determine where it is needed, and which models/terminologies need alignment in order to facilitate some otherwise hypothetical workflow. The semantic linking activity in the ENVRI environmental and Earth science RI cluster[6] was carried out in several stages:

1. We collected information from environmental and Earth science RIs and communities, regarding their requirements, adopted technologies and the current state of the art; much of the results of this process appear in Chapter 3 of this book.
2. We used the requirements gathered in the previous step to refine the ENVRI RM (described in detail in Chapter 4), which importantly (for the purpose of semantics and shared terminology) provided a common vocabulary for describing various kinds of component and activity deployed in RIs, and helped us to identify the most important interactions typically facilitated (or needed) by environmental and Earth science RIs.
3. Concurrently, we also began gathering information about the community standards, protocols, and semantic/terminological resources used by RIs and in various aspects of environmental research, data and process specification. This was performed mainly via direct interactions with technical experts involved in RI development.
4. We developed Open Information Linking for Environmental Research Infrastructures (OIL-E, described more completely in Chapter 6) to capture the stereotypical elements of environmental and Earth science RIs as identified by ENVRI RM, and define the necessary relationships between those stereotypes across different views of science, information, computation, engineering and technology. One of the roles of OIL-E, aside from allowing for various RI descriptions based on ENVRI RM to be transformed into a format that can be uploaded into an ENVRI Knowledge Base [39] and programmatically queried, was to act as a connective 'hub' ontology

---

[6] https://www.envri.eu/.

for RI architecture. This ontology allows specifications of specific concepts to be extended with other, more specific ontologies and taxonomies used by the scientific community.

5. Using the OIL-E ontology to structure the data, we began mapping the semantic landscape of environment science by encoding information about the different RIs, their component parts and their constituent processes, as well as associating standards and software to different entities where appropriate.

6. This has resulted in the creation of a knowledge base (also described in Chapter 6) to contain all the formally-encoded data, and to provide a service with which architects and developers can investigate and contribute descriptions of RIs.

7. We further investigated specific approaches for linking data encoded using OIL-E with other (meta)data sources of interest to researchers or to the RIs that support their activities. The next part of this chapter goes into these investigations in further detail.

8. Within the framework of successor projects such as ENVRI-FAIR[7], we can now focus on capturing mapping information for bridging between OIL-E and other RI knowledge representations, and on tools for semantic modelling and discovery using OIL-E and the ENVRI Knowledge Base.

Figure 1 provides a pictorial overview of the relationship between the various parts of the semantic landscape mapping in ENVRI, which was also used in various dissemination materials produced by the project.

4   Semantic Linking Scenarios

'Semantic linking' in the context of the cluster of environmental science research infrastructures is fundamentally concerned with how to contextualise (meta)data regarding research datasets, tools, methods and infrastructure such that they can be interpreted in accordance with a particular model of reality, are meaningfully comparable with similar metadata, and can be understood as part of a wider semantic landscape. This is so that (for example) we can determine the role of certain data in specific processes within a particular infrastructure. We therefore need to consider how to 'elevate' existing data semantically (by providing additional context needed to do more with the data), and how to transform those data where necessary (so that we can use them elsewhere). We need to consider what new data must be created to provide additional context to the entities we wish to model, as well as to describe the relationships between entities.

There are four semantic linking scenarios that need to be considered in the context of environmental science and environmental science research infrastructure, that we chose in the context of ENVRIplus to explore in more depth:

1. The creation of a new model for an existing artefact or process based on a formal ontology. This could be in addition to existing semantic metadata for that artefact or process, providing additional contextual information that could allow for multiple means of interaction with a given research asset, for example by creating multiple

---

[7] https://envri.eu/envri-fair/.

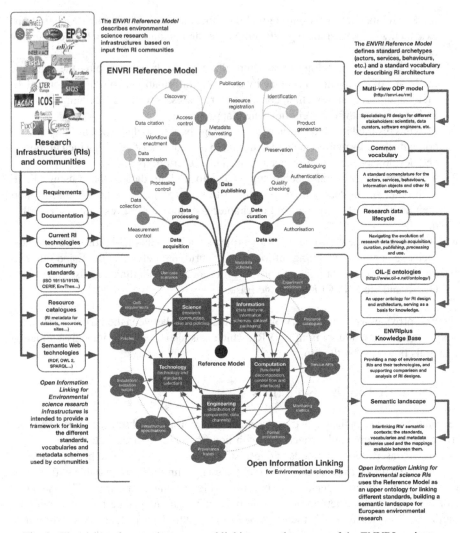

Fig. 1. The vision of semantic survey and linking over the course of the ENVRI projects.

metadata records in different schemes for the same data product for retrieval and use by different services with different protocols.

2. The enrichment of an existing model using controlled vocabulary extracted from an ontology or other formal terminological resource. In this case, the additional vocabulary provides additional metadata by which services (e.g. for search and discovery) can differentiate and classify research assets already described using a set metadata scheme and protocol.

3. The translation of an existing model from one semantic context to another. Rather than augmenting or linking to existing semantic metadata, this is the scenario where entirely new metadata is generated from existing metadata, generally for inclusion

in another metadata catalogue or repository which requires a different scheme for describing research assets.

4. The linking of two models for the same entity (or conceptually overlapping entities) by generating additional 'bridging' metadata between existing metadata records. This is the linked open data approach, whereby information existing independently in multiple contexts about the same or similar entities is somehow made connected such that an external query service can navigate between contexts and aggregate the results from each.

All of these four scenarios have overlaps in their objective and concerns to the extent that it is not always clear to which scenario a given semantic linking operation belongs (and in many cases an operation could justifiably belong to more than one), but nonetheless it is useful to consider how semantic technologies might be used to address each case in turn.

4.1 Semantic Contextualization

The most basic form of 'semantic linking' is the (re-)contextualisation of data already somehow modelled using some ontology or metadata scheme. Typically, this involves describing and classifying entities using a new ontology or other metadata scheme, which provides new metadata that can be used to discover and retrieve information about those entities (or the entities themselves if they exist as data). Doing this for multiple ontologies/schemes binds the data in question to two or more different semantic domains, and so allows the data to be examined in either context; this is most appropriate in the case of multiple systems that might want to query the data, but where each system supports a different schema. A benefit of this kind of 'multiple classification' is that it creates sample data for constructing more formal semantic mappings between two different semantic models should it later be determined that all data in one model needs to be transformed into the other. The main benefit, however, and the distinguishing factor from the scenario where the second model is simply generated from the first model automatically, is that the second model may capture information not representable by the first model, thus increasing the amount of information about a data entity available. For example, one model might not capture procedural aspects of how a dataset is created, while another does; thus, it is not possible to simply generate the metadata required by the latter model from the former model. It may be useful however to simultaneously describe the dataset using *both* models for the additional flexibility such multi-modelling grants, such as support for two different querying systems that each expect a specific model to be used.

In the context of the ENVRIplus project, different kinds of entities with semantic connotations (datasets, metadata schemes, vocabularies, etc.) were described using the OIL-E ontology and so classified in terms of ENVRI RM, where possible with direct links to their respective access points (e.g. URLs for querying and retrieving metadata) or specifications (e.g. landing pages for ontologies) as appropriate. Figure 2 provides an example of such contextualisation in data acquisition, specifically the collection of data regarding phytoplankton.

In Fig. 2, concepts from four of the five viewpoints defined in OIL-E are used (though actually only one concept is used from the technology view). In addition, a number of

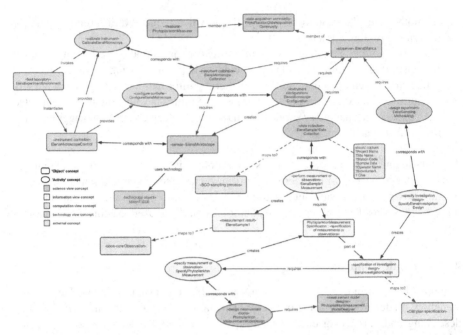

Fig. 2. Modelling the acquisition of data regarding phytoplankton across multiple views in OIL-E.

points at which entities might be further explicated using other ontologies are highlighted (using the OBOE, OBI or BCO ontologies). This allows us to describe the activity of data collection (answering who, what and how), the data being created, and the processes involved. Each of these views could be elaborated upon or linked to a larger dataset in OIL-E or, using one of the linking methods described in the following sections, translated into another ontological model.

The ENVRI Knowledge Base was the primary vehicle for exploring this kind of semantic linking within the ENVRIplus project: by collecting information about different RIs using the terminology of ENVRI RM and the framework of OIL-E, we were able to explore and visualise the resulting knowledge network and perform some fundamental comparative analyses.

4.2 Semantic Enrichment

Often, it is not necessary to create new descriptions of entity data from scratch. While some aspects of research infrastructure (particularly processes) are rarely formally described in any machine-actionable representation, other things (particularly datasets or services) already have descriptive metadata based on some formal model. The issue then becomes not that of how to (re-)model the entity in question, but how to 'plug in' the existing model into a wider semantic context such that the information within the model can be made better use of by a greater variety of knowledge-driven services. One approach is to transform the existing model into a new model that is somehow more 'semantically interoperable'; we address this in the next section. Another approach is to enrich the existing model 'externally', by creating linking data hosted outwith the model that allows an external actor to find and retrieve the information in the current model; we address this in the section on semantic bridging. A third approach is to enrich the existing model *internally*, by taking controlled vocabulary from an external ontology or thesaurus and annotating the model where it permits the insertion of such vocabulary, allowing for external services to harvest that information and thus 'comprehend' the context in which the model is applied. For example, we can take observation data from an RI such as the Integrated Carbon Observation System (ICOS)[8] and annotate the datasets with terms from the EnvThes thesaurus for ecosystem observations in order to better identify the scientific context of each observation set—e.g. that it pertains to the North Atlantic Oscillation[9], or to snow accumulation[10].

We consider here the example of CERIF (Common European Research Information Format) [18]. CERIF is a recommendation for the contextualisation of research activity, relating people to organisations, to projects, to equipment, to datasets and other research products. Investigated as a possible base scheme for cross-RI joint research asset catalogues, CERIF is notable for how it separates its semantic layer from its primary entity-relationship model. Most CERIF relations are semantically agnostic, lacking any particular interpretation beyond identifying a link. Almost every entity and relation can be assigned a classification however that indicates a particular semantic interpretation (e.g. that the relationship between a *Person* and a *Product* is that of a creator and their creation), allowing a CERIF database to be enriched with concepts from an external semantic model (or several linked models). In this respect, the vocabulary provided by OIL-E was investigated as a means to further classify objects in CERIF in terms of their role in a research infrastructure, e.g. classifying individuals and facilities by the roles they play in research activities, datasets in terms of the research data lifecycle, or computational services by the functions they enable. This can provide additional operational context for faceted search—for example to identify which processes generated a data product, or to search for quality-assured datasets only.

Some examples of classifications based on ENVRI RM stereotypes defined in OIL-E are given in Table 1. Classifying CERIF entity classes such as *Person, Facility, Result Entity* or *Service* using OIL-E concepts such as *environmental scientist, data provider, persistent dataset* and *virtual laboratory* is simple enough, but OIL-E can also be used to classify various classes of RI activity involving interactions between instances of CERIF entity in a way that is particularly suitable for describing time-bounded events involving those entities. For example, given a CERIF relation between a Person and the Result Entity that the person in question annotated, that relation can be classified using the 'annotate data' information action concept in OIL-E, with CERIF also capturing the time of annotation.

Semantic enrichment of this kind need not be limited to one particular semantic context. Providing additional information about the *scientific* context for datasets (e.g. categorising the experimental method applied to generate the data or the branch of science to which the data belong) is also important, and there exist many vocabularies to

---

[8] https://www.icos-ri.eu/.

[9] http://vocabs.lter-europe.net/EnvThes/20403.

[10] http://vocabs.lter-europe.net/EnvThes/20949.

Table 1. Example classifications of CERIF entities based on ENVRI RM stereotypes.

CERIF entity	OIL-E concept	Example classifications
'Event'	'behaviour'	'data collection [behaviour]', 'data replication [behaviour]'
'Equipment'	'resource'	'sensor network', 'storage system'
'Facility'	'resource'	'data repository', 'research infrastructure'
'Organisation Unit'	'actor'	'data publisher', 'semantic mediator'
'Person'	'actor'	'environmental scientist', 'engineer'
'Result Entity'	'persistent data'	'QA-assessed data', 'annotated data'
'Service'	'computational object'	'catalogue service', 'data broker'

do this (and indeed many are already in use for just this purpose). Aside from the pre-scribed code-lists of ISO 19115, environmental science research infrastructures such as AnaEE[11] and LTER-Europe[12] are actively developing better vocabularies for describing ecosystem and biodiversity research data, building upon existing SKOS vocabularies (such as EnvThes, referenced above).

There is no need to restrict annotation of metadata to one specific controlled vocab-ulary, especially if links between terms in different vocabularies can be established [40]. The identification of synonymous, subsuming and intersecting terms (and the publication of such links in a machine-accessible way such as on the Semantic Web) can provide the basis for better semantic search whereby a greater range of data products with similar characteristics can be retrieved on query without necessarily sharing precisely the same controlled vocabulary for their metadata. Making use of such linked vocabulary would simplify the task of integrating resource metadata from multiple catalogues as it would reduce the need to map all metadata values into a single master vocabulary (with the likely resulting loss of nuance), while still retaining the benefits of cross-RI search and discovery. A number of environmental and Earth science RIs such as AnaEE, LTER-Europe, LifeWatch and ICOS are now investigating such linking of vocabularies as part of an effort to make their respective resource catalogues more interoperable.

4.3 Semantic Mapping

Semantic mapping concerns the full mapping of data from one semantic context to another, with all the necessary structural transformation that entails. Such mapping might be applied on a targeted basis to specific metadata records, or to the results of specific queries retrieved from metadata servers, or there may be a mass translation of an entire catalogue. In general however, full semantic mapping is performed when integrating data from multiple sources into a single corpus with a single ontology and vocabulary. In the context of environmental science research infrastructure, this most typically arises

---

[11] https://www.anaee.com/.

[12] http://www.lter-europe.net/lter-europe.

in the construction of joint catalogues combining metadata records from heterogeneous sources. In order to facilitate search and discovery over the entire joint catalogue, the metadata gathered must of course be aligned to the greatest extent possible, which means one standard scheme, though cross-RI search and discovery can be further enhanced by identifying links between synonymous or related vocabulary terms, which can be seen as a kind of semantic bridging between controlled vocabularies (see next section).

A mapping agent will access the source of the data, apply the mapping, and record the mapped data in some target resource. The mapped data are then independent of the original source, but this also means that the data may need to be updated at times if the source changes, and a process is therefore needed to trigger such updates or to regularly poll the source for changes. Various tools exist for defining mappings between different ontologies or metadata schemes. An example of such a tool is the 3M Mapping Memory Manager[13], which implements the X3ML framework [19] for specifying translations from XML-based metadata schemes to RDF.

In addition to the enrichment activity described above, we have also explored mapping from various different common metadata schemes into CERIF RDF, which applies the CERIF 1.6 standard to RDF [20]. Figure 3 shows a snapshot of semantic mapping, in this case defining mapping rules from the metadata scheme used by the CKAN-based EUDAT B2FIND service[14] to CERIF RDF using 3M.

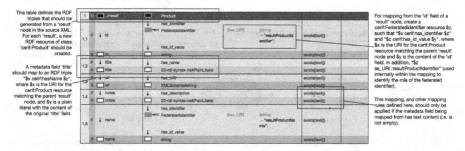

Fig. 3. Example of mapping rules generated in 3M: XML harvested from CKAN to CERIF RDF.

Mappings in 3M are described by X3ML mapping rules relating elements found in the source (XML-based) scheme to RDF triples in the target scheme, subject to various syntactic conditions (e.g. element type, parent hierarchy and internal content). Thus, 3M interprets the information fields in the source scheme based on the RDF *subject-predicate-object* model; each parent node in the source document is associated with a given RDF resource (the subject), and the content of each mapped field is used to generate triples linking that resource to other RDF resources (objects) via predicates derived from the field in question. For example in Fig. 3 above, each result node is mapped to a 'Product' resource, and the contents of each result node's 'id' field is associated with the product via a triple linking the product to a 'FederatedIdentifier' object via the predicate 'has_identifier'. 3M supports the specification of generators to

---

[13] https://github.com/isl/Mapping-Memory-Manager.
[14] https://eudat.eu/services/b2find.

produce unique identifiers for new RDF resources constructed during mapping of terms, and also provides various test and analytics facilities by which to evaluate (for example) the completeness of a given mapping. Examples of mappings into CERIF RDF, including mapping from OIL-E to CERIF have been published online [21] as part of the technical output of the VRE4EIC project[15], which the ENVRI community participated in.

Regarding the schema mapping between XML and RDF triples, we developed another work which provides insights in two folds [43]. Firstly, testify the validity of single matcher in a column based manner for the semantic data types. Secondly, testify the validity of a highly configurable framework that utilises hierarchical classification in order to construct a composable pipeline. Based on this vision, a Reconfigurable pipeline for Semi-Automatic Schema Matching (REPSASM)[16], was implemented to solve the customizability of the matching problem by providing an environment in which a user can create, configure and experiment with their own schema-matching procedure.

Other tools exist for transformation of data records, particularly between formats and models. For example, the derivation of RDF from relational database tables can be done quite naively by treating each table as the subject of an RDF triple, each column as the predicate, and each cell as the object, but this rarely creates a good representation of the source data. Instead, the use of tools such as Ontop[17], which applies the R2RML standard [22] for mapping relational database schemes to RDF, can allow relational databases to be queried as if they were RDF (see Fig. 4).

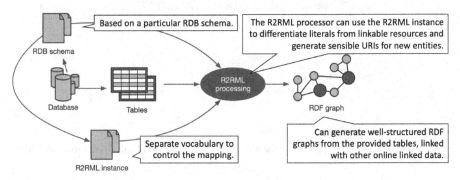

Fig. 4. Using R2RML to generate RDF from relational database tables.

This essentially performs the desired mapping on the fly, allowing for the benefits of mature relational database management systems to be retained; such an approach is being applied by RIs such as AnaEE[18] to extract metadata for semantic annotation using a standard ecosystem ontology built on the OBOE ontology. One thing to consider is that because this type of mapping can be performed at query time, it is not necessary to actually fully transform all the content of a relational database into RDF in advance, or

[15] https://www.vre4eic.eu/.
[16] https://github.com/JordyBottelier/arpsas.
[17] https://ontop.inf.unibz.it/.
[18] https://www.anaee.com/.

indeed at all. Instead, transformation can be performed on the results of queries on the database, presenting those results as native RDF without any indication that the source data exist in a different format or schema. Based on the types of query and retrieval operations performed on a data corpus, this kind of on-the-fly mapping might be more performant than transforming the entire data corpus in advance, especially if changes to the corpus might later need to be propagated to the mapped data. The decision on whether to map everything in advance to create a unified data source, or to map on demand just the information extracted from queries, is an important one to make when carrying out semantic mapping; balancing stability, performance, liveness and other concerns against one another. It is not a binary choice however. Certain key metadata (used for locating data for example) could be mapped in advance to create an 'upper' database via which to query individual data sources, or the results of recent or recurring queries for which mapping has already been performed could be cached in a central locale, reducing data retrieval time. All these approaches to mapping can be automated, and so be used 'under the surface' to improve the interoperability of RI systems and create the appearance of standardisation from the researcher perspective.

4.4 Semantic Bridging

Sometimes, the main barrier to interoperability is not the format of the metadata describing the data or service of interest, nor is it the vocabulary used within the metadata record, but simply the inability to find and access the metadata in question in an efficient, seamless way. RIs work diligently to provide portals via which researchers can find and access the data they are responsible for curating, but often this still carries the requirement to visit the RI's specific data portal and manually make the relevant request. Many RIs do contribute specific classes of data to aggregators (such as Copernicus), but often data is still kept in specific silos, retrievable yet isolated.

There are a number of ways to address this problem, including the construction of more cross-RI joint catalogues to expose RI resources to broader communities, but here we focus on a single approach, which is that of linked data [23]. The linked data approach is to leverage Semantic Web technologies to publish resource metadata in an open, retrievable way that can easily be cross-referenced by others in their own published (meta)data, so creating a wide-spanning distributed knowledge graph that can be navigated programmatically by discovery and query services. If RI resource metadata is available online as linked data of this (or a functionally-equivalent) form, then semantic linking might be reducible to simply creating more links between local knowledge graphs to build or add to a global, cross-RI knowledge graph. We refer to this approach to semantic linking as *semantic bridging*.

Semantic bridging is mainly applicable where there is a commonality of data format, but there is a need for additional semantic context for computational services to be able to infer that information from two or more sources actually relates to the same artefact. One case might involve relating entities referred to in the description of an RI process or subsystem with existing metadata regarding those entities, perhaps hosted by the very RI being described—for example the ENVRI Knowledge Base might refer to datasets provided by the ICOS RI, for which RDF information is provided by the

ICOS Carbon Portal[19]. Simply including the URI links used by the Carbon Portal in the metadata provided by the ENVRI Knowledge Base would allow any system querying the knowledge base to follow through to the Carbon Portal without manual intercession by a human investigator.

Another example involves the bridging between online provenance data structured according to the W3C PROV standard [24] with an OIL-E description of an infrastructure process such that there are direct links between a provenance dataset and a reference to that dataset in the ENVRI Knowledge Base, allowing queries to be distributed across both datasets. We can use SHACL rules [25] to describe how to generate additional RDF triples classifying entities in the provenance graph using OIL-E, and then automatically assert them in the knowledge base, with pointers back to the provenance data. Figure 5 provides a (simplified) example of such a rule, for relating a PROV activity to an OIL-E behaviour.

```
:ProvActivityMappingShape a sh:NodeShape ;
  sh:targetClass prov:Activity ;
  sh:rule [
    a sh:SPARQLRule ;
    rdfs:label "Map PROV entities onto OIL-E science view.";
    sh:prefixes prov: , oil: ;
    sh:construct """
      PREFIX oil: <http://www.oil-e.net/ontology/oil-base.owl#>
      PREFIX prov: <http://www.w3.org/ns/prov#>
      CONSTRUCT {
        $this a oil:Behaviour .
      } WHERE {
        $this a prov:Activity .
        ...
      }"""
  ] .
```

Fig. 5. Sample SHACL rule for mapping PROV-O activities to OIL-E science view behaviours.

SHACL allows us to define the conditions under which to produce new data (via SPARQL construct queries) that can be inserted into the ENVRI Knowledge Base and used by a distributed query broker to find and retrieve information from the provenance store as if it were an extension of the RI description in the knowledge base. The main challenge is the construction of 'conditional' rules that allow for the different kinds of provenance graph, as even within the PROV standard there are various ways to build a provenance trace depending on the primary concerns of the developer.

In this case, the linking of PROV data to RI specifications in OIL-E confers another benefit, which is that we can validate whether the structure of PROV traces (involving interactions between Agents, Activities and Events) matches the form of the RI provenance tracking behaviour as defined using ENVRI RM. Thus such bridging allows for possible validation of the provenance graph based on OIL-E definitions, and allows

[19] https://www.icos-cp.eu/.

for a distributed query broker to potentially access the provenance data directly via the bridging data in the knowledge base.

5 Discussion

Semantics in heterogeneous distributed systems are plagued by many of the problems of knowledge representation in general, such as how to achieve adequate computability, consistency and completeness in data coming from various sources produced in various different ways. The Semantic Web provides one means to represent and publish information in a lightweight, machine-actionable way, but it does not remove the necessity to deal with these problems, adding to them further issues of data redundancy, unreliability and limited performance versus more tightly integrated data models such as used in relational databases. Considerable attention has been given to the openness, extensibility and computability of Semantic Web standards, weighing different options (e.g. the use of SKOS over OWL [26, 27] to reduce the complexity of specifying controlled terminologies and their relationships). The use of linked data for describing resources (of all kinds) is already well-established, with research now focusing on different approaches to generating linked data from various sources and with how to navigate and query distributed information. Examples of such recent research include the generation of a navigable Graph of Things from an array of live IoT data sources [28] and the use of crowdsourcing to provide real-time transport data in rural areas [29], both topics with relevance to how RIs gather and expose field observations acquired via sensors or human experts. On the topic of distributed query, various frameworks have been proposed such as LDQL [30] and LILAC [31], which may make linked data based search over distributed metadata catalogues more practical and efficient than is currently the case.

Most geospatial technologies currently used by environmental and Earth science RIs have been developed independently of the Semantic Web, with recommendations such as INSPIRE[20] being mostly technically (albeit not conceptually) disjoint from it. Instead, bodies such as OGC have produced a number of open standards for Web access of metadata which are in common use by many RIs, usually brokered via software such as GeoNetwork[21]. This poses a barrier for integration of geospatial catalogues published via technologies such as CSW or OAI-PMH into the Semantic Web, and adaptors are still needed to query such data sources and present responses in RDF format (e.g. [32]), though there are also unifying technology proposals such as OGC's GeoSPARQL[22] to at least partially address this gap.

For mapping between a modest set of standards, manual mapping with tool support remains most practical, but automation may help to accelerate the construction of new mappings, provided that the precision and recall of such mappings can be made sufficient (most likely at present by mixing machine learning techniques with expert supervision and refinement). While how best to map metadata between different terminologies and

---

[20] https://inspire.ec.europa.eu/.

[21] https://geonetwork-opensource.org/.

[22] http://www.opengeospatial.org/standards/geosparql.

models remains an open question, automated mapping techniques can at least be (somewhat) objectively evaluated by comparing performance against human-crafted ontology sets covering the same domain (e.g. OntoFarm for conference organisation [33]). Given that syntactic mapping is still a big part in building semantic mapping, it is necessary to consider not only synonymous and otherwise-related terms in English, but also multilingual support; Bella et al. [34] provide an example of how to conduct mapping not rooted solely in measuring against a base English syntax.

Metadata descriptions of research assets are not limited to 'characteristic' information; provenance data (which might be structured according to a standard such as PROV-O) for data products and processes are also an important target for semantic linking, especially for creating unified (or at least *unifiable*) records of how research assets are used and where they came from; such records may be generated from scientific workflow management systems with provenance support [35, 42]. Such systems remain important for reproducible data science; most scientific investigations must follow a clear workflow, and there have been a number of workflow management systems developed with different characteristics and target applications [36], several of which have been applied to data science [37]. The use of ontologies for verification and validation of workflows has already been explored (e.g. [38]), and the ability to construct and validate such workflow specifications using metadata from service catalogues demonstrates that the cataloguing problem is not wholly centred on datasets.

The need to use controlled vocabulary within scientific datasets is self-evident, as is the need for standard schemes to describe such datasets, but it is still difficult for researchers, particularly researchers working independently, to even identify the best terminologies to use with their data (e.g. to use in particular data fields or to annotate their data), let alone to apply them in order to make it easy to integrate and interpret as part of a larger data corpus. For example, various repository services now exist that host controlled vocabularies and ontologies for use by researchers (e.g. BioPortal[23] and AgroPortal[24]), but there is a lack of standard tools for discovering these terminological resources and evaluating their appropriateness to researchers' own needs and those of their communities. This represents a fundamental problem that must also be addressed when considering approaches to semantic linking—there is not much value in harmonising standards that researchers themselves are not fully aware of, nor is it useful if the mappings, translation services and other products of harmonisation are themselves invisible to the scientific community. This is another area in which community-driven initiatives such as ENVRI and RDA might prove invaluable.

6   Conclusion

Semantic linking is a topic of considerable importance for the effective realisation of seamless interoperability between research infrastructure, needed to achieve the kind of open data and open science research commons being now promoted by initiatives such as

---

[23] https://bioportal.bioontology.org/.
[24] http://agroportal.lirmm.fr/.

DataONE[25] and EOSC. While standardisation of metadata schemes, protocols and terminology across different areas of domain science can and does enhance interoperability between different data and resource providers (and can be considered the main driver of such interoperability in practice), it is clear that there will remain necessary disparities between communities driven by their need to attend to the specific requirements of their own researchers and as a byproduct of legacy technology choices. As long as these disparities exist, there will be a need for some kind of translation of data between two or more data models, executed at the intersection between different services operating in different semantic domains. Thus, the examination of different techniques and the adoption of specific technologies to perform these translations on demand remains an important facet in the promotion of interoperability within and across research infrastructure.

There are various ways to enhance the semantic interoperability of data and services provided by RIs. In this chapter we have provided an overview of some techniques that were investigated in the context of the Horizon 2020 ENVRIplus project:

- **Semantic contextualisation**, where we increase the body of contextual information available about the resources and data that already exists by applying ontologies and other meta-models to describe those resources and data in different ways, increasing the number of facets by which we can explore them.
- **Semantic enrichment**, where we use controlled vocabularies to further classify and annotate existing metadata records to make search and discovery easier.
- **Semantic mapping**, where we develop transformation models by which to fully convert information described in one data model into another, minimising information loss.
- **Semantic bridging**, where we generate additional linking data to 'bridge' between two online data sources, leveraging the power of linked data to permit distributed querying of a wider network of knowledge.

Our overview of these techniques only scratches the surface of what is required to improve semantic interoperability and what is currently being done by various communities and community initiatives. Practical semantic alignment requires considerable attention on the part of semantic modellers and RI developers. In particular, it is necessary to identify where such attention should be focused: the specific standards, protocols, models and terminologies that would provide the greatest benefit if linked; as well as the specific intermediary transformation services which, if deployed in the right place, would expedite data integration and service composition for the most relevant scientific use-cases. To make these judgements, it's important to understand exactly how these semantic resources are being used already by RIs and research communities, and where interdisciplinary research is being stymied by a lack of standardisation or interoperability.

Acknowledgements. This work was supported by the European Union's Horizon 2020 research and innovation programme via the ENVRIplus project under grant agreement No 654182.

---

[25] https://www.dataone.org/.

References

1. Zhao, Z., et al.: Reference model guided system design and implementation for interoperable environmental research infrastructures. In: 2015 IEEE 11th International Conference on e-Science, Munich, Germany, pp. 551–556. IEEE (2015). https://doi.org/10.1109/eScience.2015.41

2. Nieva de la Hidalga, A., et al.: The ENVRI Reference Model (ENVRI RM) version 2.2, November 2017. https://doi.org/10.5281/zenodo.1050349

3. Martin, P., et al.: Open information linking for environmental research infrastructures. In: 2015 IEEE 11th International Conference on e-Science (e-Science), pp. 513–520. IEEE (2015). https://doi.org/10.1109/eScience.2015.66

4. Martin, P., Chen, Y., Hardisty, A., Jeffery, K., Zhao, Z.: Computational challenges in global environmental research infrastructures, chap. 12. In: Chabbi, A., Loescher, H.W. (eds.) Terrestrial Ecosystem Research Infrastructures: Challenges and Opportunities, pp. 305–340. CRC Press (2017). https://zenodo.org/record/3361569

5. Wilkinson, M.D., et al.: The FAIR guiding principles for scientific data management and stewardship. Sci. Data **3**, 1–9 (2016)

6. ISO 19115-1:2014: Geographic information—Metadata—Part 1: Fundamentals. ISO standard, International Organization for Standardization (2014)

7. ISO 19139:2007: Geographic information—Metadata—XML schema implementation. ISO/TS standard, International Organization for Standardization (2007)

8. Erickson, J., Maali, F.: Data catalogue vocabulary (DCAT). W3C recommendation. W3C (2014). http://www.w3.org/TR/2014/REC-vocab-dcat-20140116/

9. Nebert, D., Voges, U., Bigagli, L.: OGC catalogue services 3.0—general model. OGC implementation standard. Open Geospatial Consortium (2016). http://docs.opengeospatial.org/is/12-168r6/12-168r6.html

10. Lagoze, C., Van de Sompel, H.: The making of the open archives initiative protocol for metadata harvesting. Libr. Hi-tech **21**(2), 118–128 (2003)

11. Berners-Lee, T., Hendler, J., Lassila, O., et al.: The semantic web. Sci. Am. **284**(5), 28–37 (2001)

12. W3C OWL Working Group: OWL 2 web ontology language. W3C recommendation. W3C (2012). https://www.w3.org/TR/2012/REC-owl2-overview-20121211/

13. Bechhofer, S., Miles, A.: SKOS simple knowledge organization system reference. W3C recommendation. W3C (2009). http://www.w3.org/TR/2009/REC-SKOS-reference-20090818/

14. W3C SPARQL Working Group: SPARQL overview. W3C recommendation. W3C (2103). http://www.w3.org/TR/2013/REC-sparql11-overview-20130321/

15. Madin, J., Bowers, S., Schildhauer, M., Krivov, S., Pennington, D., Villa, F.: An ontology for describing and synthesizing ecological observation data. Ecol. Inform. **2**(3), 279–296 (2007)

16. Schentz, H., Peterseil, J., Bertrand, N.: EnvThes—interlinked thesaurus for long term ecological research, monitoring, and experiments. In: EnviroInfo, pp. 824–832 (2013)

17. The European Commission: Realising the European Open Science Cloud. The European Commission (2016). https://ec.europa.eu/research/openscience/pdf/realising_the_european_open_science_cloud_2016.pdf

18. Jörg, B.: CERIF: the common European research information format model. Data Sci. J. **9**, 24–31 (2010)

19. Marketakis, Y., et al.: X3ML mapping framework for information integration in cultural heritage and beyond. Int. J. Digital Libr. **18**(4), 301–319 (2016). https://doi.org/10.1007/s00799-016-0179-1

20. Martin, P., Remy, L., Theodoridou, M., Jeffery, K., Zhao, Z.: Mapping heterogeneous research infrastructure metadata into a unified catalogue for use in a generic virtual research environment. Future Gener. Comput. Syst. **101**, 1–13 (2019). https://doi.org/10.1016/j.future.2019. 05.076

21. Theodoridou, M., Ivanovic, D., Martin, P., Remy, L., Muckensturm, M.: X3ML mappings from common metadata schemes to CERIF RDF (2019). https://doi.org/10.5281/zenodo.254 8732

22. Sundara, S., Das, S., Cyganiak, R.: R2RML: RDB to RDF mapping language. W3C recommendation. W3C (2012). http://www.w3.org/TR/2012/REC-r2rml-20120927/

23. Berners-Lee, T.: Linked data. W3C Design Issues (2006). https://www.w3.org/DesignIssues/ LinkedData.html. Accessed 26 Feb 2018

24. Groth, P., Moreau, L.: PROV-overview. W3C note. W3C (2013). http://www.w3.org/TR/2013/ NOTE-prov-overview-20130430/

25. Kontokostas, D., Knublauch, H.: Shapes constraint language (SHACL). W3C recommendation. W3C, July 2017. https://www.w3.org/TR/2017/REC-shacl-20170720/

26. Stellato, A.: Dictionary, thesaurus or ontology? Disentangling our choices in the semantic web jungle. J. Integr. Agric. **11**(5), 710–719 (2012)

27. Baker, T., Bechhofer, S., Isaac, A., Miles, A., Schreiber, G., Summers, E.: Key choices in the design of simple knowledge organization system (SKOS). Web Semant.: Sci. Serv. Agents World Wide Web **20**, 35–49 (2013)

28. Le-Phuoc, D., Quoc, H.N.M., Quoc, H.N., Nhat, T.T., Hauswirth, M.: The graph of things: a step towards the live knowledge graph of connected things. Web Semant.: Sci. Serv. Agents World Wide Web **37**, 25–35 (2016)

29. Corsar, D., Edwards, P., Nelson, J., Baillie, C., Papangelis, K., Velaga, N.: Linking open data and the crowd for real-time passenger information. Web Semant.: Sci. Serv. Agents World Wide Web **43**, 18–24 (2017)

30. Hartig, O., Pérez, J.: LDQL: a query language for the web of linked data. Web Semant.: Sci. Serv. Agents World Wide Web **41**, 9–29 (2016)

31. Montoya, G., Skaf-Molli, H., Molli, P., Vidal, M.E.: Decomposing federated queries in presence of replicated fragments. Web Semant.: Sci. Serv. Agents World Wide Web **42**, 1–18 (2017)

32. Patroumpas, K., Georgomanolis, N., Stratiotis, T., Alexakis, M., Athanasiou, S.: Exposing INSPIRE on the semantic web. Web Semant.: Sci. Serv. Agents World Wide Web **35**, 53–62 (2015)

33. Zamazal, O., Svátek, V.: The ten-year OntoFarm and its fertilization within the onto-sphere. Web Semant.: Sci. Serv. Agents World Wide Web **43**, 46–53 (2017)

34. Bella, G., Giunchiglia, F., McNeill, F.: Language and domain aware lightweight ontology matching. Web Semant.: Sci. Serv. Agents World Wide Web **43**, 1–17 (2017)

35. Altintas, I., Barney, O., Jaeger-Frank, E.: Provenance collection support in the Kepler scientific workflow system. In: Moreau, L., Foster, I. (eds.) IPAW 2006. LNCS, vol. 4145, pp. 118–132. Springer, Heidelberg (2006). https://doi.org/10.1007/11890850_14

36. Liew, C.S., Atkinson, M.P., Galea, M., Ang, T.F., Martin, P., Hemert, J.I.V.: Scientific workflows: moving across paradigms. ACM Comput. Surv. **49**(4), 66:1–66:39 (2016). https://doi. org/10.1145/3012429, http://doi.acm.org/10.1145/3012429

37. Mork, R., Martin, P., Zhao, Z.: Contemporary challenges for data-intensive scientific workflow management systems. In: Proceedings of the 10th Workshop on Workflows in Support of Large-Scale Science. ACM (2015). https://doi.org/10.1145/2822332.2822336

38. Miksa, T., Rauber, A.: Using ontologies for verification and validation of workflow-based experiments. Web Semant.: Sci. Serv. Agents World Wide Web **43**, 25–45 (2017)

39. Zhao, Z., et al.: Knowledge-as-a-Service: a community knowledge base for research infrastructures in environmental and earth sciences. In: 2019 IEEE World Congress on Services (SERVICES), Milan, Italy, pp. 127–132. IEEE (2019). https://doi.org/10.1109/SERVICES.2019.00041
40. Liao, X., Zhao, Z.: Unsupervised approaches for textual semantic annotation, a survey. ACM Comput. Surv. **52**, 1–45 (2019). https://doi.org/10.1145/3324473
41. Zhao, Z., et al.: Scientific workflow management: between generality and applicability. In: Fifth International Conference on Quality Software (QSIC 2005), Melbourne, Australia, pp. 357–364. IEEE (2005). https://doi.org/10.1109/QSIC.2005.56
42. el Khaldi Ahanach, E., Koulouzis, S., Zhao, Z.: Contextual linking between workflow provenance and system performance logs. In: 2019 15th International Conference on eScience (eScience), San Diego, CA, USA, pp. 634–635. IEEE (2019). https://doi.org/10.1109/eScience.2019.00093
43. Liao, X., Bottelier, J., Zhao, Z.: A column styled composable schema matcher for semantic data-types. Data Sci. J. 18–25 (2019). https://doi.org/10.5334/dsj-2019-025

Authentication, Authorization, and Accounting

Alessandro Paolini[1]([✉]) [ID], Diego Scardaci[1], Nicolas Liampotis[2] [ID],
Vincenzo Spinoso[1] [ID], Baptiste Grenier[1] [ID], and Yin Chen[1] [ID]

[1] EGI Foundation, Amsterdam, The Netherlands
{alessandro.paolini,diego.scardaci,vincenzo.spinoso,
baptiste.grenier,yin.chen}@egi.eu
[2] GRNET, Athens, Greece
nliam@grnet.gr

Abstract. Environmental research infrastructures and data providers are often required to authenticate researchers and manage their access rights to scientific data, sensor instruments or online computing resources. It is widely acknowledged that Authentication, Authorization and Accounting (AAA) play a crucial role in providing a secure distributed digital environment. This chapter reviews the advanced AAA technology and best practices in the existing pan-European e-Infrastructures. It also discusses the challenging issues of interoperability in federated access and presents state-of-the-art solutions.

Keywords: Authentication · Authorization · Accounting

1 Introduction

A challenge that any operational Research Infrastructure (RI) has to deal with is controlling the access to these services and resources: the identity of the users needs to be verified, and once this is done successfully, the proper rights have to be granted to the users to perform the operations they are supposed to do. It is widely acknowledged that Authentication, Authorization and Accounting (AAA) play a crucial role in providing a secure distributed digital environment. In the rest of the chapter, we will discuss the first two 'AA's – Authentication and Authorization; then, address the issues for the last 'A' – Accounting, separately. We will review the state-of-the-art of AAA, and discuss the best practice in EGI e-Infrastructure [1][1].

All the procedures, policies, and technologies used to implement such basic activities are part of the so-called Authentication and Authorization Infrastructure (AAI): it is a key service meant to ensure that services and resources are accessed by the users in a secure way and that at the same time the users personal data are stored in a safe manner.

---

[1] EGI is the first European-wide publicly-funded e-Infrastructure. It currently federates 237 computing centres across Europe and world-wide, providing high-throughput (grid and cloud) computing, storage and data resources to support European research, at the moment, having over 1 Million CPU cores and almost 700 Petabytes (disk and tape) storages.

© The Author(s) 2020
Z. Zhao and M. Hellström (Eds.): Towards Interoperable Research
Infrastructures for Environmental and Earth Sciences, LNCS 12003, pp. 247–271, 2020.
https://doi.org/10.1007/978-3-030-52829-4_14

Different AAI technologies have been using over the years to secure access to digital devices, the most common practice in e-Infrastructures such as EGI is the Public Key Infrastructure (PKI), which will be outlined in Sect. 2.

While PKI technology has been generally used for the access mainly to non-web based services, there was also the necessity to assign a digital (and single) identity to the users in order to regulate the access (generally) to web-based services. The need for user identity to cross borders between organisations, domain and services, lead to the creation of federated identity environments. Home organisations (e.g. a university, library or research institute.), who operate an Identity Provider (IdP)[2], register users by assigning a digital identity – in this way, they are able to authenticate their users and provide a limited set of attributes that characterise the user in a given context. Resource owners (Service Providers) delegate the authentication to Identity Providers in order to control access to the provided resources. An Identity federation is a group of Identity and Service Providers that sign up to an agreed set of policies for exchanging information about users and resources to enable access to and use of the resources. There are many Research and Education identity federations around the globe and they commonly have a national coverage[3]: for example, eduGAIN [2] interconnects identity federations around the world, simplifying access to content, services and resources for the global research and education community. We are reporting about issues, challenges and requirements for operating interoperable AAIs in Sect. 3.

In Sect. 4 we will discuss the solution to the identify federations and depict the "AARC Blueprint Architecture (BPA)[4]", created with the purpose to provide a set of interoperable architectural building blocks for software architects and technical decision-makers, who are designing and implementing access management solutions for international research collaborations.

In order to provide an implementation example of AARC BPA, in Sect. 5, we will describe the Check-in service, the AAI platform for the EGI Infrastructure.

Section 6 is about Accounting (the third "A" of AAA). Accounting provides the method for collecting and sending user activity information used for billing, auditing, and reporting, such as user identifies, start and stop times, executed commands, number of packets, and number of bytes. The accounting tool used by EGI is called APEL, originally created for the LHC Computing Grid (LCG). APEL parses batch, system and gatekeeper logs generated by a site and builds accounting records, which provide a summary of the resources consumed based on the attributes, such as CPU time, Wall Clock Time, Memory and EGI user DN (Domain Name). APEL is the underpinning technology of the EGI accounting portal [4] that supports the daily operation of e-Infrastructure.

Finally, we will conclude this chapter in Sect. 7.

---

[2] By definition, an IdP is a system that creates, maintains, and manages identity information for principals (users, services, or systems) and provides principal authentication to other service providers (applications) within a federation or distributed network.

[3] The REFEDs map to discover worldwide identity federations https://refeds.org/federations/fed erations-map.

[4] AARC Blueprint Architecture (BPA) [16] is produced by the AARC project [3], an EC H2020 project, aims to address the increased need for federated access and for authentication and authorisation mechanisms by research and e-infrastructures.

2 Public Key Infrastructure and Digital Certificates

The purpose of this section is giving an overview of the basilar concepts of the Public Key Infrastructure, providing the required information useful for the context of this document: for a more exhaustive description, the reader can have a look at the references linked to this section.

A Public Key Infrastructure (PKI) is a set of roles, policies, procedures and technologies to authenticate electronic users and devices: by using a cryptographic technique it enables entities to securely communicate on an insecure public network, and reliably verify the identity of an entity via digital signatures [5]. The identity of each entity is bound to a key pair to encrypt and decrypt messages: a public and a private key which are mathematically related. Therefore, this model makes use of asymmetric cryptology algorithms with the following properties:

1. It is impossible to derive the private key from the public one.
2. The public key can be distributed to other entities in the system to encrypt messages which can be decrypted only by the corresponding private key, which therefore must be kept secret.

Let's assume that two entities in the system, John and Beth, need to exchange some digital content between them. The following basilar steps will be accomplished:

1. Both John and Beth have their own key pairs. They are safely storing their private key and they have sent their public key to each other.
2. Before sending a message to Beth, John encrypts it using her public key.
3. To decrypt the message, Beth uses her private key.

Anyway, this simplified process highlights an important concern: how can John be sure that the public key used to encrypt the message for Beth really belongs to her?

To address this issue, one way to do is to introduce in the model a trusted third party that certifies the integrity and the ownership of the public keys: this new entity is called Certification Authority (CA) and has the important role of storing, issuing and signing the digital certificates used to verify that a particular public key belongs to a certain entity. The CA certificate could be self-signed or signed by another CA, and it is used to sign all the Certificate Signing Request (CSR) containing public keys. At the same time, a CA periodically publishes the so-called Certificate Revocation List (CRL) which contains a list of all the revoked certificates: this is to constantly make aware all the entities about the validity of the issued certificates.

When a CA is used, the previous example can be modified in the following way:

1. Assume that the CA has issued a digital certificate that contains its public key. This certificate is signed with the CA private key.
2. Beth and John agree to use the CA to verify their identities.
3. Beth requests a certificate to the CA, by sending it a CSR containing her public key.
4. The CA verifies her identity and issues the certificate making it publicly available.

5. John retrieves the certificate, verifies it, so he can assume that the public key in the certificate does indeed belong to Beth.
6. John uses Beth's verified public key to encrypt a message to her.
7. Beth uses her private key to decrypt the message from Bob.

Another use case is when John wants to send a message to Beth allowing Beth to verify that the message has really been sent by John. In such a case, John should digitally sign the message with his private key and Beth can verify his identity validating the signature with John's public key. This case is particularly relevant for distributed infrastructure, including EGI, because it enables EGI services to verify the identity (the authentication process) of a user submitting a task (e.g. run a workflow in the EGI infrastructure).

In summary, digital certificates are a way to perform mutual authentication between two parties: in this process, the two parties authenticate each other through verifying the provided digital certificate so that both parties are assured of the others' identity. In technology terms, it refers to a client (web browser or client application) authenticating themselves to a server (website or server application) and that server also authenticating itself to the client through verifying the public key certificate/digital certificate issued by the trusted CAs. The process is therefore called "certificate-based mutual authentication" [8]: since the users can securely access a server by exchanging a digital certificate instead of a username and password, this helps in preventing phishing, keystroke logging and man-in-the-middle (MITM) attacks among other common problems with password-based authentication.

The current standard that defines digital certificates is called X.509 Version 3 [7]. An X.509 v3 certificate includes the following elements:

- Certificate serial number
- The digital signature of the CA
- The public key of the user to whom the certificate is issued
- Identity of the owner
- Date of expiration
- Name of the CA that has issued the certificate

In the digital world, one single CA usually covers a predefined geographic region or administrative domain (if not all the world), such as an organization, a country, or a set of countries, therefore the identity vetting process would not scale-up if done by the CA itself. This is the reason why the task of verifying the identity (and personal data) of entities requesting their digital certificates is usually delegated to a network of subordinated Registration Authorities (RAs) who act on behalf of the parent CA in their assigned sub-domain.

In a world where large scale distributed computing is deployed on a production scale, across organisations, across countries, and across continents, a common trust domain for distributed computing has been created to join the several existing certification authorities into a single authentication domain and thus enabling sharing of computing and resources worldwide: the Interoperable Global Trust Federation (IGTF) [6] has been created to coordinate and manage this trust domain. The IGTF is a body to establish common policies and guidelines that help establish interoperable, global trust relations between providers of e-Infrastructures and cyber-infrastructures, identity providers, and other qualified relying parties. It is divided in three Policy Management Authorities (PMAs) covering the Asia Pacific, the Americas and Europe, Middle-East and Africa.

2.1 Proxy Delegation

In order to use large scale distributed computing infrastructure, such as EGI, a user need a way to copy its own credentials to the machines where its workflows/jobs are going to be executed: this is necessary to allow remote sub-processes or other services in the e-infrastructure to perform on behalf of the user (delegation), particular operations needed to successfully complete the workflow, like the access of data belonging to the user stored on different resource providers, or start sub-jobs on other resources. If the user really utilises their own long-living personal certificate to do so, this would lead to security problems (in case of stolen credentials), and it would also make difficult arranging in advance the several delegations: that is why the X.509 Proxy Certificates [9] have been created.

Proxy Certificates allow an entity holding a standard X.509 public-key certificate to delegate some or all of its privileges to another entity which may not hold X.509 credentials at the time of delegation. This delegation can be performed dynamically, without the assistance of a third party, and can be limited to arbitrary subsets of the delegating entity's privileges. Once acquired, a Proxy Certificate is used by its bearer to authenticate and establish secure connections with other parties in the same manner as a normal X.509 end-entity certificate. Moreover, Proxy Certificates usually have a very limited lifetime (generally a few hours) to mitigate the impact of an eventual security breach.

In the EGI Federation [1], the users need to be members of a Virtual Organization (VO) [10] in order to access the resources: a VO is a way of grouping users usually working on the same project and using the same application software. After an agreement with resources providers, VOs have been granted usage of a specific set of resources and services in the infrastructure.

When creating a proxy, an Attribute Authority (AA) can be contacted to release and attach to the proxy the attributes required (if allowed) to access the resources. In the EGI e-Infrastructure, the AA is implemented by VOMS [11]: several VOMS servers, hosting the VOs and information about the enrolled users, are operated by the Resource Centres members of the infrastructure. This kind of proxy carrying the VO attributes is therefore commonly named "VOMS proxy".

2.2 Robot Certificates

Rather than accessing e-infrastructure services directly, users can access them via a portal (or a "Science Gateway"), which can provide a more accessible interface to the services. Quite often user portals provide users with the capability of using institutional credentials to authenticate themselves; then the portal authenticates to the e-infrastructure services by mapping these credentials to the so-called robot certificates [12]. The robot certificates[5] are owned by an individual (often the VO manager) who is accountable for the robot operations. In this way, it is not necessary for a user to request a personal X.509 certificate and the registration to a VO, often perceived as a burden due to the bureaucracy: this contributes to increase the user-friendliness of the platforms. Use of robot certificates is then internally accounted for by the portals in compliance to the VO Portal policy.

3   Issues and Challenges for Interoperable AAI

Controlling access to research-related resources and collaborative tools is challenging, particularly when dealing with research communities that can be geographically dispersed across Europe and the globe. The growth of identity federations at the national and international level has proved to be a successful model to efficiently increase scientific collaboration. An identity federation is intended as any number of organizations agreeing to interoperate under a certain rule, a federation policy, set to authenticate and authorise users. Federations are usually circles of trust in which each organisation agrees to trust the Identity Management of the other members.

In this section, we give a quick depiction on the barriers that communities usually face to adopt and use federated access, and what are the common requirements for implementing an interoperable AAI framework.

The AARC project [3], an EC-funded H2020 project, made a survey with 14 European scientific research communities and conducted interviews with a selected but broad representation of user communities in order to understand the most common issues and to classify the several requirements [13].

According to the AARC report [13] and the FIM4R (Federated Identity Management for Research) whitepaper [14], the communities, in general, viewed the Federated Identity Management as an important mean to enable access to shared resources, but the most perceived barrier was the lack of adequate information about it (this highlighted the need to provide guidelines and training, as well as online resources and material for management and decision-makers to facilitate AAI appropriation by each community). Other important barriers were the lack of funding[6], the excessive bureaucracy when joining a federation, and the lack of clarity on benefits within the organization. Moreover, the survey confirmed that the web-based authentication method cannot solve alone the AAI challenge for VOs: many users still prefer non-web-based authentication, as well as protocol translation and delegation. Most communities reported also that Identity Federations' coverage for their collaboration is poor.

From the interviews and the discussion about the requirements, it was clear that, besides the need to cover functional gaps between the communities, building a federated AAI requires the definition of common policies that cover the necessary legal and operational practices for all the entities involved in the AAI ecosystem. The outcome

---

[5] Since the portal is an automated entity, the e-infrastructure services consider it to be a "robot". The portal operator obtains a "robot certificate" that enables the portal to authenticate to e-infrastructure services.

[6] Often, institutes do not have enough funding for paying the necessary resources and full-time staff to manage them.

of the analysis and prioritization of these requirements was a fundamental input for the high-level AAI Blueprint Architecture, which will be described in the next section.

The requirements were classified into two categories: (A) architectural and technical, and (B) policies and best practices.

In the first category we have:

- **User and Service Provider friendliness:** the Federated AAI framework should provide simple and intuitive tools that are able to address the needs of users with different levels of computer literacy and enable more Service Providers (SPs) (commercial and non-commercial) to connect.
- **Homeless users:** the Federated AAI framework should support users without a federated institutional Identity Provider (IdP), such as citizen scientists and researchers without formal association to research laboratories or universities.
- **Different Levels of Assurance:** credentials issued under different policies and procedures should include the provenance of the level under which they were issued.
- **Community-based authorisation:** the Federated AAI framework should enable communities to manage the assignment of attributes to their members for authorisation purposes.
- **Attribute aggregation/Account linking:** the Federated AAI framework should support the aggregation of identity attributes originating from different sources of authority, including federated IdPs and community-based attribute authorities.
- **Federation solutions based on open and standards-based technologies:** open and standards-based AAI technologies should be used by the different communities to allow for interoperability by means of suitable translation services
- **Persistent user identifiers:** the Federated AAI framework should reference the digital identities of users through long-lasting identifiers.
- **Unique user identities:** Each user should have a single digital identity to allow SPs to uniquely identify their users.
- **User-managed identity information:** A user should be able to self-manage some of their attributes, e.g. through a web-based User Interface (UI). Depending on the attribute type, update restrictions should be imposed.
- **User groups and roles:** the Federated AAI framework should support the assignment of groups to users, as well as the assignment of roles to users within their groups.
- **Step-up authentication:** the Federated AAI framework should provide an additional factor or procedure that validates a user's identity for high-risk transactions or according to policy rules.
- **Browser & non-browser based federated access:** the Federated AAI framework should provide federated access to both web-based and non-web-based services/applications.
- **Delegation:** the Federated AAI framework should provide the capability for the users to delegate to third parties, mostly computational tasks or services, to act on their behalf. This allows users to run thousands of actions in parallel without the need for interactive access, for example, to save output data (as described in Sect. 2.1).
- **Social media identities:** the Federated AAI framework should support common social media providers, such as Google and LinkedIn, but also the researcher identity providers, such as ORCID, to act as authentication providers and/or attribute authorities.

- **Integration with e-Government infrastructures:** the Federated AAI framework should support broader cross-domain collaboration including e-Government infrastructures.
- **Effective accounting:** the Federated AAI framework should support effective accounting across distributed, heterogeneous data infrastructures.

In the second category, policies and best practices, the requirements are the following:

- **Policy harmonisation:** all participating entities in the AAI ecosystem (IdPs, AAs, SPs) should commit to a common policy framework regarding the processing of personal data. This framework should incorporate at least the GÉANT Data Protection Code of Conduct [15].
- **Federated incident report handling:** A common procedure should be adopted for reporting security incidents that involve federations spreading across multiple administrative domains.
- **Sufficient attribute release:** the set of attributes released to SPs should be extended, primarily, to allow consuming services to operate and, also, to allow for more advanced features, such as personalisation of services.
- **Awareness about identity federations:** the benefits offered by identity federations should be promoted to all stakeholders, such as (commercial) service providers and identity providers that have not joined a federation yet.
- **Semantically harmonised identity attributes:** a common set of vocabularies should be used by the different communities to denote identity attributes managed by identity providers and attribute authorities.
- **Simplified process for joining identity federations:** the bureaucracy involved in joining identity federations should be reduced.
- **Best practises for terms and conditions:** AARC could offer guidelines for describing the terms and conditions that service providers (operated in the R&E) should use.

4 A General Solution: The AARC Blueprint Architecture

The way researchers collaborate can vary significantly between different scientific communities. Some are highly structured, with thousands of researchers who could be located virtually anywhere in the world. Typically, these are communities that have been working together for a long time, that want to share and have access to a wide range of resources, and have had to put in place practical solutions to make the collaborations work. On the other hand, there are also a number of smaller, more diverse research communities working within specific or across multiple scientific disciplines. Typically, these are either nascent communities being established around new scientific domains or communities in specific domains that do not need to promote widespread and close collaboration among researchers. In between these two extremes are scientific communities of all varieties in terms of size, structure, history, etc.

Over the past few years, the AARC project [3] has been working together with e-infrastructures, research infrastructures, research communities, AAI architects, and implementers to get a better understanding of their experiences and needs regarding sharing and accessing resources within research collaborations. The goal has been to collectively define a set of architectural building blocks and implementation patterns, the "AARC Blueprint Architecture" (BPA), that will allow the development of interoperable technical solutions for international intra- and interdisciplinary research collaborations.

Research infrastructures and e-infrastructures can already rely on eduGAIN [2] and the underlying identity federations to authenticate their users: the AARC BPA builds on top of eduGAIN and adds the functionality required to support common use cases within research collaborations, such as access to resources based on community membership.

While previous versions of the BPA [16] provide a blueprint for implementing an AAI, the latest iteration of the BPA (AARC-BPA-2019) [17] focuses on the interoperability aspects, to address an increasing number of use cases from research communities requiring access to federated resources offered by different infrastructure providers. Hence the "community-first" approach, which introduces the Community AAI. The purpose of the Community AAI is to streamline researchers' access to services, both those provided by their own infrastructure as well as services shared by other infrastructures. User authentication to the Community AAI uses primarily institutional credentials from national identity federations in eduGAIN, but, if permitted by the community, can also use other IdPs.

Specifically, in the community-first approach, we can distinguish among three types of services that can be connected to the Community AAI:

1. community services - provided only to members of a given community
2. generic services - provided to members of different communities
3. infrastructure services - provided by a given research infrastructure or e-Infrastructure to one or more Community AAI (typically through a dedicated infrastructure proxy).

AARC-BPA-2019 [17] is accompanied by a set of guidelines and informational documents that provide guidance on the interoperable expression of information, including

- community user identifiers [18]
- group membership and role information [19]
- resource-specific capabilities [20]
- affiliation information [21].

4.1 The AARC Blueprint Architecture Building Blocks

The current BPA version champions a proxy[7] service architecture in which services in a research collaboration can connect to a single point, the SP-IdP-Proxy (hereafter termed "proxy"), which itself takes the responsibility for providing the connection to the identity federations in eduGAIN, thus reducing the need for each service having to separately

---

[7] Not to confuse with the proxy certificate mentioned in Sect. 2. Hereafter the word "proxy" is meant as "proxy server": a computer server or an application that acts as an intermediary for requests from clients seeking resources from other servers.

connect to a federation (eduGAIN). As shown in Fig. 1, the latest iteration of the AARC Blueprint Architecture (AARC-BPA-2019) [17] defines five-component layers: User Identity, Access Protocol Translation, Community Attribute Services, Authorisation and End Services. Each layer groups one or more components based on their functional role.

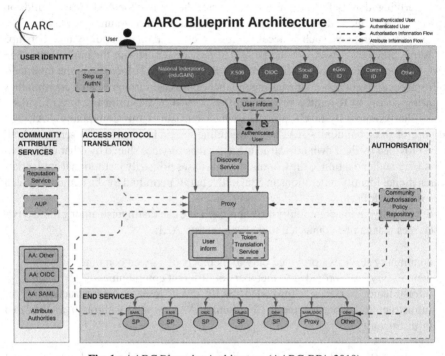

Fig. 1. AARC Blueprint Architecture (AARC-BPA-2019).

The *User Identity Layer* contains services for the identification and authentication of users. In existing implementations in the research and education space, these services typically include Security Assertion Markup Language (SAML) identity providers, certification authorities, and OpenID Connect (OIDC) or OAuth2 Providers (OPs). Although the focus of the services in this layer is to provide user authentication, often some end-user profile information is released as part of the authentication process.

The *Community Attribute Services Layer* groups services related to managing and providing information (attributes) about users. Typically, they provide additional information about the users, such as community group membership and roles, on top of the information that might be provided by services from the User Identity Layer.

The *Access Protocol Translation Layer* addresses the requirement for supporting multiple authentication technologies. It includes the following services:

- SP-IdP-Proxy (proxy), which serves as a single integration point between the Identity Providers from the User Identity Layer and the Service Providers in the End Services

Layer. Thus, the proxy acts as an SP towards the Identity Federations for which this proxy looks like any other SP, while towards the internal SPs it acts as an IdP.

- Token Translation Services, which translate identity tokens between different technologies.
- Discovery Service, which enables the selection of the user's authenticating IdP.
- User inform, which allows users to be informed regarding the processing of their personal data.

The *Authorisation Layer* controls access to the End Services Layer. The AARC BPA allows the implementers to delegate many of the complex authorisation decisions to central components, which can significantly reduce the complexity of managing authorisation policies, and their evaluation for each service individually.

The *End Services Layer* contains the services users want to use. Access to these services is protected (using different technologies). These services can range from simple web-browser-based services, such as wikis or portals for accessing computing and storage resources, to non-web-browser-based resources such as APIs, login shells, or workload management systems.

4.2 The "Community-First" Approach

As mentioned above, the latest BPA iteration fosters the interoperability among AARC BPA compliant AAIs that are operated by different research and e-Infrastructures by introducing the so-called Community AAI, which follows the proxy-based architecture shown in Fig. 1. It is therefore responsible for dealing with the complexity of using different identity providers with the *community services*. Furthermore, the Community AAI can add attributes to the federated identity that in turn can enable service providers to control access to their resources. These community-specific services only need to connect to a single identity provider, i.e. their Community AAI.

Apart from the community-specific services, there are *generic services*, such as the RCauth.eu Online CA, which serve the needs of several communities and are thus connected to more than one Community AAI. Being connected to multiple Community AAIs requires generic services to provide some form of IdP discovery, in order to be able to redirect the user to the relevant Community AAI[8]. Additionally, the generic services should support some means of doing "IdP hinting" (see [22]), thereby allowing "community branding" of the service and automatically redirecting the user to the corresponding Community AAI.

Communities may also require access to various services which themselves are behind (another) proxy, as often is the case with resources offered by e-Infrastructures or Research Infrastructures (Infrastructures hereafter). These *Infrastructure Proxies*[9]

---

[8] Primarily to get the user's identity via the community IdP, but also potentially to obtain attributes from community attribute authorities.

[9] An AAI service of a research infrastructure or e-Infrastructure (hereafter termed infrastructure) that enables access to resources offered by Service Providers connected to that infrastructure. This AAI service does not provide community membership management. Specifically, the infrastructure proxy comprises two AARC BPA component layers: the Access Protocol Translation and the Authorisation.

can be connected to different Community AAIs - see Fig. 2. So, just as for the generic services, Infrastructure services should be able to hint to the Infrastructure Proxy which Community AAI to use (see [22]).

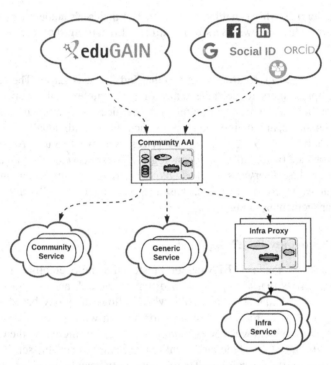

Fig. 2. A Community-first approach based on the AARC Blueprint Architecture. Researchers access services/resources using their institutional (eduGAIN), social or community-managed IdP via their Community AAI. Community services are connected to a single Community AAI, whereas generic services can be connected to more than one Community AAIs. e-Infrastructure services can be connected to different Community AAIs through a single e-infrastructure SP proxy. (A community-managed IdP is useful when there is a collaboration that wants to release attributes at IdP level for its members. This would allow to streamline the authentication process at community level. It can also be useful when a large part of the collaboration members does not have their own identity provider.)

It should be noted that the "community-first" approach does not impose a requirement on communities to deploy and operate a Community AAI on their own. Communities could make use of either dedicated or multi-tenant deployments of AAI services operated by a third-party, typically a generic e-Infrastructure. A multi-tenant AAI service deployment supports different communities, as depicted in Fig. 3. It typically appears as a single entity to its connected IdPs and SPs. Such multi-tenant deployments are aimed at medium-to-small research communities/groups or individual researchers. Yet it should be emphasised that also in the multi-tenant AAI scenario, the community managers are responsible for managing their community members, groups and authorisation attributes.

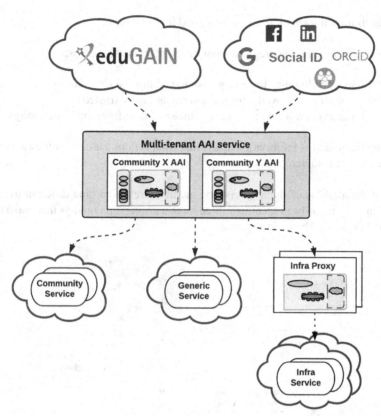

Fig. 3. Multi-tenant deployment of AAI services in "community-first" approach to the AARC Blueprint Architecture.

4.3 Authorisation Models

Authorisation models describe the organisational flow of authorisation information. Any other information needed by the service to fulfil actions such as personalisation, accounting, traceability, is out of the scope of this chapter. The organisational flow of authorisation information follows this lifecycle:

- Definition of authorisation information at one or more Attribute Authorities (AA)
- Aggregation of authorisation information
- Use of authorisation information for making an authorisation decision
- Enforcement of the authorisation decision

260   A. Paolini et al.

Authorisation information can be classified into two types:

1. User-attributes (often aggregated from different sources) such as:

   - Affiliation within the Home Organisation and/or the Community
   - Assurance, i.e. how well attribute assertions can be trusted
   - Group and role information (these primarily come from the Community)

2. Capabilities such as information describing what actions a user is entitled to perform
   on a specific resource.

Based on the analysis of the authorisation architectures from nine different use cases
detailed in [23], it has been identified three main authorisation models that make use of
an SP-IdP-Proxy, as shown in Fig. 4:

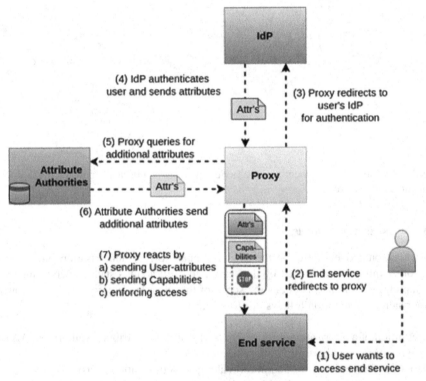

Fig. 4. The flow of authorisation information for a user who wants to access an end service in a
BPA-compliant infrastructure.

1. Centralised Policy Information Point (step 7a in Fig. 4): the proxy aggregates
   user attributes, such as group membership information and roles, and makes them
   available to the end-services

2. Centralised Policy Management and Decision Making (step 7b in Fig. 4): the proxy conveys the authorisation decision to the end-services in the form of capabilities
3. Centralised Policy Management and Decision Making and Enforcement (step 7c in Fig. 4): the proxy enforces the decision directly at the proxy.

Centralised Policy Information Point. In this model, the proxy aggregates the information and makes it available to the end services so they can make the authorisation decision. This allows the service to perform fine-grained access control because all information necessary for an informed decision is available. However, scalability may become an issue for large deployments. For example, it may become non-trivial to consistently update authorisation across a large number of services, as the authorisation policy needs to be replicated to every service. Additionally, services may see user-specific authorisation data, such as group membership, that might be intended for other services. This may be problematic with regard to the "data minimisation principle". Furthermore, this puts the onus on the services to correctly interpret and act on the obtained authorisation information.

Centralised Policy Management and Decision Making. In this model, the proxy makes the authorisation decision and encodes this decision into resource-specific authorisation information, typically in the form of capabilities. This allows the decision at the proxy to be based on additional information which the proxy might prefer not to send to the services. This is generally simpler for the end services to implement since the complexity of interpretation of the authorisation information is handled by the proxy. In contrast to the approach described in the previous model, this puts the onus on the proxy to correctly interpret and act on the authorisation information. Note that in this model:

1. the proxy is creating and/or translating authorisation statements
2. the proxy may need to make a mix of capabilities and user attributes available for the service to be able to properly enforce the authorisation decision.

Centralised Policy Management and Decision Making and Enforcement. In this model, the proxy makes the authorisation decision, as in the case of Centralised Policy Management and Decision Making. Furthermore, the proxy is responsible for enforcing that decision. This allows the integration of services that might not be capable of doing any authorisation, with only little modification. However, it requires the proxy to understand the authorisation policy of the end services. Often this type of authorisation enforcement is only used for certain parts (e.g. a global black- or whitelist) while using the other models for the rest of the authorisation. For example, in case the proxy grants the user access to the end service, this model may be followed by either of the other two models described.

Considerations on the Different Models
1. Authorisation implementations SHOULD support the Centralised Policy Information Point model for end services that require full control over the authorisation process. Authorisation implementations MUST be aware that in this model it is easy to send more data than required to end service. Filtering MAY be a solution.

2. Authorisation implementations SHOULD support the Centralised Policy Management and Decision Making model for simplifying the authorisation process for
   the end services. Authorisation implementations MUST be aware that the onus for
   correctly interpreting and acting upon authorisation information is put on the proxy.
3. Authorisation implementations SHOULD only use the Centralised Policy Management, Decision Making and Enforcement model for a partial authorisation decision
   (e.g. central suspension), and combine it with one of the two models above.
4. Depending on the requirements of the Service Providers reached through the proxy,
   it is possible to use a hybrid approach, combining any of the three models above,
   in a single authorisation flow. In all these flows the proxy can supplement the
   attributes from the authenticating IdP with information from AAs. The three different approaches address whether and how this information is passed on to the end
   services.

5   The EGI AAI Platform

The Check-in service is the AAI Platform for the EGI infrastructure [24] that implemented the AARC Blueprint Architecture. The Check-in service enables the integration
of external Identity Providers (e.g. from eduGAIN [2] and individual organisations) with
the EGI services through the Check-in Identity/Service Provider Proxy component, so
that users are able to access the EGI services (web and non-web based) using existing
credentials from their home organisations. To this end, Check-in has been published in
eduGAIN as a Service Provider. Through eduGAIN, EGI operational tools and services
that are connected to Check-in can become available to more than 3000 Universities and
Institutes from the 60 eduGAIN Federations with little or no administrative involvement.

Compliance with the REFEDS Research and Scholarship (R&S[10]) entity category
and the Sirtfi[11] framework, the Check-in service ensures sufficient attribute release, as
well as operational security, incident response, and traceability for 170 Identity Providers
from 25 identity federations that support R&S and Sirtfi. Complementary to this, users
without an account on a federated Identity Provider are still able to use social media or
other external authentication providers for accessing EGI Services that do not require
substantial level of assurance [25].

The adoption of standards and open technologies by Check-in, including SAML
2.0[12], OpenID Connect[13] and X.509 v3, has facilitated interoperability and integration
with the existing AAIs of other eInfrastructures and research communities, such as

---

[10] The REFEDS Research and Scholarship Entity Category (R&S) is one of the Entity Categories
defined by REFEEDS, https://refeds.org/category/research-and-scholarship.

[11] Sirtfi - A Security Incident Response Trust Framework for Federated Identity, defined by
REFEDS https://refeds.org/wp-content/uploads/2016/01/Sirtfi-1.0.pdf.

[12] SAML 2.0 standard is produced by the SSTC on 1 May 2012: https://wiki.oasis-open.org/sec
urity/FrontPage#SAML_V2.0_Standard.

[13] OpenID Connect 1.0 is a simple identity layer on top of the OAuth 2.0 protocol. It allows Client
to verify the identity of the End-User based on the authentication performed by an Authorisation
Service. The OpenID specification is at http://openid.net/developers/specs/.

ELIXIR[14] and LToS[15]. The Check-in service enables users to manage their accounts from a single interface, to link multiple accounts/identities together and to access the EGI services based on their roles and Virtual Organisation (VO) membership rights. For VOs that do not operate their own Group/VO management system, the Check-in service provides an intuitive interface to manage their users and their respective roles and group rights. For VOs that operate their own Group/VO management system, the Check-in service has a comprehensive list of connectors that allows integrating their systems as externally managed Attribute Authorities (AA).

In summary, user communities have several options to integrate with Check-in in order to access the EGI resources:

- Users authenticate using their institutional identity provider, which is part of an identity federation and eduGAIN;
- Users authenticate using their ORCID, social media a community-specific identity provider, for example, in the case of ELIXIR;
- Users authenticate using their Community AAI (see also Sect. 4.2), for example, in the case of ELIXIR;
- Authorisation information about the users (VO/group memberships and roles) is managed by the community's group management service, which is connected to Check-in as an external attribute authority;
- Communities that do not operate their own Group/VO management service can leverage the group management capabilities of the Check-in platform.

EGI Check-in is a contribution towards the development of Single Sign On (SSO) to e-infrastructures for European researchers. It lowers the barriers to use of EGI resources today and has been designed with an eye to the integration with other planned and probable developments. Check-in service can be accessed at https://aai.egi.eu/.

5.1 EGI Check-in Architecture

Figure 5 illustrates a high-level view of the Check-in architectural elements that deliver the system's functionality. It depicts the system's functional structure, including the key functional components, their responsibilities, the interfaces they expose, and the interactions between them.

The core of EGI AAI Check-in service is the **IdP/SP Proxy** component, which acts as a bridge between the EGI services and external authentication sources and identity providers. This decoupling of the internal services and the external authentication

---

[14] ELIXIR: A Europe leading life science organisations in managing and safeguarding the data being generated by publicly funded research. https://www.elixir-europe.org/.

[15] LToS: The long-tail of science refers to the individual researchers and small laboratories who - opposed to large, expensive collaborations - do not have access to computational resources and online services to manage and analyse large amounts of data. EGI provides the Application on Demand (AoD) service, which is a platform allows individual researchers and small research teams to perform compute and data-intensive simulations on large, distributed networks of computers in a user-friendly way. https://wiki.egi.eu/wiki/Long-tail_of_science.

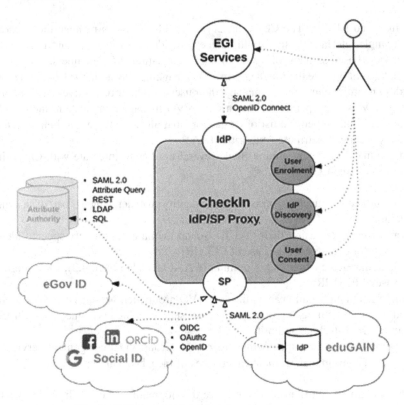

Fig. 5. EGI checkIn high-level functional architecture.

sources/identity providers reduce the complexity of the service implementation as it removes dependencies on the heterogeneity of multiple IdPs, Federations, Attributes, Authorities and different authentication and authorization technologies. This complexity is handled centrally by the proxy.

The introduction of an IdP/SP Proxy entity brings additional benefits. Specifically, as illustrated in Fig. 5, services only need to establish trust with one entity, the IdP/SP proxy. Typically, services will have one static configuration for the IdP/SP proxy. Having one configured IdP also removes the requirement from the service providers to operate their own IdP Discovery Service (a common requirement for services supporting federated access). Furthermore, all internal services will get consistent and harmonised user identifiers and attributes, regardless of the home organisation or the research community the authenticating user belongs to. Finally, this separation simplifies change management processes, as the internal services are independent of the IdPs run by the home organisations. Similarly, IdPs establish trust with one entity, the operator of the IdP/SP proxy, and they are not impacted by the operational changes introduced by each individual service.

The **User Enrolment and VO Management** service supports the management of the full life cycle of user accounts in the Check-in service. This includes the initial user registration, the acceptance of the terms of use of EGI, account linking, group and VO management, delegation of administration of VOs/Groups to authorised users and the configuration of custom enrolment flows for VOs/Groups via an intuitive web interface.

5.2 Token Translation: Integration with RCAuth.Eu Online CA

For various use cases, a user might need to use different types of credentials: for example, the user has an institutional account but she needs to access a storage element that requires an X.509 (proxy) certificate. So it is necessary to translate those institutional credentials into the precise format allowing the access to that particular service. In order to provide such functionality, the EGI Check-in service has been connected to the new RCauth.eu Online CA [26].

The RCauth Online CA issues certificates to end-entities based on a successful authentication to a Federated Identity Management System (FIMS) operated by an eligible Registration Authority – typically a FIMS Identity Provider (IdP) operated by an academic or research organisation.

When a certain web-flow requires a X.509 credential, the user will be redirected via a component, a so-called Master Portal[16], to the Online CA[17]. There the user will log in again transparently (due to SSO) to the Check-in service and will have to give consent for the management of user credentials. It will then be redirected to the originating service. In the process, a new credential is cached in the Master Portal which subsequently will be retrieved by whichever service initiated this flow, typically a Science Gateway.

When it is needed, a VOMS proxy (as seen in Sect. 2.1) can be requested initially. When the user is already enrolled in the VOMS server, this can be done completely transparently; otherwise a form of provisioning is needed.

The components of the service, as shown in Fig. 6, can be categorised in the following way:

- The **blue component** [many] represents the Service Provider Portal which the user wants to use. These are usually the Science Gateways (VO Portals) run by VOs. Given the wide variety of scientific disciplines in EGI, this scenario may include many such portals.
- **Red components** [few] correspond to the Master Portal. The scenario may comprise a few of these services, each one corresponding to the e-Infrastructures (like EGI) using RCAuth.eu.

  - *Master Portal*: acts as a caching service for user credentials (proxy certificates), taking some load of the RCauth.eu backend. Moreover, it also intermediates between two separate trust domains: the domain (single) of the Delegation Server and the domains (many) of connecting Service Provider Portals (Science Gateways). This improves the scalability of the model since instead of registering ALL Portals to the single Delegation Server directly, now registered Portals can be split between a few Master Portals running in front of the Delegation Server.

---

[16] The architecture design of the Master Portal is at: https://wiki.nikhef.nl/grid/AARC_Pilot_-_Architecture.

[17] The architecture of the RCAuth Online CA is at: https://wiki.nikhef.nl/grid/AARC_Pilot_-_RCAuth.eu.

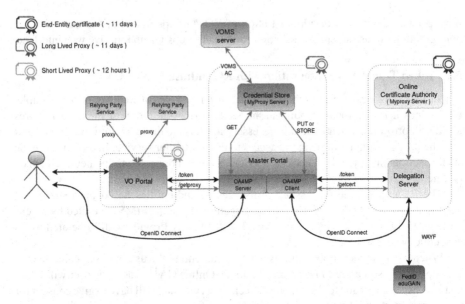

Fig. 6. RCAuth.eu Online CA scenario [27]. (Color figure online)

- *Credential Store*: is a MyProxy server used by the Master Portal to actually store the user proxies.

- **Yellow components** [one] represent an Online CA with a web frontend, and it's what we call RCauth.eu. Given the hardware security module (HSM) cost and high-security requirement, there is only one Online CA component.

  - *Delegation Server*: is the web frontend service which talks to the Online CA to generate certificates for authenticated users.
  - *Online CA*: is a Certificate Authority running on an HSM. This service (although called online) is only directly accessible to the Delegation Server in front of it.
  - *WAYF*: an IdP/ SP Proxy with an internal filter for accepting authentication sources directly. This gives full control for RCauth.eu over the eligibility of IdPs.

- **Purple components** [many] represent the different authentication sources that RCauth.eu is accepting (or planning to accept in the near future).

The addition of the Master Portal component to the schema, based on a replication of the CILogon software[18], moves all the complexity of caching the user credentials and of interacting with the Online CA away from the VO-run science gateways. The net result is that it makes easy for the VO portals to securely obtain credentials, based on the OpenID Connect protocol (see footnote 9) (acting as a client). The Master Portal takes

---

[18] CILogon is an integrated identity and access management platform that enables researchers to log on to cyberinfrastructure (CI): https://www.cilogon.org/.

care of obtaining the longer-lived end-entity certificates, caching them in the form of a proxy certificate and handling the additions of the VO-based attributes. Due to the more modular setup, having this extra component in the middle also makes it easier to reuse the same online CA for different e-Infrastructures.

6 Accounting

In previous sections, we have discussed two aspects of AAA. The final plank in the AAA framework is accounting, which measures the resources a user consumes during access. This can include the amount of system time or the amount of data a user has sent and/or received during a session. Accounting is carried out by logging of session statistics and usage information and is used for authorization control, billing, trend analysis, resource utilization, and capacity planning activities [28]. Accounting is fundamental in measuring the resource usage for each VO and verifying it's in line with the SLAs and the corresponding requirements/pledges negotiated with it. Moreover, EGI's "pay for use" model, which supposes that resources are paid by the customer periodically as they are consumed, will surely make use of accounting data in the future, as soon as the volumes of usage will be high enough.

For the purpose of this book, we are going to provide an overview of the accounting implementation in EGI.

The EGI Accounting Infrastructure (Portal and Repository) supports the daily operations of EGI and it is useful for assessing the real usage of the computing, cloud, and storage resources.

It is a complex system that involves various sensors in different regions, all publishing data to a central repository. The data are processed, summarised and displayed in the accounting portal, which acts as a common interface to the different accounting record providers and presents a homogeneous view of the data gathered and a user-friendly access. There are dedicated views for different types of users, for example national resources managers, Virtual Organisation (VO) Managers, resource centres administrators and the general public.

The Accounting Repository is based on APEL [29], a tool that collects accounting data from sites participating in the EGI. The accounting information is gathered from different sensors into a central accounting database where it is processed to generate statistical summaries that are available through the EGI Accounting Portal[19] [4]. Statistics are available for view in different detail by users, VO Managers, site administrators and anonymous users according to well-defined access rights.

The Accounting Portal is a web application based on Apache, and MySQL, which has as its primary function to provide users with customised accounting reports, containing tables and graphs, as web pages. It also offers RESTful web services to allow external entities to gather accounting data.

---

[19] EGI Accounting Portal is one of EGI core services that provide data accounting information for EGI users: https://accounting.egi.eu/.

The Accounting Portal consists of a backend (Fig. 7), which aggregates both data and metadata in a MySQL database, using the APEL SSM (Secure Stomp Messenger)[20] messaging system to interact with the Accounting Repository and several scripts, which periodically gather the data and metadata. It relies on a model that allows the representation of the data in several ways, focusing on different views (grid, cloud, storage, multicore, user statistics etc.) and integrating metadata (topology, geographical data, site status, nodes, VO users and admins, site admins etc.). Secure Stomp Messenger (SSM) is based on Apache ActiveMQ[21].

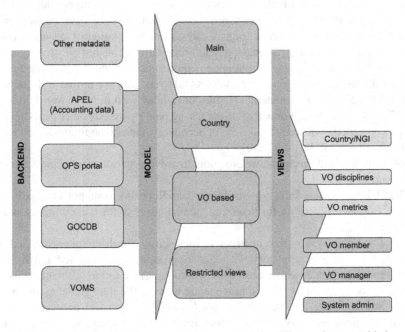

Fig. 7. Accounting Portal information sources and the different views provided.

A set of specific views exposes the data to the user. These views contain a form to set the parameters and metric of the report, a number of tables showing the data parametrised by two selectable dimensions and filtered by several parameters, a line graph showing the table data, and pie charts showing the percentage distribution on each dimension.

The Accounting Portal has to refresh its database periodically with data from the Accounting Repository to ensure that information published are up-to-date.

Metadata is a category of data that complements the raw accounting data and allows the portal to organise, categorise and import new meaning to it. This metadata includes:

---

[20] APEL SSM is the messaging system used by APEL to transmit messages: https://wiki.egi.eu/wiki/APEL/SSM.

[21] Apache ActiveMQ is a multi-protocol, java-based messaging server, http://activemq.apache.org/.

- Geographical Metadata: Country and Operations Centre affiliation of sites. Generally, this follows current borders, but there are important exceptions.
- Topological Metadata: Sites are presented in trees, there are Country and Operations Centres trees that correspond to geographical classifications.
- Role Metadata: VO members and managers, and the site admins records. This metadata controls the access to restricted views.
- Country affiliation data: Each user record contains a user identifier that has his/her user name and membership data. These data are used in anonymised statistics per country, like how much resources from other countries are used by a given country and the distribution of its resources used by other countries.
- VO Data: To make possible VO selection in the user interface, the portal stores the list of all the VOs. They are also used to filter incorrect VO names, provide access to VO managers, and arrange to account by VO discipline (such as "High Energy Physics", "Biomedicine", "Earth Sciences", etc.). Information is gathered from the Operations portal using its XML based APIs.
- Site status metadata: Sites must be filtered to exclude those that are not in production (due to being closed or being in test mode). There must be also metadata to aggregate the accounting history of sites whose name has been changed.
- Other metadata: There are also other metadata like local privileges, SpecInt calculations, publication status, VO activities and more. Some of these metadata is calculated internally using other types of metadata and published for other EGI operational tools, like VO activity data.

Views in the portal differ in the type of showed accounting data, the site organization or the restricted nature of data. The Cloud view is a view of the sites that are part of EGI Federated cloud platform, which uses Cloud middleware. Some relevant views to be considered are:

- The main grid/cloud view, showing metrics like "Sum elapsed CPU time" or the "Number of jobs". An important metric to evaluate the performance of a cloud Resource Centre is the "Elapsed time * Number of Processors (hours)", together with the "Number of VMs" running at that RC.
- Operations centres and Countries view: similar to the main view, but showing data per country or per "Operations Centre": the generic Operations Centre, in general, is mapped to a country, but there are cases where it's a group of countries (e.g. NGI_IBERGRID), or a fraction of a country (e.g. CERN).
- Disciplines View: A view that provides accounting data per VO scientific disciplines defined by EGI.

7 Conclusion

In this chapter, we started with a review of the advanced AAI technology and discussed them with the best practices in the EGI e-Infrastructure. We also brought up the interoperable AAIs and the identity federation issues that challenging today's science collaborations. We presented the AARC Blueprint Architecture as one of the sound solutions and provided an implementation example of EGI Check-in service. We finally addressed the last 'A' of AAA – Accounting, and described the technology and services used by EGI.

AAAI solutions have been rarely implemented in the ENVRI Research Infrastructures. The experience described here are generic AAA solutions and have been implemented and used by EGI e-Infrastructure to support daily operations. These solutions can be easily extended and adopted by ENVRI RIs.

Acknowledgements. This work was supported by the European Union's Horizon 2020 research and innovation programme via the ENVRIplus project under grant agreement No 654182.

The work presented in this paper was also supported by the EGI-Engage H2020 project (grant no. 654142), the EOSC-hub project (grant no. 777536), the AARC project (grant no. 653965) in particular for Sect. 3, and the AARC2 project (grant number 730941) in particular for Sect. 4.

References

1. EGI e-Infrastructure Homepage. www.egi.eu. Accessed 30 Apr 2019
2. eduGAIN Homepage. https://edugain.org/. Accessed 30 Apr 2019
3. The AARC project Homepage. https://aarc-project.eu. Accessed 30 Apr 2019
4. EGI Accounting Portal. https://accounting.egi.eu/. Accessed 30 Apr 2019
5. Adams, C., Lloyd, S.: Understanding PKI: Concepts, Standards, and Deployment Considerations, pp. 11–15. Addison-Wesley Professional (2003). ISBN 978-0-672-32391-1
6. The Interoperable Grid Trust Federation (IGTF). https://www.igtf.net/. Accessed 30 Apr 2019
7. X.509 Version 3. https://tools.ietf.org/html/rfc5280. Accessed 30 Apr 2019
8. The Transport Layer Security (TLS) Protocol Version 1.3. https://tools.ietf.org/html/rfc8446. Accessed 30 Apr 2019
9. Proxy Certificate Profile. https://tools.ietf.org/html/rfc3820. Accessed 30 Apr 2019
10. Virtual Organization (VO). https://wiki.egi.eu/wiki/Glossary_V3#Virtual_Organisation. Accessed 30 Apr 2019
11. VOMS. http://italiangrid.github.io/voms/documentation.html. Accessed 30 Apr 2019
12. Robot Certificates. https://www.eugridpma.org/guidelines/robot/. Accessed 30 Apr 2019
13. Kanellopoulos, C., Liampotis, N., van Dijk, N., Solagna, P.: Analysis of user community and service provider requirements. The AARC project Deliverable DJRA1.1 (2015). https://aarc-project.eu/wp-content/uploads/2015/10/AARC-DJRA1.1.pdf
14. Atherton, C.J., et al.: Federated Identity Management for Research Collaborations (Version 2.0) (2018). http://doi.org/10.5281/zenodo.1307551
15. The GÉANT Data Protection Code of Conduct. https://www.geant.org/uri/Pages/dataprotection-code-of-conduct.aspx. Accessed 30 Apr 2019
16. Kanellopoulos, C., Stevanovic, U., Hardt, M., the rest of the JRA1 Team: AARC Blueprint Architectures. The AARC project Deliverable DJRA1.2 (2017). https://aarc-project.eu/wp-content/uploads/2017/05/DJRA1.2-AARC-Blueprint-Architectures-1.pdf
17. AARC Consortium Partners, AppInt Members, Liampotis, N. (ed.): Evolution of the AARC Blueprint Architecture. AARC2 project Deliverable DJRA1.4 (2019). https://aarc-project.eu/wp-content/uploads/2019/05/AARC2-DJRA1.4_v2-FINAL.pdf
18. AARC-G025: Exchange of affiliation information between infrastructure. https://aarc-project.eu/guidelines/aarc-g025/. Accessed 30 Apr 2019
19. AARC-G002: Expressing group membership and role information. https://aarc-project.eu/guidelines/aarc-g002/. Accessed 30 Apr 2019

20. AARC Consortium Partners, AppInt Members: AARC-G027: Guidelines for expressing resource capabilities (2018). https://doi.org/10.5281/zenodo.2247446
21. AARC-G025: Exchange of affiliation information between infrastructures. https://aarc-project.eu/guidelines/aarc-g025/. Accessed 30 Apr 2019
22. AARC-G049: A specification for IdP hinting. https://aarc-project.eu/guidelines/aarc-g025/. Accessed 30 Apr 2019
23. Hardt, M., et al.: AARC2-DJRA1.2: scalable, integrated authorisation models for SPs. The AARC2 project Deliverable DJRA1.2 (2018). https://aarc-project.eu/wp-content/uploads/2018/07/AARC2-DJRA1.2_V4-FINAL.pdf
24. Kanellopoulos, C., Liampotis, N., Solagna, P., Salle, M.: Identity Management for Distributed User Communities. The EGI-Engage project Deliverable, D3.9 (2017). https://documents.egi.eu/public/ShowDocument?docid=3017
25. Groep, D., Jensen, J., Linden, M., Stevanovic, U., Vaghetti, D.: Expression of REFEDS RAF assurance components for identities derived from social media accounts. The AARC2 project Deliverable, AARC_G041 (2018). https://aarc-project.eu/wp-content/uploads/2018/03/AARC-G041-Expression-of-REFEDS-RAF-assurance-components-for-social-media-accounts.pdf
26. RCauth.eu Homepage. https://rcauth.eu/. Accessed 30 Apr 2019
27. The AARC RCAuth.eu pilot Homepage. https://wiki.nikhef.nl/grid/AARC_Pilot_-_RCAuth.eu. Accessed 30 Apr 2019
28. Ren, J., Tongtong, L.: Enterprise Security Architecture, Handbook of Technology Management. Wiley, Hoboken (2010)
29. EGI APEL WiKi page: https://wiki.egi.eu/wiki/APEL. Accessed 30 Apr 2019

Virtual Research Environments for Environmental and Earth Sciences: Approaches and Experiences

Keith Jeffery[1]([✉]) [iD], Leonardo Candela[2] [iD], and Helen Glaves[3] [iD]

[1] Keith G Jeffery Consultants, Faringdon, UK
keith.jeffery@keithgjefferyconsultants.co.uk
[2] National Research Council of Italy,
Istituto di Scienza e Tecnologia dell'Informazione "A. Faedo", Pisa, Italy
leonardo.candela@isti.cnr.it
[3] British Geological Survey, Nottingham, UK
hmg@bgs.ac.uk

Abstract. Virtual Research Environments (VREs) are playing an increasingly important role in data centric sciences. Also, the concept is known as Science Gateways in North America where generally the functionality is portal plus workflow deployment and Virtual Laboratories in Australia where the end-user can compose a complete system from the user interface to use of e-Infrastructures by a 'pick and mix' process from the offered assets. The key aspect is to provide an environment wherein the end-user - researcher, policymaker, commercial enterprise or citizen scientist - has available with an integrating interface all the assets needed to achieve their objectives. These aspects are explored through different approaches related to ENVRI.

Keywords: VRE · Science gateway · Virtual laboratory · Workflow management

1 Introduction

Research has increasingly become specialised into communities such as oceanography, ecology, geology, materials science. However, many phenomena can only be understood by bringing together the research activities of several communities. Examples include the relationship between shellfish pollution, algal blooms and agricultural use of nitrates or the relationship between ill-health, climate and social conditions. Over the last few years, many communities have developed pan-European research infrastructures (RIs) bringing together several national research teams and assets such as datasets, software, publications, expert staff, sensors and equipment. One way to assist and encourage interdisciplinary research is to bring together the communities and assets of the RIs.

However, this poses a problem. Each community has developed its own standards and practices for research methods, data formats, software to be used, etc. This makes

© The Author(s) 2020
Z. Zhao and M. Hellström (Eds.): Towards Interoperable Research
Infrastructures for Environmental and Earth Sciences, LNCS 12003, pp. 272–289, 2020.
https://doi.org/10.1007/978-3-030-52829-4_15

it difficult for e.g. an ecologist to utilise oceanographic data. The heterogeneity is represented especially in digital representations of data, software, persons, organisations, workflows and equipment. However, many of these assets are represented digitally by metadata providing a succinct description of the asset. The metadata standard chosen varies from community to community. On the other hand, there is a limited set of basic things (entities or objects) that are involved in research (like data, persons and samples) and so the various metadata standards have some commonality in the things they represent– although they do so in different ways.

Thus, the 'line of attack' to provide multidisciplinary challenges for researchers is to try to harmonise the metadata and thus gain access to – and (re-)utilisation of – the assets. There are two basic approaches: the software broker approach provides mapping and conversion between pairs of metadata standards. This results in n(n-1) converter pairs. The alternative approach is to choose a canonical superset metadata standard and convert each metadata standard to/from that. This results in n converter pairs. This metadata-driven brokering is now regarded as the best approach [1, 8]. However, again we have two choices; the canonical superset may be realised physically – so providing an 'umbrella' consistent metadata resource or catalogue over all the participating RIs or the superset metadata may just be a reference syntax (structure) and semantics (meaning) and each RI provides its pair of converters. The latter approach leads to architecture with peer RI to RI communication, requiring quite some software at each RI to interact with the other RIs and generate appropriate workflows. The former leads to a system over the RIs – linked to them via APIs (Application Programming Interfaces) - has the advantage of a 'helicopter view' over the participating RIs and so can generate workflows optimally. Either way, the core of a VRE is the superset catalogue (whether conceptual or physical) [12].

In fact, a VRE provides more than access to the assets of RIs; it also provides researcher intercommunication through various means and software to generate work-flows to harness the available analytics, visualisation and simulation capabilities of the RIs. Ideally the VRE workflow should be optimised to ensure co-location of data and software which means moving data to the software from the various RIs participating or – especially as datasets become larger – moving the software to the data. This has implications in terms of access rights, privacy and security and in finding an equitable method of 'payment' for use of the RI assets. The VRE may also use e-Is (e-Infrastructures) such as external curated storage or supercomputing services with the requirement to manage the deployment of (parts of) the workflow to these e-Is. The VRE should assist the researcher with research management; assisting in finding relevant research, assisting in research proposals, tracking research portfolio and cataloguing research outputs (such as scholarly publications, patents, datasets, software) since increasingly funding organisations utilise such information in planning future research programmes and in evaluating the quality of research proposals.

Recognition of the importance and utility of VREs is increasing. Similar concepts exist in North America (Science Gateways) [2] and in Australia (Virtual Laboratories). The RDA (Research Data Alliance) VRE Interest Group[1] was initiated by the leaders of

---

[1] https://www.rd-alliance.org/groups/vre-ig.html.

VRE4EIC and EVER-EST (this was very much a European initiative) but now includes key experts from Science Gateways and Virtual Laboratories.

This chapter discusses three initiatives dealing with the development of Virtual Research Environments: (i) the D4Science experience, an infrastructure enacting the development of several instances of Virtual Research Environments serving the needs of various communities of practice; (ii) the EVER-EST project, an EU project supporting the development of one Virtual Research Environment for the Earth science community; and (iii) the VRE4EIC project, an EU project proposing a reference architecture for Virtual Research Environments where metadata-based interoperability plays a key role.

2   The D4Science Approach and Experiences

D4Science is a hybrid infrastructure specifically conceived to support the development and operation of Virtual Research Environments by the as-a-Service provisioning mode.

The D4Science VRE Manager is a service enacting the definition, deployment and operation of *Virtual Research Environments on demand* (on D4Science infrastructure premises [3]. D4Science-based Virtual Research Environments (VREs) are *web-based*, *community-oriented*, *collaborative*, *user-friendly*, *open-science-enabler* working environments for scientists and practitioners willing to work together to undertake a certain (research) task [4, 5]. From the end-user perspective, each VRE manifests in a web application (a) comprising several components and (b) running in a plain web browser. Every component is aiming at providing VRE users with facilities implemented by relying on one or more services provisioned by diverse providers. In fact, every VRE is conceived to play the role of a *gateway* giving seamless access to the *datasets* and *services* of interest for the designated community and their tasks while hiding the diversities originating from the multiplicity of resource providers.

The following *key features* characterise the service:

- *Wizard-based VRE characterisation:* the service offers a wizard-based mechanism enabling authorised users (aka VRE designers) to easily select the features (e.g. datasets, facilities and policies) characterising the needed VRE;
- *Dynamic context management:* the service automatically creates the security context needed by the service instances contributing to the VRE to work in a secure and organised manner;
- *Open and extensible resource model:* the service relies on a resource model to know what are the available features to be proposed at *VRE definition time* (i.e. what are the features and the capabilities supportable by the currently available services) and how these features have to be deployed at *VRE creation time* (i.e. what are the services to be configured and how they should be instructed to support the requested features);
- *Per VRE customisable UI:* the service offer facilities enacting authorised users to customise the UI of the VRE, e.g. to define the pages it should be structured in, to allocate the VRE UI components per page and to add web content. Moreover, it provides VRE users with the web app needed to use the working environment and its facilities;

- *Ready to use basic services:* the service equips every VRE with key services enacting the VRE members to cooperate by common facilities, i.e. (a) a *shared workspace* to store and organise items of interest; (b) a *social networking area* to, e.g. post messages, have discussions, express opinions; (c) a *catalogue* to publish artefacts resulting from the VRE activity; (d) a *user management area* to deal with VRE membership (e.g. invite new members), create groups, assign roles.

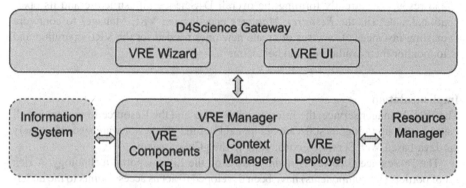

Fig. 1. The overall architecture of the D4Science VRE manager service.

Design

The D4Science VRE Manager service architecture is depicted (Fig. 1)

The main components are:

- the *VRE Manager service*, i.e. the component implementing the entire business logic related with VRE management. It comprises three subcomponents: (i) the *VRE Components KB*, i.e. the component called to build the knowledge base consisting of potential features (and accompanying services) that can be instantiated at VRE definition time. These features will be built by exploiting the information stored into the *Information System*, namely services and their capabilities, datasets, software components, hosting nodes; (ii) the *VRE Deployer*, i.e. the component called to transform the VRE specification produced by the *VRE Wizard* into a concrete deployment plan consisting of services (and their accompanying configurations) to be deployed to satisfy the specified features. This service is also responsible to implement the deployment plan by either instructing/configuring existing service instances or creating new ones to serve the VRE application; (iii) the *Context Manager*, i.e. the component called to interact with the *Resource Manager* to create the application context needed by the target services to work together and behave as expected;
- the *D4Science Gateway*, i.e. the front-end of the service. It hosts two sets of portlet: (i) the *VRE Wizard*, i.e. the portlet supporting authorised users to specify the features a new VRE should have by selecting them from an ever updated list of possible one resulting from the D4Science offering captured by the Information System; and (ii) the

VRE UI, i.e. the set of portlets forming the specific VRE working environment. These portlets include those providing access to the basic facilities (e.g. user management and shared workspace) as well as those providing access to specific services deployed in the VRE;

• additional services enacting the VRE Manager to implement the VRE in the overall D4Science infrastructure settings. These services include: (i) the *Information System* providing the VRE Manager with a comprehensive and ever updated list of services and resources currently forming the overall D4Science infrastructure and its operational state; (ii) the *Resource Manager* enacting the VRE Manager to configure existing instances of services or create new ones needed for the VRE operation and to monitor their availability and behaviour.

Implementation

The VRE Manager service, the Information System and the Resource Managers are all based on the homologous software components of the gCube software system, namely are Java-based Web Services contributing to the gCube system.

The D4Science Gateway is mainly based on the Liferay portal technology. A rich set of portlets (UI components) have been developed to act as access points to the underlying services as well as portal has been equipped with additional software components integrating it with the rest of D4Science services, e.g. components dealing with AuthN and AuthZ, and components interfacing with the Information System.

Deployment

The components presented above are designed to be allocable on many nodes and to exist in multiple instances.

In particular, the *VRE Manager* service can be deployed on a machine other than that hosting the D4Science Gateway. Moreover, many VRE Managers can be deployed in the infrastructure each serving a specific *virtual organisation*. This deployment option is key for multi-tenancy scenarios where diverse communities are provided with their own features set at VRE definition phase.

The *D4Science Gateway* is conceived to be deployed on a cluster with an instance per node plus a proxy acting as a unifying access point. Every instance can be configured to give access to a number of VREs (e.g. a community gateway contains all the VREs created for the needs of such a community) and to host the VRE Wizard enacting the creation of new VREs. Every VRE consists of a number of portlets organised according to the VRE specification.

The *Information System* is a conceptually centralised service yet its architecture is highly distributed and scalable thus to be able to serve many communities and cases. The resources are registered per virtual organisation and per virtual research environment (thus implementing the "application context" created by the VRE Manager).

The *Resource Manager* is a conceptually centralised service having actuators on every node hosting a D4Science service. A hierarchy of interoperating instances can be built thus having instances taking care of coordinating the management of services at the level of virtual organisation with instances taking care of resources management at the level of every VRE.

Use Cases

The D4Science VRE Manager service has been used to deploy and operate hundreds of VREs on D4Science premises. These VREs have been deployed to serve very diverse scenarios stemming from application contexts ranging from agri-food (AGINFRA+) to social sciences and humanities (PARTHENOS), environmental science (ENVRIplus), fisheries and conservation, aquafarming (iMarine and BlueBRIDGE), social mining (SoBigData.eu). A comprehensive list of currently supported VREs is available online[2].

3 The EVER-EST Approach and Experiences

3.1 The Challenge

Vast amounts of data about our planet are now available to researchers and it is important that this data is easily discoverable, accessible and properly exploited, preserved and shared in order to provide information for a whole spectrum of stakeholders: from scientists and researchers to decision and policy makers at the highest level.

Virtual Research Environments (VREs) provide the IT infrastructure to enable researchers to collaborate, share, analyse and visualise data over the internet. The development of a number of e-infrastructures within Europe and other areas to support activities such as Data Discovery and access has provided the foundations for the development of VREs.

The EVER-EST project (European Virtual Environment for Research – Earth science Themes: a solution) aimed to create a virtual research environment (VRE) focused on the requirements of the Earth science community. Within the earth sciences there are major challenges such as climate change research and ensuring the secure and sustainable availability of natural resources and understanding natural hazards which require interdisciplinary working and sharing of large amounts of data across diverse geographic locations and science disciplines to work towards a solution.

The project includes a major work stream to develop a virtual research environment, and this builds on a number of e-infrastructures which have been created under European Commission funding in recent years. Other work packages test this emerging infrastructure using appropriate use cases.

3.2 Creating a Virtual Research Environment

Scientific research in the Earth Sciences is conducted on many different scales from the local to the global. Much of this research is becoming increasingly multidisciplinary and being conducted by researchers who are not necessarily co-located. To support this increasingly distributed approach to Earth science research there is a demand for virtualised collaborative working environments where researchers can share resources e.g. data, workflows, ideas, knowledge and results.

Key objectives of the EVER-EST project are:

---

[2] https://services.d4science.org/explore.

- Creation of a virtual research environment (VRE) that provides a platform and suite of generic services to support collaborative research in the Earth Sciences;
- Validation of the EVER-EST VRE by the four pre-selected Virtual Research Communities (VRCs) that bring unique use cases in terms of their data, workflows, working practices, and desired outcomes;
- Validation of the novel use of the Research Objects concept for application in the Earth science domain. The concept of Research Objects has previously been validated by other disciplines such as astrophysics;

Engaging with the wider Earth Sciences community to promote adoption of the EVER-EST VRE as a solution for dynamic and potentially cross-disciplinary collaborative research.

3.3   Validate the Virtual Research Environment with Four Main Virtual Research Communities

The VRE was validated and evaluated through these four real-world use cases which are provided by existing communities of practice from the Earth science domain. The EVEREST consortium includes a key representative for each of the four Virtual Research Communities who is responsible for the tailoring and validation phase of the EVER-EST VRE for the specific use case and must also ensure the involvement and engagement of additional members of the community outside the EVER-EST project. The VRCs are:

- Sea Monitoring VRC – led by CNR-ISMAR
- Natural Hazards VRC (floods, geological, weather, wildfires) –led by NERC
- Land Monitoring VRC – led by SatCen
- Supersites VRC (volcanoes and seismic) – led by INGV

3.4   Implement and Validate the Use of "Research Objects" in Earth Science

The EVER-EST project defined, implemented and validated the use of "Research Objects" concepts and technologies in the Earth science domain as a mean to establish more effective collaboration.

Modern scientists are calling for mechanisms that go beyond the publication of datasets. They increasingly need to systematically capture the life cycle of scientific investigations and provide a single-entry point to access the information about the hypothesis investigated, the datasets used, the computations and experiments carried out, their outcomes, the people involved in the research, etc. Research Objects (RO) provide a structured container to encapsulate research data and the associated methodologies along with essential metadata descriptions.

3.5   Definition of EVER-EST Building Blocks

During the initial phase of the EVER-EST project, the technical activities focused on the assessment and definition of the main interface between the EVER-EST building blocks and the integration activities of the core infrastructure.

A study on the novel use of Research Objects in the Earth Sciences was carried out in consultation with the virtual research communities (VRCs). This was combined with an in-depth discussion to identify the requirements for the individual use cases provided by the VRCs including a definition of the data that needs to be integrated into the EVER-EST infrastructure (Fig. 2).

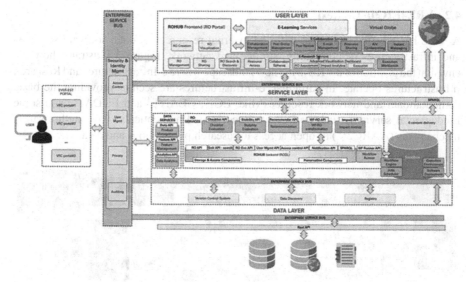

Fig. 2. The overall architecture of the EVER-EST VRE.

4 The VRE4EIC Approach and Experiences

4.1 Introduction

VRE4EIC aims at providing a model for Virtual Research Environments, which includes requirements, reference architecture and implementation on two use cases to demonstrate its feasibility and innovative impact. VRE4EIC has chosen CERIF[3] (Common European Research Information Format: an EU recommendation to Member States) to denote the superset catalogue.

VRE4EIC has undertaken a considerable amount of requirements collection and analysis, and has characterised many RIs to understand their available interfaces. The architecture has been designed and constructed. The prototype has been evaluated by the RIs that are in the project (ENVRI and EPOS) first, and then other RIs will be invited to evaluate the system.

In parallel, VRE4IC has been cooperating with other VRE projects, notably EVER-EST in Europe but also – via the VRE Interest Group of RDA (Research Data Alliance)[4] -

---

[3] http://www.eurocris.org/cerif/main-features-cerif.
[4] https://rd-alliance.org/groups/vre-ig.html.

SGs (Science Gateways) in North America[5] and VLs (Virtual Laboratories)[6] in Australia. In parallel, the various metadata groups in RDA, coordinated by Metadata Interest Group (MIG), are working on a standard set of metadata elements - to be used to describe RI assets in catalogues - which are not simple attributes with values but will have internal syntax and semantics [6].

4.2   VRE4EIC in Context

A VRE has to effectively deal with the external resources of data, software services, and infrastructures of computing, storage and network. Figure 3 illustrates how we envision the position of a VRE in the new landscape where e-Infrastructures and Research Infrastructures operate. In particular, e-Infrastructures are seen as providing the basic computational and network resources (like EGI[7], GEANT[8] and EUDAT[9]) and some fundamental services, such as federated access and authentication and authorisation mechanisms (AARC2[10]) or open access to research publications and data.

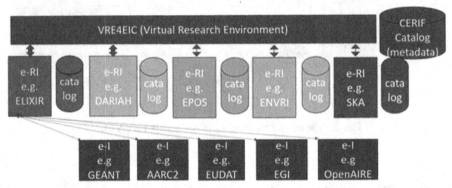

Fig. 3. Positioning of a Virtual Research Environment in relation to e-Research Infrastructures (e-RIs) and e-Infrastructures (e-Is).

Research-Infrastructures, on the other hand, employ the services and resources of e-Infrastructures to provide resources for their research communities. Each RI is devoted to a specific discipline, or cluster of related disciplines (e.g. DARIAH[11] is for the Humanities, EPOS for the Earth sciences). A VRE, in turn, sits on top of RIs to enable scientific communities to access data, services and tools from and, above all, across RIs. The CERIF-based Catalogue is central to achieve the VRE functionality, as it copes with the interoperability issues implied in that functionality, as described in the rest of the paper.

---

[5] https://sciencegateways.org/.

[6] https://nectar.org.au/labs/.

[7] https://www.egi.eu/.

[8] https://www.geant.org/.

[9] https://eudat.eu/.

[10] https://aarc-project.eu/.

[11] https://www.dariah.eu/.

There is an alternative architecture where the e-VRE components are built into each e-RI. However, this means that each e-RI has to maintain in its catalogue the catalogue content of all other e-RIs for interoperability with the usual problems of currency and integrity, especially if the native catalogue of an e-RI uses an insufficiently rich metadata format.

4.3 The VRE4EIC e-VRE Reference Architecture

At the general level, the Reference Architecture conforms to the multi-tiers view paradigm used in the design of distributed information systems [7]. Following this paradigm, we can individuate three logical tiers in the e-VRE:

- The *Application* tier, which provides functionalities to manage the system, to operate on it, and to *expand* it, by enabling administrators to plug new tools and services into the e-VRE.
- The *Interoperability* tier, which deals with interoperability aspects by providing functionalities for: i) enabling application components to discover, access and use e-VRE resources independently from their location, data model and interaction protocol; ii) publishing e-VRE functionalities via a Web Service API; and iii) enabling e-VRE applications to interact with each other.
- The *Resource Access* tier, which implements functionalities that enable e-VRE components to interact with e-RIs resources. It provides synchronous and asynchronous communication facilities.

Figure 4 depicts the logical tiers of e-VRE and shows their placement in an ideal space between the e-researchers that use the e-VRE and the e-RIs that provide the basic resources to the e-VRE. Based on the analysis of the requirements, a set of basic functionalities have been individuated and grouped into six conceptual components:

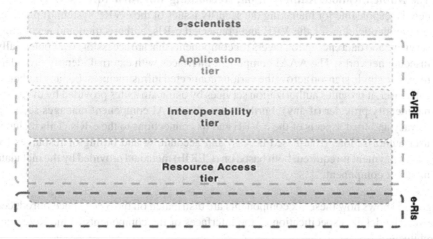

Fig. 4. Architectural tiers in a Virtual Research Environment.

- The e-VRE management is implemented in the **System Manager** component. The System Manager can be viewed as the component enabling Users to use the *core* functionalities of the e-VRE: access, create and manage resource descriptions, query the e-VRE information space, configure the e-VRE, plug and deploy new tools in the e-VRE and more.
- The **Workflow Manager** enables users to create, execute and store business processes and scientific workflows.
- The **Linked Data (LD) Manager** is the component that uses the LOD (Linked Open Data) paradigm, based on the RDF (Resource Description Framework) data model, to publish the e-VRE information space - i.e. the metadata concerning the e-VRE and the e-RIs in a form suitable for end-user browsing in a SM (Semantic Web)-enabled ecosystem.
- The **Metadata Manager** (MM) is the component responsible for storing and managing resource catalogues, user profiles, provenance information, preservation metadata used by all the components using extended entity-relational conceptual and object-relational logical representation for efficiency.
- The **Interoperability Manager** provides functionalities to implement interactions with e-RIs resources in a transparent way. It can be viewed as the interface of e-VRE towards e-RIs. It implements services and algorithms to enable e-VRE to: communicate synchronously or asynchronously with e-RIs resources, query the e-RIs catalogues and storages, map the data models. The Interoperability Manager is also responsible for efficiently managing the integration of third-party software, enabling the RA to virtually acquire any desired functionality that is not directly offered by any component of the RA. A case in point is the functionality required to assist researchers in communication with peers and in the administrative processes that are implied by research management. In general, this is the strategy chosen by the project to cope with all those aspects that are under standardization and, as such, do not tolerate formalisation at this stage.
- The **Authentication, Authorisation, Accounting Infrastructure (AAAI)** component is responsible for managing the security issues of the e-VRE system. It provides user authentication for the VRE and connected e-RIs, authorisation and accounting services, and data encryption layers for components that are accessible over potentially insecure networks. The AAAI component interfaces with external identity providers to enable single sign-on across the various connected infrastructures. For any authenticated user, it provides authorization services by using attributes provided by the external identity provider (if any). Furthermore, the AAAI component manages security, privacy and trust aspects of the e-VRE and its connections to the e-RIs. This includes user authorisations (role-based access) and accounting and billing of resources for which payment is required, both based on (CERIF) metadata provided by the metadata manager component.

Figure 5 shows how these six components are distributed on the 3-tier space introduced above. The detailed specification of the interfaces of the components of the Reference Architecture.

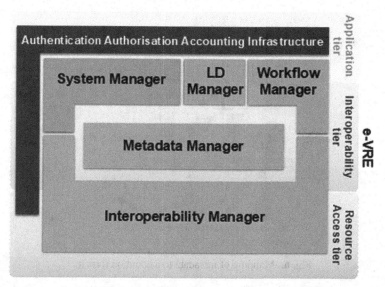

Fig. 5. Conceptual components and logical tiers.

Catalogue and Mapping

The metadata catalogue describes, provides access to and records actions on the assets of the RIs addressed by e-VRE. Mapping is required to represent the inhomogeneities [8, 12] of each RI in a homogeneous way to permit interoperability using the catalogue and thus is core to the reference model. The 3 M web application[12] is an open source application suite which supports schema mapping, Unique Resource Identifier (URI) definition and generation, (meta)data transformation, provision and aggregation. 3 M is based on the X3 ML mapping definition language for describing the schema mappings. 3 M is used to define mappings between various metadata formats used in existing VREs/RIs and the e-VRE. 3 M allows data experts to transform their internal structured data and other associated contextual knowledge to other formats. Fields or elements from a source database are aligned with one or more entities described in the target format. The purpose of this is typically for integration with other (meta)data also transformed to the same target format.

The process of mapping (meta)data using the 3 M tool is shown (Fig. 6). The first step is to define the mapping between two formats using the 3 M tool. This step needs at least two resources: the source schema (or an XML sample) and the target schema, in this case CERIF expressed in RDF. This step produces an X3 ML document describing the mapping that has been realised in 3M.

This result is used by the X3ML engine to apply the transformation defined in the mapping to a set of data. This data is harvested from a source repository through a harvester to get a set of data that has exactly the same format as the source schema (or XML sample). The X3ML engine is then able to transform the data to the target schema using the rules defined in the X3ML file resulting from step 1. The result of this second

---

[12] https://www.ics.forth.gr/isl/index_main.php?l=e&c=721.

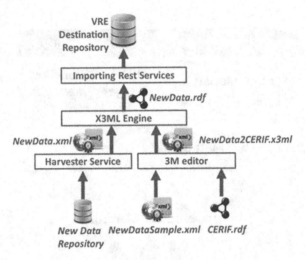

Fig. 6. Mapping of metadata to a common format.

step is an RDF file containing the data harvested in the target schema. This last result can finally be imported in the destination repository using a REST service.

Development

During the VRE4EIC project, we have conducted a Gap Analysis to identify the most needed components in existing e-RIs and VREs, with a special attention to the EPOS and ENVRIplus Research Infrastructures [11]. The analysis highlighted the heterogeneity of approaches and technologies adopted by current VRE and e-RI systems, especially in relation to the management of resource catalogues; additionally, the security infrastructure technologies adopted by most of VREs/e-RIs have limitations when executing operations on a distributed workspace. The components that have been selected by the Gap Analysis are the Metadata Manager, the AAAI Manager and the Node Manager. In order to implement these components and fit them into the EPOS and the ENVRIplus architectures, the VRE4EIC Consortium has made a plan that is illustrated next [10].

Figure 7 shows an overview of the Reference Architecture, including the subcomponents in which every component has been structured for modularity reasons. For instance, the Workflow Manager has three sub-components: the WF Configurator that implements workflows definition functionalities, the WF executor implementing execution functionalities and the WF repository component implementing storage management for workflows. For the same modularity reasons, the Query Manager has been elevated to the role of independent component. Thus, overall the Reference Architecture includes seven main components, each corresponding to a functional area:

- Identity, access and logging Management (AAAI Manager)
- System management (System Manager)
- Catalogues management including interoperability conversions (Metadata Manager)
- Metadata publishing using Linked Data Paradigm (LD Manager)
- Workflows management (Workflow Manager)

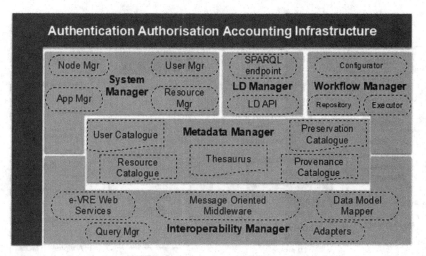

Fig. 7. The developed reference architecture.

- External application integration management (Interoperability Manager)
- Distributed query management (Query Manager).

Virtual Research Environments are dynamic systems; when new tools or technologies emerge a VRE should be able to integrate them. This means that the e-VRE architecture should be easily expandable by adding new software modules or replacing existing software components. Additionally, a component should be replaced or evolved (for instance using new software libraries) without affecting other components. The e-VRE should be potentially used in every research domain; for every domain it should be able to adopt the right technology to implement its functionalities. Deep integration (i.e. integration via Adapters) should be exposed as services in a standardised way to enable users to build clients not depending from the particular integration technology. The e-VRE system must be scalable to meet dynamic changes in the load of research computing processes at component level and independently deployable since they can be reused in other VREs.

In order to meet these requirements, an approach based on Microservices[13] has been chosen. As a result, the building blocks of the TA are autonomous services cooperating with each other to implement the above functional areas. The interaction between the TA services is mainly implemented using an asynchronous paradigm, based on the concept of event. The result is an event driven architecture [9]. Figure 8 shows the resulting micro-services, highlighting the components included, each characterised by the colour relative to the tier where the component belongs.

A repository has been created on GitHub to host the codebases of e-VRE services (VRE4EIC project, 2018). The e-VRE Services will be developed independently and the integration will be done using the APIs published by the Node service. A server has

---

[13] https://martinfowler.com/articles/microservices.html.

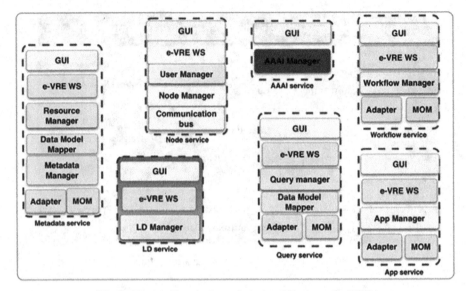

Fig. 8. The services in the technical architecture of e-VRE.

been created on the CNR ISTI cloud (v4e-lab.isti.cnr.it), hosting a continuous integration framework and a number of services used in the development of integration tests.

A repository has been created on GitHub to host the codebases of e-VRE services (VRE4EIC project, 2018). The e-VRE Services will be developed independently and the integration will be done using the APIs published by the Node service. A server has been created on the CNR ISTI cloud (v4e-lab.isti.cnr.it), hosting a continuous integration framework and a number of services used in the development of integration tests.

Use of e-VRE
Novel elements of the proposed reference architecture for an enhanced Virtual Research Environment include the metadata mapping, the microservice architecture and the co-development (i.e. evaluation on the architecture via workshops, and keeping developers and end-users in the feedback loop). In this section we will briefly demonstrate these novel elements by presenting two scenarios.

The first scenario demonstrates the integration between an external application and the e-VRE system (see Fig. 9). The proposed use case is to use the e-VRE Taverna plugin to enable users to create workflows, using resources from, for example, the European Plate Observing System.

The three boxes represent the EPOS, the e-VRE and the "user system" (e.g. laptop). Initially an EPOS user launches the TAVERNA workbench application in order to execute some scientific workflow (step 1 in the figure) on his/her own laptop. In order to access to workflows provided by the e-VRE system, the user installs a plugin that automatically connects to the workflow configurator component (in the e-VRE system) and fetches web services descriptions managed by the e-VRE metadata manager (step 2 and step 3 in Fig. 9). The metadata manager, in turn, accesses web services descriptions in the EPOS workflows catalogue (step 4, which can be executed at runtime or off-line by

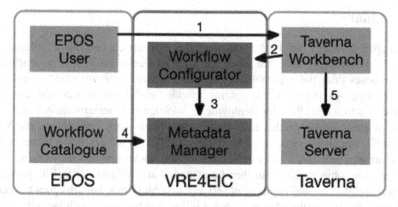

Fig. 9. Illustration of the first scenario.

ingesting information in advance). The so created workflow is then saved into the Taverna repository and executed on the Taverna Server (step 5). The description of the Workflow is also saved in the storage of the Metadata Manager, so that it can be launched later or re-used in the context of another workflow. This enables any non-skilled user to take advantage of workflows and web services from EPOS domain (potentially, from any domain) just by installing a plugin on its workflow application (in this case the Taverna Workbench).

The second scenario demonstrates the use of the e-VRE metadata catalogue to discover assets across RIs (see Fig. 10). Once the descriptions have been acquired and transformed into the CERIF format via the 3M technology described in Sect. 6.2, the user authenticates (step 1), implying his credentials being verified (2) and passed on the involved services (3). He then executes a catalogue search (4) which returns metadata records relative to resources belonging to multiple domains. The described assets can then be accessed (5) to be viewed on the appropriate viewer (e.g. for geological maps) or to be given as input to some simple local processing engine, such as waveform plotting, matlab, and the like (6).

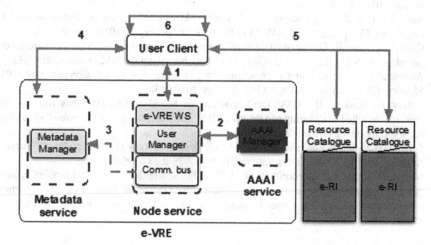

Fig. 10. Discovery of and access to assets.

5 Summary

This chapter reviewed briefly the origins of the VRE concept and then covered three recent EC-funded VRE research projects with relevance to ENVRI. It is clear that all the approaches share the same objective of enabling users to discover, access and re-use assets for their own purposes. All systems provide capabilities for accessing assets and composing into a workflow for deployment. D4Science concentrates on a vertically-integrated architecture but this approach has provided many domain-specific VREs. EVER-EST has also concentrated on domain-specific examples but achieves a more general architecture by the use of research objects, encapsulating the 'working set' of assets into one object which can then be managed and utilised. VRE4EIC provides a reference architecture and component services to achieve this, but also goes further.

Other support is required to approach a full researcher workbench including access to a communications system, office system and systems related to the management of research. The reference architecture of VRE4EIC has appropriate interfaces to achieve this.

The global recognition of the need for VREs (and similar SGs in North America and VLs in Australia) promises a vibrant future research and development activity in this area leading to better offerings for the researchers (and other user) community.

Acknowledgements. This work was supported by the European Union's Horizon 2020 research and innovation programme via the ENVRIplus project under grant agreement No 654182.

References

1. Nativi, S., Jeffery, K., Koskela, R.: Brokering with Metadata. ERCIM News 100, Special theme: Scientific Data Sharing and Re-use (2015). http://ercim-news.ercim.eu/en100/special/rda-brokering-with-metadata
2. Barker, M., et al.: The global impact of science gateways, virtual research environments and virtual laboratories. Fut. Gener. Comput. Syst. **95**, 240–248 (2019). https://doi.org/10.1016/j.future.2018.12.026
3. Assante, M., et al.: The gCube system: delivering virtual research environments as-a-service. Fut. Gener. Comput. Syst. **95**, 445–453 (2019). https://doi.org/10.1016/j.future.2018.10.035
4. Candela, L., Castelli, D., Pagano, P.: Virtual research environments: an overview and a research agenda. Data Sci. J. **12**, GRDI75–GRDI81 (2013). https://doi.org/10.2481/dsj.GRDI-013
5. Assante, M., et al.: Enacting open science by d4science. Fut. Gener. Comput. Syst. **101**, 555–563 (2019). https://doi.org/10.1016/j.future.2019.05.063
6. Jeffery, K., Koskela, R.: RDA: The Importance of Metadata. ERCIM News 100, Special theme: Scientific Data Sharing and Re-use (2015). http://ercim-news.ercim.eu/en100/special/rda-the-importance-of-metadata
7. Schuldt, H.: Multi-tier architecture. In: Liu, L., Ozsu, M.T. (eds.) Encyclopedia of Database Systems. Springer (2009). https://doi.org/10.1007/978-0-387-39940-9_652
8. Martin, P., Remy, L., Theodoridou, M., Jeffery, K., Sbarra, M., Zhao, Z.: Mapping heterogeneous research infrastructure metadata into a unified catalogue for use in a generic virtual research environment. Fut. Gener. Comput. Syst. **101**, 1–13 (2019). https://doi.org/10.1016/j.future.2019.05.076

9. Michelson, B.M.: Event-driven architecture overview - event-driven SOA is just part of the EDA story: Patricia Seybold. Group (2006). https://doi.org/10.1571/bda2-2-06cc
10. Bailo, D., Jeffery, K.G., Spinuso, A., Fiameni, G.: Interoperability oriented architecture: the approach of EPOS for solid earth e-Infrastructures. In: 2015 IEEE 11th International Conference on e-Science, pp. 529–534. IEEE, Munich (2015). https://doi.org/10.1109/eScience.2015.22
11. Zhao, Z., et al.: Reference model guided system design and implementation for interoperable environmental research infrastructures. In: 2015 IEEE 11th International Conference on e-Science, pp. 551–556. IEEE, Munich (2015). https://doi.org/10.1109/eScience.2015.41
12. Remy, L., et al.: Building an integrated enhanced virtual research environment metadata catalogue. J. Electron. Libr. (2019). https://zenodo.org/record/3497056

Case Studies

Case Study: Data Subscriptions Using Elastic Cloud Services

Spiros Koulouzis[1] ⓘ, Thierry Carval[2] ⓘ, Jani Heikkinen[3] ⓘ, Antti Pursula[3] ⓘ, and Zhiming Zhao[1](✉) ⓘ

[1] Multiscale Networked Systems, University of Amsterdam, 1098XH Amsterdam, The Netherlands
{s.koulouzis,z.zhao}@uva.nl
[2] Ifremer, Brest, France
thierry.carval@ifremer.fr
[3] CSC - IT Center for Science, Espoo, Finland
{jani.heikkinen,antti.pursula}@csc.fi

Abstract. To perform data-centric research in environmental and earth sciences, researchers need effectively query, select and access data products from different research infrastructures. When providing observation data continuously, infrastructure is expected to create and deliver customised data products, e.g. for specific geo-regions, time durations or observation parameters, to enhance its ability to serve the research communities. Such kind of services often have time-critical requirements; some tasks need to be carried out within specific time windows when the data products are needed for real-time modelling or simulation frameworks.

Keywords: Research infrastructure · Data subscription · Cloud computing

1 Introduction

Many environmental and Earth science Research Infrastructures (RIs) act as data hubs and publishers of scientific data and serve their user communities via an integrated data portal [6]. The Euro-Argo RI [8] is a typical example of a long-established, distributed RI from the marine domain and is the European contribution to the Argo programme. Argo monitors the world's oceans measuring temperature, salinity, pressure, etc. via the deployment of robotic floats to create a roughly even network of data collecting nodes across the marine surface of the earth. These floats periodically send data back via satellite to data assembly centres, which provide integrated, cleaned data products to various regional centres, archives and research teams; all data is then made publicly available via a common portal within 24 h of acquisition.

Due to the maturity in data acquisition, Euro-Argo seeks improved publishing methods for accessing existing curated data collections, and thus, prototypes a subscription service for their data. In contrast, to merely providing collected data freely for download and requiring researchers to monitor the core Argo dataset for updates manually,

Z. Zhao and M. Hellström (Eds.): Towards Interoperable Research
Infrastructures for Environmental and Earth Sciences, LNCS 12003, pp. 293–306, 2020.
https://doi.org/10.1007/978-3-030-52829-4_16

researchers are instead allowed to subscribe to specific subsets of Argo data and have updates pushed to their cloud storage, thus streamlining data delivery and accelerating data science workflows involving those data.

In this chapter, we will demonstrate how the Dynamic Real-time Infrastructure Planner developed in the project can be used for optimising virtual infrastructures for the EuroArgo research infrastructure to realise its data subscription service. The use case is prototyped based on EGI FedCloud and EUDAT's B2SAFE. This chapter is an extension of the work published in [1].

2   Data Subscription in RIs

2.1   A Data Subscription Scenario in EuroArgo

In the Euro-Argo data subscription scenario, investigators subscribe to customised views (e.g. specific regions, time durations, and observation parameters) on the Argo data using a data subscription service. Euro-Argo provides the infrastructure services needed for computing data products to match each subscription and then dispatches those products to their destinations; the subscription service can then distribute the tailored updates to investigators' private storage.

A typical subscription task can be made up of a set of input parameters:

1. An area expressed as a bounding box (geospatial data are widespread in environmental and earth science).
2. A time range (typically investigators want the most recent data, but updates to past readings due to quality control or restoration of missing data may also be of interest).
3. A list of parameters required in the data products (e.g. temperature or salinity; in advanced cases, this may be a derivative parameter which must itself be computed from some base parameters).
4. Optionally, a deadline (deadlines may be expressed in terms of maximum accepted time for delivery of the data product).

To deliver the data subscription service, a distributed infrastructure is needed for computing data products and delivering subscriptions to users. The subscription scenario is often time-critical where a number of subscriptions must be fulfilled on a deadline to receive the data products. Different products may require different degrees of processing at different times and place differing levels of load on the processing infrastructure.

Such a data subscription scenario serves both end-users and application workflows for which the retrieval of subscribed-to data is a crucial input. Frequently these workflows require specific data to be delivered within a specific time window and often have firm or soft real-time requirements [9]. The type of real-time requirement is specified by the end-user or the workflow developer.

As the volume of subscriptions and the customizability of subscriptions increases, so too does the pressure on the underlying infrastructure providing the data, the bandwidth for transport and the processing capacity. At the same time, there will be periods of low activity between rounds of updates. Thus, we need a scalable infrastructure to support the data subscription processing pipeline so as to not unnecessarily tie up resources while still permitting acceptable quality of service during peak periods.

2.2   Generalising the Service to Different RIs

In addition to Euro-Argo, other RIs are now looking into the data subscription scenario as an approach to better serve their communities. RIs differ in several aspects, such as in their maturity in various data life cycle phases, in their internal diversity, and in collaboration between RIs. Projects such as ENVRIplus seek approaches to enable convergence through reference modelling [7, 10], helping the RIs to identify common processes and structures and to adopt best practices, and integrate to common infrastructural environment (eInfra) services.

However, several of the problems encountered in traditional data curation and publishing still exist and can be summarised as follows:

- accumulating, large, and complex datasets can only be disseminated with extensive effort
- frequently changing datasets can only be monitored with extensive effort
- unpublished, confidential data can only be disseminated to the designated audience

Moreover, as discussed in Chapter 12, other encountered challenges include lack of accounting information, lack of data provenance information, and the complexity involved in using and integrating distributed systems [5]. For example, the latter challenge emerges when data flows can exist between curation, processing, and publishing subsystems, each which can be provided by different RIs. As a result, the total number of data flows can increase to an extent not easily managed by an investigator.

To answer these problems and challenges in general, a data subscription model was proposed to change the way how subsets of frequently changing data collections are published/disseminated to designated investigators.

2.3   Data Subscription Model

Data subscriptions are built upon the well-known Publish/Subscribe messaging pattern providing advantages such as loose coupling of publishers and subscribers in time, space, and synchronization [3]. The pattern typically consists of three types of entities: publishers, subscribers, and a message broker or topologies of brokers forming a communication infrastructure. The broker can implement a messaging matching scheme through which subscribers receive only a subset of the total published messages. The two most common schemes are topic-based and content-based matching. In contrast to the latter, the topic-based matching let the publisher decide the classes of messages to which subscribers can register/subscribe to.

In this model, subscribers register to topics, more specifically to globally unique persistent data identifiers (PID) [see chapter 9]. There are several reasons for this. First, the data discovery process [see chapter 4] can be independent of creating a subscription. Second, the subscriber can create a subscription even before there is published events on the data identifier. Thus, the subscribers are truly uncoupled from the publishers in time. Third, persistent identifiers are seen as the widely accepted approach to support research data re-use and sharing, while also enabling provenance tracking, and consequently, enabling micro-attribution. Fourth, in order to provide a common and robust model, the semantically immutable persistent identifiers provide several valuable characteristics to build on, such as in direction and option for data granularity levels.

In terms of data life cycle stages, Fig. 1 shows the primary flows between the stages and a matching process of the data subscription model.

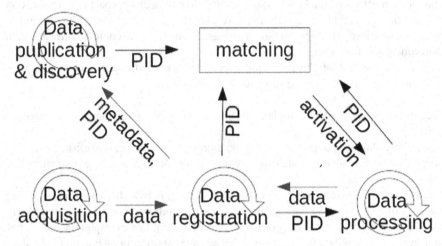

Fig. 1. The data life cycle stages in the Data Subscription Mode.

3   Architectural Design and Prototype

Figure 2 illustrates a functional depiction of the Euro-Argo data subscription scenario.

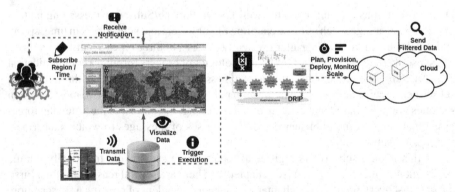

Fig. 2. The data subscription scenario is one where researchers can subscribe to the specific data they are interested in (e.g. marine data from floats in the Mediterranean) via a simple community portal, and have updates pushed to their workspaces.

3.1 Architecture Design

We applied a prototype of the data subscription service in the scenario depicted in Fig. 3. Currently, the resources from e-infrastructures such as EUDAT [4] and EGI FedCloud are used. Figure 3 shows the use-case scenario based on the use of EUDAT and EGI services. In this case, EUDAT provides services for data subscription, storage, and data transfer, while EGI FedCloud provides the services for the computing of data products for each subscription.

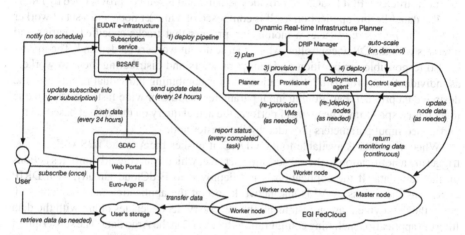

Fig. 3. A context diagram showing interactions between components in the Euro-Argo data subscription scenario. The subscription service invokes DRIP to plan, provision and deploy the subscription data processing pipeline. Subscriptions and processing are event-driven, triggered by updates pushed to the B2SAFE data repository. The deployment is scaled with demand [1].

The data subscription service scenario thus involves the following basic components:

1. A data selection community portal serving as the front-end;
2. The global data assembly centre of Euro-Argo [8], providing the source research dataset;
3. B2SAFE data repository [2] provided by EUDAT;
4. A deployment of DRIP [1] (deployed within EGI FedCloud);
5. A data filtering application. This is the software that actually takes the input parameters and composes the requested data product from the raw source research dataset;
6. EGI FedCloud virtual resources, forming the fundamental infrastructure for data processing and transportation;
7. EUDAT data subscription service (which maintains and matches the subscriptions defined via the data selection community portal).

Users interact with the subscription service via a portal, registering to receive updates for specific areas and time ranges for selected parameters such as temperature, salinity, and oxygen levels and optionally set a deadline for receiving the requested results. The global data assembly centre (GDAC) of Euro-Argo receives new datasets from regional centres and pushes them to the EUDAT B2SAFE data service. The subscription service itself maintains records of subscriptions including references to selected parameters and associated actions. The subscription service is based on well-defined APIs that allow connecting it with various community front-ends and infrastructure platforms. The role of DRIP then is to plan and provision a customised infrastructure dynamically with demand, and to deploy, scale and control the data filtering application to be hosted on that infrastructure. EGI FedCloud provides actual cloud resources provisioned by DRIP.

The data filtering application itself is composed of a master node and a set of worker nodes. The *master node* uses a monitoring process that tracks specified metrics and interacts with the DRIP controller, which can scale out workers on demand. The master is also responsible for partitioning input parameters and distributing them to workers as individual tasks for parallel execution and for combining individual results into the desired data product. Partitioning input parameters should provide faster execution due to increased speed-up. The *workers* perform the actual query on the dataset based on the partitioned input parameters provided by the master node.

When new data is available to the GDAC, it pushes them to the B2SAFE service, triggering a notification to the subscription service, which consequently initiates actions on the new data. If the application is not deployed to FedCloud already, then DRIP provisions the necessary VMs and network so that the application may be deployed. Next, the deployment agent installs all the necessary dependencies along with the data filtering application including configurations to access on the Argo data. The subscription service signals to the application master node the availability of the input parameters to be processed, whereupon it partitions the input tasks into sub-tasks and distributes them to the workers. If the input parameters include deadlines then the master will prioritise them accordingly. The monitoring process keeps track of each running task and passes that information to the DRIP controller. If the programmed threshold is passed, then the controller will request more resources from the provisioner. Finally, the results of each task are pushed back to the B2SAFE service triggering a notification to the subscription service, after which it notifies the user[1].

3.2 Infrastructure Customisation and Performance Optimisation

To meet the time-critical constraints of the data subscription service, data products for all subscriptions should be processed and distributed within a certain time window. Resources need to be elastic to support all tasks without wasting significant resources during less active periods. To this end, DRIP provides an auto-scaling option to ensure on-time delivery of the requested data, based on the total budget available for conscripting resources (not that this budget need be monetary; it could also be tied to other metrics such as energy use). However, simply adding resources is not always enough to provide the best possible performance for an application—to fully take advantage of the available resources it is often necessary to change the invocation parameters of an application and partition them in a manner that will achieve good scalability and efficiency.

---

[1] The use case online demo: https://www.youtube.com/watch?v=PKU_JcmSskw&t=12s.

Two basic optimisation strategies have been investigated for partitioning and scheduling subscription tasks in order to minimise resource usage while meeting all necessary deadlines.

Input Partitioning. We investigated two types of input partitioning: *linear* and *logarithmic*. With linear partitioning, we simply divide the input range into equal parts for parallel processing. With logarithmic partitioning, we split the range into larger sections at the beginning of the range (accounting for the sparser data recorded early in the Euro-Argo dataset) and smaller sections towards the end (when observations become more detailed).

Deadline-Aware Auto-Scaling. The user has the option to specify a deadline for obtaining the requested data. To ensure on-time data delivery, the application master calculates the 'importance' of each task based on its deadline and input parameters:

$$Imp(task) = (|P| \cdot w_p) + (ttd \cdot w_p) + (tr \cdot w_d) + (\alpha \cdot w_a) \tag{1}$$

In Eq. 1, P is the parameter list, ttd is the time-to-deadline, tr is the time range, α is the area and w_p, w_d, w_t, w_α are the respective weights that determine each parameter's importance.

Ascertaining the prioritisation of tasks allows for smarter scaling behaviour on the part of the provisioning system by determining which parameters thresholds should be placed to trigger scaling. Figure 4 illustrates how the process of the deadline-aware auto-scheduling proceeds.

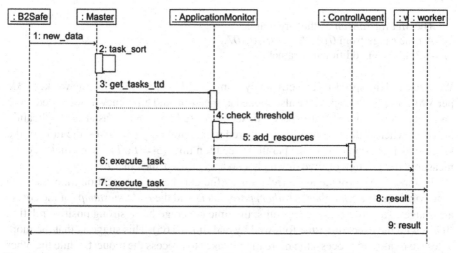

Fig. 4. Deadline-aware auto-scheduling flow. As soon as the GDAC pushes out new data, the process begins. All tasks are sorted according to Eq. 1, then the application monitor constantly evaluates the next task's time-to-deadline. If it is greater than the chosen threshold, then the controller provisions more resources.

4 Experimental Results

In this section, we present the results of the experiments described in the previous section.

4.1 Input Partitioning

Before attempting to partition input, parameters and distribute them to worker nodes, we must first identify which parameter is responsible for the most computing time when generating the data products. To do this, we generated a set of tasks on a region of randomly selected raw data requiring computing of all parameters. We performed 550 tasks spanning the Mediterranean Sea while requesting data in a time window from 1999 to 2007 and covering more than 400 possible parameters in the data products. We executed these tasks on identical VMs and measured their execution time to determine the correlation between area, time range and the number of parameters with execution time. Additionally, we investigated the effect of the end date on execution time, e.g. whether execution time changes when processing three months of data leading up to 1999 rather than leading up to 2007 (indicating a general shift in the typical volume of data collected at different points in time). We use this correlation analysis to select a suitable partitioning strategy. For our experiments we used EGI's FedCloud as our test-bed; all VMs in these experiments were identical, with two cores and two GBs of RAM.

We tested the logarithmic partitioning strategy under the assumption that input data are not always equally distributed, and therefore the load balance on the worker nodes would not be the same. For both strategies we applied the same task with the following input parameters:

1. The *Mediterranean* as the target area.
2. A time range from *01/01/99* to *01/01/07*.
3. 412 different additional parameters.

We measured the speed-up and efficiency using 1, 2, 4, and 8 VMs with one worker node per VM for both strategies. We also looked at speed-up and efficiency as we added more tasks per worker node. With speed-up, we measured how much faster an application becomes when adding more VMs compared with using only one VM—the ratio of the sequential execution time to the parallel execution time ($S = T_s/T_p$). For efficiency, we measured the fraction of time in which a node is utilised such that $E = S/p$.

In Table 1, we provide the correlation coefficients between execution time and each time *coverage, area, number of (other) parameters* and the *end timestamp* (of the coverage range). According to these results, the time coverage has a strong positive relation (0.93) with the execution time followed by end time (0.65). This suggests that the more dates we request to process, the more time it takes to process the request, while the other variables do not indicate any particular strong relationship with the execution time.

Figure 5 and Fig. 6 show the speedup and efficiency results. The lines indicated as 'log1' and 'log4' indicate speed-up for logarithmic partitioning while assigning 1 and 4 tasks per worker respectively. The lines indicated as 'lin1' and 'lin4' represent linear partitioning with the same assignments. These results indicate that the logarithmic partitioning with 4 tasks per worker performs best (log4), which aligns with Table 1.

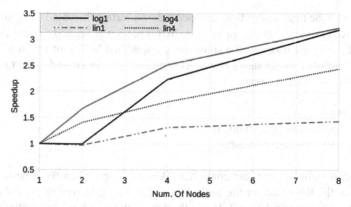

Fig. 5. Speed-up for linear and logarithmic partitioning strategies assigning 1 and 4 tasks per worker.

Table 1. Correlations with the execution time of various parameterisation options.

Correlations	Execution time
Execution time	1.00
Time coverage	0.93
Area	0.03
Num. of parameters	0.02
End timestamp	0.65

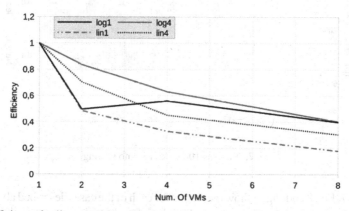

Fig. 6. Efficiency for linear and logarithmic partitioning strategies assigning 1 and 4 tasks per worker.

4.2 Deadline-Aware Auto-Scaling

Using Eq. 1 we ranked 100 tasks each with the same deadline but varying areas and time ranges. After ranking these tasks, we set the time-to-deadline as a metric for a monitoring process. When the time-to-deadline dropped below a certain threshold, a signal was sent to the controller to scale up the application. In this particular setup the controller started a new VM each time it received a signal until a specified VM limit was reached, after which the controller would start a new worker on each VM in a round-robin fashion. We examined three different cases:

1. no scaling,
2. scaling with a static threshold, and
3. scaling with a dynamic threshold.

In the case of static scaling, the controller takes no action when the time-to-deadline drops below the threshold. In the second case, the threshold was set to a static value (chosen after an empirical study). In the third case, the threshold was initially set to a specific value, but as soon as the time-to-deadline dropped below the threshold a signal was sent to the controller to scale the application, and the new threshold value was set to the current time-to-deadline minus a selected factor. For the third case, we tried to avoid aggressive scaling in an attempt to provision only as many VMs as necessary so that we could finish all tasks in time. For this experimental setup we specified a limit to the number of VMs to eight with two workers per VM, meaning that the maximum number of workers at any time was 16—this represented the budget limit that might be imposed by the application developer to prevent 'run-away' scheduling of VMs.

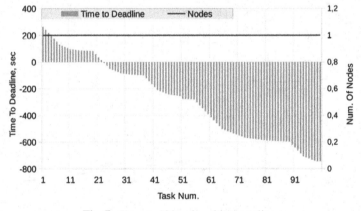

Fig. 7. Process 100 tasks with no scaling.

Figure 7, Fig. 8 and Fig. 9 show the results for each of the cases described above. In all figures the -axis represents the task number, the left-side -axis the time to the deadline (in seconds) and the right-side -axis the number of nodes used for each execution. Also, in Fig. 8 and Fig. 9 we show the threshold for triggering the addition of more resources. In Fig. 7, although the cost of the application is minimal (only one VM) after approximately 22 tasks are initiated, all deadlines are missed. In Fig. 8, we observe all tasks are processed within their deadline, but the controller over-provisions VMs for the task, reaching the specified limit of 16 workers (two workers per VM) very quickly.

Finally, in Fig. 9, we see that the controller provisions just enough workers to complete all tasks on time with the exception of the last, which overshoots its deadline by two seconds (which may or may not be unacceptable given the strictness of the deadline imposed—in this particular instance, however, we deem it acceptable given the overall high quality of service provided).

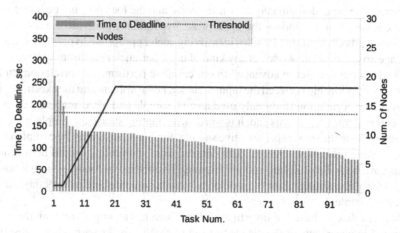

Fig. 8. Process 100 tasks with a static threshold.

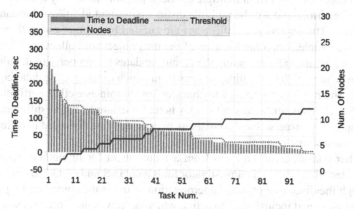

Fig. 9. Process 100 tasks with a dynamic threshold.

5   Discussion

The results presented here demonstrate that a linear partitioning strategy can provide non-linear variations in speed and efficiency. This can be attributed to an unequal load distribution, where some workers were assigned far smaller loads than others, despite the data being split 'evenly' across a certain dimension. In the case of the Euro-Argo dataset used for this experiment, this is because more recent data samples contain more data than older samples (due to improvements in data acquisition over time), which explains why the logarithmic partitioning performed better. However, the recorded speed-up can be improved further if the partitioning is calibrated based on the actual end date selected for a sample. Moreover, a more linear speed-up could be achieved if the partitioning was performed based on all, rather than just one, input dimensions. This is not a trivial task; however, as the input domain may be -dimensional and the load may not be linear across all dimensions, making finding the appropriate hyperplanes to divide the domain into equal task loads challenging. Besides identifying such appropriate hyperplanes, another challenge arises: how can we select any kind of input parameter partitioning if we cannot analyse the input data-set in advance? In our case, we performed a correlation study to identify the relationship between the input parameter of a problem and the execution time. However, that correlation study only used a small sample and often analysing the entire data-set is not practical. To this end, it is worth investigating statistical sampling methods that may provide the most representative sample. Such a process may be complemented by an iterative process where real data coming from monitoring would help evaluate and improve both the sampling and partitioning. Historical observations on the same or similar (for a given judgement of 'similarity') can also contribute to selecting the best partitioning strategy.

One area that we have not investigated, but which has an impact on both the performance and requirements of the data subscription pipeline is the case where subscribers subscribe not just to one custom view on a single dataset (albeit a very rich one), but to a view that combines data from multiple datasets, possibly hosted by multiple RIs.

In this scenario, there will be multiple distinct persistent identifiers for data objects or collections. The objects and collections are curated by another entity than the subscriber. For example, when the location of the data objects or collections changes, the subscription remains valid assuming the curator updates the property of the identifier.

Moreover, there will be multiple sources from which to retrieve the data required for processing, and it will be necessary to consider how to join as well as partition the data in a way that accounts for factors not in play here; for example, where different datasets are geographically dispersed and so workers may actually be deployed in different data centres to ensure performance.

A further consideration emerges from a requirement for immutable semantics of persistent identifiers. When the investigator finds a data set and metadata describing the data through the discovery process, interpretation of using the corresponding identifier in a subscription, and inclusion of an action which perceives the structural semantics of the data, need to be considered.

6   Conclusion and Future Work

In this paper, we have presented a service prototype to define data subscriptions to Euro-Argo data, connected to invoking a processing pipeline optimised with Dynamic Real Time Infrastructure Planner (DRIP) solution. We demonstrated how DRIP could be used to automatically select and provision infrastructure resources, deploy services, and optimise the runtime quality for the EUDAT data subscription service based on a study case involving the Euro-Argo research infrastructure.

The DRIP microservice suite optimises the runtime quality of service provided by a data service deployed dynamically on a virtualised e-infrastructure, with a particular focus on time-critical constraints such as deadlines for delivering data to a distributed set of targets. We demonstrated how to select an optimal strategy for partitioning the input tasks into workers using a modicum of expert knowledge concerning the specifics of an application. The results clearly show the value of integrated systems such as DRIP for dynamic optimisation of data services in research support environments, and how with further investigation and development they might be used for a number of similar applications cases involving distributed services and large, dynamic datasets. Furthermore, we showed subscription matching in the presence of existing data life cycle stages through which investigators can free up time and be notified of significant events in subscribed data.

The demonstrated Data subscription service prototype is part of the EUDAT innovation portfolio. The starting point for developing the subscription service has been to provide a generic component that can be connected to different community frontends, and that can utilise different e-infrastructure platforms for automated processing. This paper presents the successful Euro-Argo pilot case within ENVRIplus project that demonstrates the interaction of several service components from several providers: the Euro-Argo data portal, EUDAT B2SAFE data storage, the EUDAT data subscription service, the DRIP solution, and EGI FedCloud. The EUDAT Collaborative Data Infrastructure has identified the potential in the subscription model and considers it as a possible new addition to the data management services, depending on the interest of user communities and availability of development resources.

Regarding the DRIP solution, it is necessary to acknowledge the difficulty still inherent in building generic solutions for fully automated optimisation of infrastructure for arbitrary data services. Some degree of application-specific customisation is still necessary when applying infrastructure-level optimisation. However, further investigation and classification of different kinds of data service will assist in identifying the best mechanisms and heuristics for optimisation.

In this light, an important future work will be deploying DRIP as an optimisation engine for a broader range of services provided on behalf of environmental RI—by doing this, we will be able to explore a wider range of usage scenarios and so identify new optimisation strategies for input partitioning and dynamic provisioning of infrastructure. For example, DRIP could consider how resource failures would have an impact on deadlines and the strategies for swiftly reacting to such events. Moreover, integrating DRIP with data processing frameworks from specific research domains will also be important for refining our approach, allowing us to work in complement with established and new frameworks for scientific data handling. For example, automated data quality of distributed data streams is an important aspect of many disciplines including environmental science. Challenges such as in-time resource scaling and optimal resource placement will be studied in the context of DRIP. This will add to continuing global efforts to consolidate research infrastructure and other research support environments.

Acknowledgements. This work was supported by the European Union's Horizon 2020 research and innovation programme via the ENVRIplus project under grant agreement No 654182.

References

1. Koulouzis, S., et al.: Time-critical data management in clouds: challenges and a dynamic real-time infrastructure planner (DRIP) solution. Concurr. Comput. Pract. Exp. e5269 (2019). https://doi.org/10.1002/cpe.5269
2. Cacciari, C., Fares, M., Fiameni, G., Michelini, A., Danecek, P., Wittenburg, P.: Adoption of the B2SAFE EUDAT replication service by the epos community. In: EGU General Assembly Conference Abstracts 16 (2014)
3. Eugster, P.T., Felber, P.A., Guerraoui, R., Kermarrec, A.M.: The many faces of publish/subscribe. ACM Comput. Surv. (CSUR) 35(2), 114–131 (2003)
4. Gentzsch, W., Lecarpentier, D., Wittenburg, P.: Big data in science and the EUDAT project. In: Global Conference (SRII) 2014, pp. 191–194 (2014), http://doi.org/10.1109/SRII.2014.34
5. Ahanach, E. el K., Koulouzis, S., Zhao, Z.: Contextual linking between workflow provenance and system performance logs. In: 2019 15th International Conference on eScience (eScience), pp. 634–635. IEEE, San Diego (2019). https://doi.org/10.1109/eScience.2019.00093
6. Hu, Y., et al.: Deadline-aware deployment for time critical applications in clouds. In: Rivera, F.F., Pena, T.F., Cabaleiro, J.C. (eds.) Euro-Par 2017. LNCS, vol. 10417, pp. 345–357. Springer, Cham (2017). https://doi.org/10.1007/978-3-319-64203-1_25
7. Atkinson, M., Hardisty, A., Filgueira, R., Alexandru, C., Vermeulen, A., Jeffery, K., Loubrieu, T., Candela, L., Magagna, B., Martin, P., et al.: A consistent characterisation of existing and planned RIs. ENVRIplus deliverable 5.1, submitted on 30 April 2016
8. Wong, A., Keeley, R., Carval, T.: The Argo data management team (2013) Argo quality control manual, version 2.8. Argo Data Management (2010)
9. Evans, K., et al.: Dynamically reconfigurable workflows for time-critical applications. In: Proceedings of the 10th Workshop on Workflows in Support of Large-Scale Science - WORKS 2015, pp. 1–10. ACM Press, Austin, Texas (2015). https://doi.org/10.1145/2822332.2822339
10. Zhao, Z., et al.: Reference model guided system design and implementation for interoperable environmental research infrastructures. In: 2015 IEEE 11th International Conference on e-Science, pp. 551–556. IEEE, Munich (2015). https://doi.org/10.1109/eScience.2015.41

Case Study: ENVRI Science Demonstrators with D4Science

Leonardo Candela[1]([✉]) [ID], Markus Stocker[2,3] [ID], Ingemar Häggström[4] [ID],
Carl-Fredrik Enell[4] [ID], Domenico Vitale[5] [ID], Dario Papale[5] [ID], Baptiste Grenier[6] [ID],
Yin Chen[6] [ID], and Matthias Obst[7] [ID]

[1] National Research Council of Italy, Istituto di Scienza e
Tecnologie dell'Informazione "A. Faedo", Via G. Moruzzi, 1, 56124 Pisa, Italy
leonardo.candela@isti.cnr.it
[2] TIB Leibniz Information Centre for Science and Technology, Welfengarten 1 B,
30167 Hannover, Germany
markus.stocker@tib.eu
[3] MARUM Center for Marine Environmental Sciences, PANGAEA Data Publisher for Earth
and Environmental Science, Leobener Strasse 8, 28359 Bremen, Germany
[4] EISCAT Scientific Association, Box 812, 981 28 Kiruna, Sweden
{ingemar.haggstrom,carl-fredrik.enell}@eiscat.se
[5] Department for Innovation in Biological, Agro-Food and Forest Systems (DIBAF),
University of Tuscia, Via San Camillo de Lellis, 01100 Viterbo, Italy
{domvit,darpap}@unitus.it
[6] EGI Foundation, Science Park 140, Amsterdam, The Netherlands
{baptiste.grenier,yin.chen}@egi.eu
[7] Swedish Lifewatch, Gothenburg, Sweden
matthias.obst@marine.gu.se

Abstract. Whenever a community of practice starts developing an IT solution for its use case(s) it has to face the issue of carefully selecting "the platform" to use. Such a platform should match the requirements and the overall settings resulting from the specific application context (including legacy technologies and solutions to be integrated and reused, costs of adoption and operation, easiness in acquiring skills and competencies). There is no one-size-fits-all solution that is suitable for all application context, and this is particularly true for scientific communities and their cases because of the wide heterogeneity characterising them. However, there is a large consensus that solutions from scratch are inefficient and services that facilitate the development and maintenance of scientific community-specific solutions do exist. This chapter describes how a set of diverse communities of practice efficiently developed their science demonstrators (on analysing and producing user-defined atmosphere data products, greenhouse gases fluxes, particle formation, mosquito diseases) by leveraging the services offered by the D4Science infrastructure. It shows that the D4Science design decisions aiming at streamlining implementations are effective. The chapter discusses the added value injected in the science demonstrators and resulting from the reuse of D4Science services, especially regarding Open Science practices and overall quality of service.

© The Author(s) 2020
Z. Zhao and M. Hellström (Eds.): Towards Interoperable Research
Infrastructures for Environmental and Earth Sciences, LNCS 12003, pp. 307–323, 2020.
https://doi.org/10.1007/978-3-030-52829-4_17

Keywords: Virtual research environment · Open science · D4science · Science demonstrators · Science communities

1   Introduction

Science is highly digital, collaborative and multidisciplinary and science practices have been changed in recent decades [6]. These changes are induced by the opportunities offered by the developments in information technologies and infrastructures and are impacting the whole research lifecycle – from data collection and curation to analysis, visualisation and publishing. Research communities are dynamically aggregated, working environments conceived to support research tasks are virtual, heterogeneous and networked across the boundaries of research performing organisations. Scientists are thus asking for integrated environments providing themselves with seamless access to data, software, services and computing resources they need in performing their research activities independently of organisational and technical barriers [5]. In these settings, approaches based on ad-hoc and "from scratch" development of the envisaged supporting environments are neither viable (e.g. high "time to market") nor sustainable (e.g. technological obsolescence risk).

Environmental science is not eluding these changes, rather it is fully affected by them. A rich array of environmental research infrastructures is being organised and developed to provide their designated communities with computing resources, services and facilities for data collection and collation, processing, analytics, and publishing. Initiatives such as ENVRIplus [1] and the European Open Science Cloud [10] have been launched to make available state-of-the-art solutions for data management thus making the development and operation of environmental research infrastructures more efficient.

In spite of these developments, researchers and scientists are still struggling with the lack of working environments tailored to their specific needs, especially when operating in multidisciplinary contexts.

In this chapter, we present the D4Science-based solution for developing and operating *virtual research environments* for different communities of practice identified in selected science demonstrators. D4Science [2, 3] enacted virtual research environments promote the re-use of domain-specific existing data and services, the co-creation and co-development of the envisaged working environment, and the use of state-of-the-art solutions for collaboration, communication and Open Science.

This chapter presents four concrete and diverse science demonstrators. These cases concern (a) providing scientists willing to analyse data collected by EISCAT radars with a collaborative working environment, (b) implementing shared, standardised and reproducible data processing and quality control (QC) procedures for long-term eddy covariance (EC) flux datasets, (c) providing scientists involved in atmospheric new particle formation event analysis with computational environments for event identification and classification with built-in analysis (derivative) data FAIRification, and (d) providing scientists seeking to increase our knowledge of biodiversity organisation and ecosystem functions with a working environment to test models. In particular, the challenges and the resulting prototypical working solutions (with their pros and cons) community of practices managed to develop are presented and discussed.

2 The Collaborative Working Environment for Data Analysis

EISCAT_3D will differ from other environmental research infrastructures with respect to its configurability and data volumes. A typical environmental RI measures well-defined parameters and stores the data in a specified way. EISCAT_3D, on the other hand, will be a flexible, multi-purpose instrument. Archived data can be reanalysed to extract parameters in complementary research domains, typically for example both electron and ion densities and temperatures in the ionosphere and the influx of meteors from space. Data access rules also apply according to the agreement between EISCAT members, including embargo times for PIs of experiments. This means that users must be allowed to upload and run their own analysis software on archived EISCAT_3D datasets to which they are granted access.

The use of big data and supercomputing systems will be unfamiliar to typical EIS-CAT_3D scientists, so authentication, search and analysis should be handled by a portal. This portal should have a search GUI as well as APIs for script-based access. This line of work is also further developed by EGI and in the European Open Science Cloud Hub Competence Centre (EOSC-Hub CC) for EISCAT_3D.

In the framework of ENVRIplus, EISCAT_3D has been a pilot case in using D4Science. An advantage of the D4Science portal is that it allows uploading user software in many languages. The online R studio is well developed, but GNU/Octave, Python and several other languages are also available. Like in the DIRAC portal development in the EOSC-Hub CC, the science demonstrator had to work on existing data from the present EISCAT radars. The realtime graph plotting routine was selected as a common analysis case. This is Matlab software but runs also in Octave, which eliminates the need for a software license. File format conversion software written in Python with the HDF5 library has also been tested.

Figure 1 shows the D4Science file management GUI, which presents a familiar interface to the user. Here, program and data files were uploaded.

Fig. 1. The D4Science file GUI.

Figure 2 shows the DataMiner interface, where selected jobs are submitted for execution with selected input. Finally, Fig. 3 shows results being listed.

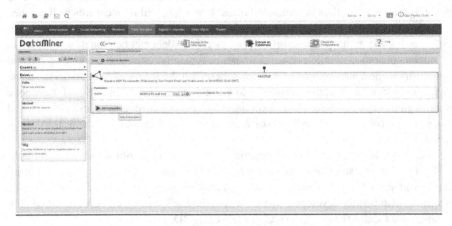

Fig. 2. The D4Science Data Miner job submission interface.

Fig. 3. The D4Science Data Miner list of completed jobs.

In EOSC CC, the development of the DIRAC system[1], originating from the LHCb detector at CERN, is ongoing. DIRAC is a job submission system that could benefit from a frontend such as D4Science. At this stage, EGI Check in AAI has been added to DIRAC to accommodate EISCAT access rules. The DIRAC metadata catalogue will also be extended for our purposes. Data will be stored using a replicating file management system such as Rucio, which is a development for the LHC ATLAS detector at CERN[2]. As for file storage backends, the project has been considering dCache or a parallel system such as Lustre or IBM GPFS.

---

[1] http://www.diracgrid.org.
[2] http://rucio.cern.ch.

Another pilot project was granted by EUDAT where EISCAT collaborated with CSC, Finland. Here, metadata constructed from EISCAT's existing Level 2 and 3 data systems were uploaded to B2SHARE. Figure 4 shows a sample B2SHARE entry. This would provide a common search interface for the two data levels. The existing data server of EISCAT has only basic search functions for listing the two levels together (namely the online schedule, ordered by date). We also foresee that all metadata will be provided for harvesting into B2FIND.

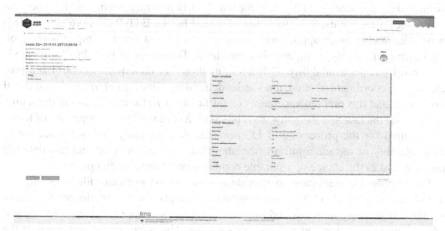

Fig. 4. A sample B2SHARE entry for existing EISCAT data, created using the B2SHARE REST API in Python software.

3   The Eddy Covariance of GHGs Fluxes Use Case

The eddy-covariance (EC) technique is considered the most direct and reliable method to calculate flux exchanges of the main greenhouse gases (GHG) over natural ecosystems and agricultural fields. The resulting measurements are extremely important to characterise ecosystem exchanges of carbon, water, energy and other trace gases and are widely used to validate or constrain parameter of land surface models via data assimilation techniques.

EC fluxes calculation involves a complex set of data processing steps that, beyond the knowledge of the technique, requires a considerable amount of computational resources. This might constitute a constraint for RIs (e.g. ICOS) that aim to simultaneously process large raw dataset sampled at multiple sites in Near Real Time (NRT) mode (i.e. provide each day fluxes estimates relative to the previous day).

The ambitious goal of this pilot investigation is to provide a computationally efficient tool able to process EC raw data and offer users the possibility to calculate fluxes according to the multiple processing scheme [9]. The ultimate aim is to establish a service that can be used by RIs that use this micrometeorological technique to measure exchanges of greenhouse gases and energy between terrestrial ecosystems and atmosphere (e.g. ICOS, LTER and ANAEE).

3.1 Virtual Research Environment

The EC technique involves high-frequency sampling (e.g. 10 or 20 Hz) of wind speed and scalar atmospheric concentration data and yields vertical turbulent fluxes. EC fluxes are computed within a finite averaging time (normally 30 min) from the covariance estimates between instantaneous deviations in vertical wind speed and gas concentration (e.g. CO_2) from their respective mean values, multiplied by the mean air density [4].

Despite the simplicity of this idea, a number of practical difficulties arise in transforming high-frequency data into reliable half-hourly flux measurements. To cope with these issues, here we used the tools implemented by the EddyPro® Fortran code [8]) an open source software application available for free download at https://www.licor.com/env/products/eddy_covariance/eddypro.html. The choice of EddyPro® software is motivated by *i*) the availability of different methods for data quality control and processing (e.g. coordinate rotation, time-series detrending, time lag determination, spectral corrections, and flux random uncertainty quantification), *ii*) the availability of the source code and *iii*) the fact that the software is based on a community-developed set of tools.

Required for the processing of EC raw data through EddyPro® software, are 1) the availability of metadata information about the EC system setup and raw data file structure, and 2) the choice of a suitable combination of processing options.

Concerning 1), users have to provide a standardised metadata file in.csv format (metadata.csv, see Table 1). This file constitutes the input of an R script that automatically builds the mandatory files ingested into the EddyPro® software (i.e. the *.metadata* and *.eddypro* files) developed ad hoc for this exercise. The organization and name of the metadata variables is based on an international standard (BADM) used also in the USA network AmeriFlux. The format of the.csv has been instead designed in order to develop a template easy to prepare by individual scientists and organised RIs.

It is important to note that in the current implementation only a few sensors are supported (the one used in ICOS) but the structure has been prepared in order to be ready to add new sensors and new processing methods, options and combinations.

In case of NRT data processing, in order to perform part of the flux corrections (i.e. spectral corrections and planar fit), 5 additional configuration files are needed: planar_fit.txt, spectral_assessment_badr.txt, spectral_assessment_lddr.txt, spectral_assessment_bapf.txt, spectral_assessment_ldpf.txt. They can be obtained by specific EddyPro® runs based on long periods of data (at least one month of data is usually required for a consistent parameter estimation). The above files have to be placed together with EC raw-data files in an archive folder (data.zip) which will constitute the input file of the current implementation (see Fig. 5).

The use of different processing options leads to different flux estimates. Discrepancies in flux estimates are caused by systematic errors introduced by the methods used in the raw-data processing stage. Since there are not tools to establish a priori which is the best combination of processing options providing unbiased flux estimates, the viable solution, proposed by [11] and implemented here, involves a multiple processing scheme where EC flux data are calculated according to different combinations of methods.

In particular, EC fluxes are calculated according to four different processing schemes resulting from a combination of block average (ba) or linear detrending (ld) and double rotation (dr) or planar fit [12] (pf) processing options (for details see [4]). All other processing options remain unchanged: maximum cross-covariance method for time lag determination, spectral correction method proposed by Fratini et al. (2012), statistical tests by Vickers and Mahrt (1997) and by Foken and Wichura (1996) for data quality control, method by Finkelstein and Sims (2001) to random uncertainty quantification.

Table 1. Description of Metadata to provide in the metadata.csv file.

Column	Variable Label (File Header)	Description (Units)
1	SITEID	Official EC station code following the FLUXNET standards (CC-Xxx)
2	LATITUDE	Geographic latitude ([-90,90] from S to N, decimal)
3	LONGITUDE	Geographic longitude ([-180,180] from W to E, decimal)
4	ALTITUDE	The altitude of ecosystem under study (m)
5	CANOPY_HEIGHT	Distance between the ground and the top of the plant canopy (m)
6	SA_MANUFACTURER	Manufacturer of the sonic anemometer (currently only gill)
7	SA_MODEL	Model of the SA (currently only SA-Gill HS-50 or -100)
8	SA_SW_VERSION	The embedded software version of the SA
9	SA_WIND_DATA_FORMAT	The format of wind data (currently only uvw)
10	SA_NORTH_ALIGNEMENT	Specify whether the SA's axes are aligned to transducers (axis) or spars (spar)
11	SA_HEIGHT	The vertical distance between the ground and the centre of the device sampling volume (m)
12	SA_NORTH_OFFSET	Specify the SA's yaw offset with respect to local magnetic north (degree positive eastward)
13	GA_MANUFACTURER	Manufacturer of the gas analyser (currently only licor)
14	GA_MODEL	Model of the GA (currently only GA_CP-LI-COR LI7200)
15	GA_SW_VERSION	The embedded software version of the GA
16	GA_NORTHWARD_SEPARATION	The distance between the centre of the sample volume of the GA and the SA as measured horizontally along the north-south axis (cm)
17	GA_EASTWARD_SEPARATION	The distance between the centre of the sample volume of the GA and the SA as measured horizontally along the east-west axis (cm)
18	GA_VERTICAL_SEPARATION	The distance between the centre of the sample volume of the GA and the SA as measured along the vertical axis (cm)
19	GA_TUBE_DIAMETER	The inside diameter of the intake tube (mm)
20	GA_FLOWRATE	The flow rate of the intake tube (l/min)
21	GA_TUBE_LENGTH	The length of the intake tube (cm)
22	FILE_DURATION	The time span covered by each raw file (min)
23	ACQUISITION_FREQUENCY	The number of records per second in raw files (10 or 20 Hz)
24	FILE_FORMAT	Specify the format of raw files (ASCII or BIN)
25	FILE_EXTENSION	Specify the raw files extension (e.g..csv,.txt or.dat)
26	LN	Logger number (from 1 to 10)
27	FN	Number of the file generated by the logger (from 1 to 10)
28	EXTERNAL_TIMESTAMP	0 or 1 if the timestamp in the file name refers to the beginning or the end of the averaging period, respectively.
29	INTERNAL_TIMESTAMP	1 if there is a timestamp internal to raw files, otherwise 0.
30	EOL	Specify the end of the line of raw files (e.g. lf)
31	SEPARATOR	The character that separates individual values in raw files
32	MISSING_DATA_STRING	Specify the character string used for missing data in raw files (e.g. NA, NaN, -9999)
33	NROW_HEADER	The number of rows in the header of the raw file
33+1	COLNAMES_1	Variable name in the first column of the raw data file
33+j	COLNAMES_j	Variable name in the j-th column of the raw data file
33+N	COLNAMES_N	Variable name in the last column of the raw data file

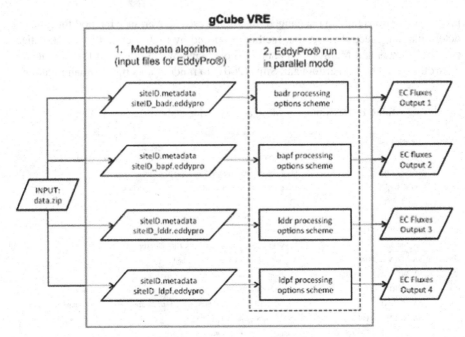

Fig. 5. EC data processing path.

To reduce the computational runtime, the implementation of the four processing schemes aforementioned is performed in parallel mode in the gCube Virtual Research Environment (VRE). The processing path is defined as in Fig. 5 (for an illustrative example see https://www.youtube.com/watch?v=ssHAfwXVF0A).

3.2 Benefits

The implementation of a multiple processing scheme as illustrated above and the direct management and use of metadata according to international standard in the eddy covariance community constitutes a novelty in the context of EC data analysis. The main advantage of multiple processing is twofold. On one hand, it offers the possibility of an extensive evaluation of the effect each method has on flux data estimation. On the other hand, by combining the output results as described by [11], it is possible to obtain more consistent estimates of the uncertainty associated with EC fluxes. The direct use of metadata instead ensures the needed flexibility for a large use of the tool if the new sensors are added in the system.

The efficiency of parallel computing implemented in the VRE drastically reduces the computational runtime required to obtain flux estimates from different processing options schemes. When using EC raw data from a single observation tower, the estimated computational time required for an NRT run is about 4 min, similar to those required for the run of a single processing scheme. This constitutes a clear advantage for any user and in particular, for RIs aiming at analyzing routinely large amounts of data. Although, here we selected only four processing option schemes, the efficiency of parallel computing implemented in the VRE offers the possibility to increase the number of processing schemes suitable for the EC data processing and post-processing steps. This might considerably improve our understanding of the performance of methods developed for EC raw-data processing and the interpretation of resulting fluxes.

4 New Particle Formation Event Analysis

Atmospheric new particle formation (NPF) is a worldwide observed phenomenon that affects human respiratory health and the global climate [7]. NPF is studied by analysing (specifically, interpreting) the particle size distribution of polydisperse aerosol as measured by a differential mobility particle sizer at specific spatio-temporal locations (thereafter observational or primary data). The Finnish Station for Measuring Ecosystem-Atmosphere Relations[3] (SMEAR) research infrastructure operates such instruments at multiple spatial locations, including at Hyytiälä in southern Finland. The research infrastructure systematically publishes the collected observational data using SmartSMEAR. The observational data is thus accessible to researchers, worldwide. With the SmartSMEAR API, the data is also accessible programmatically.

To study NPF, atmospheric physicists analyse observational data to detect and characterise NPF events. During events, new particles initially form and then grow in diameter size, typically over the course of a few hours during the daytime. The detection of such events is typically performed manually by visualizing observational data for specific spatio-temporal locations (Fig. 6). Atmospheric physicists utilise such visual primary data products to determine whether or not an event occurred at the specific day and place. Events are then characterised (i.e., described) for their attributes, such as event start and end times, classification, or growth rate. With such primary data interpretation activity, atmospheric physicists generate derivative secondary data (here data about NPF events). Secondary data are subsequently used in statistical analysis, e.g. to compute descriptive statistics, thus resulting in derivative tertiary data which are sometimes published in the scholarly literature.

The FAIRification[4] of secondary and tertiary data is an important challenge for this research community, which consists of some hundreds of researchers organised in dozens of research groups (personal communication). While primary data are relatively FAIR, the derivative data generated by the numerous researchers and research communities fare very poorly along the FAIR Data Principles. Indeed, secondary data are hardly findable and accessible, not to speak of interoperable. Tertiary data such as descriptive statistics may be found and accessed in the scholarly literature. Being printed in PDF documents, they are, however, hardly interoperable. As a result, secondary data generated by the numerous research groups and researchers in the community cannot be reused (e.g. integrated); information systems underperform in search, retrieval or processing of tertiary data; and the reproducibility of tertiary data is generally impossible.

The New Particle Formation Event Analysis VRE[5] prototyped how infrastructure can ensure FAIR secondary and tertiary data by design. FAIRification is built into the

---

[3] https://www.atm.helsinki.fi/SMEAR/.

[4] https://www.go-fair.org/fields-of-action/go-build/fairification-process/.

[5] https://marketplace.eosc-portal.eu/services/new-particle-formation-event-analysis.

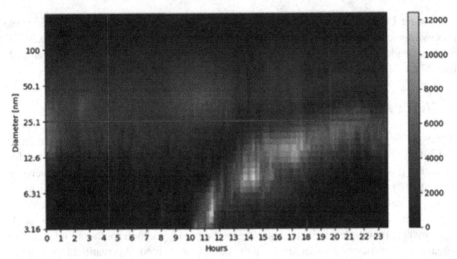

Fig. 6. Visualization of observational data for specific spatio-temporal locations. The data product shows an NPF event starting at approximately 10 am and ending shortly before 12 pm (noon). The high concentration (yellow) of particles with initially small but growing diameter size forms the typical "fingerprint" of an NPF event in observational data. (Color figure online)

infrastructure which thus ensures that data are born FAIR and frees researchers or data curators from having to FAIRify data retrospectively. Most importantly, we move data analysis into the VRE and thus harmonise data analysis across research groups; systematically catalogue secondary and tertiary data to ensure their findability and accessibility; and use languages for knowledge representation to ensure data interoperability.

4.1 Virtual Research Environment

Building on D4Science and EGI Jupyter e-Infrastructures, we developed a VRE that demonstrates how the NPF research community could perform event classification and statistical computation while the infrastructure ensures FAIR derivative (secondary and tertiary) data.

Figure 7 illustrates the VRE system architecture, its components and interactions. The NPF research community, its research groups and individual researchers access the VRE via D4Science authentication and authorization. In addition to standard VRE functionality, e.g. document management, this VRE leverages EGI Jupyter and D4Science Data Miner to provide an NPF data analysis environment with FAIR derivative data. Specialised Python functions backed by Data Miner algorithm implementations support the following operations:

- Via SmartSMEAR API, fetch primary data published by the SMEAR research infrastructure;
- Plot primary data to generate and visualise the data product required to determine whether or not an event occurred at a specified spatio-temporal location (Fig. 6);

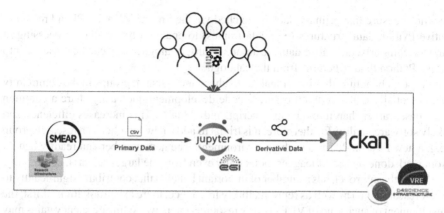

Fig. 7. VRE system architecture, its components and interactions.

- Using languages for knowledge representation (specifically, RDFS and OWL), represent derivative data (e.g. event descriptions with their attributes) richly described with a plurality of accurate and relevant attributes using vocabularies that meet domain-relevant community standards and follow FAIR principles (e.g. http://purl.obolibrary.org/obo/ENVO_01001359);
- Catalogue derivative data using the CKAN powered D4Science Catalogue;
- Retrieve catalogued secondary data (i.e., event descriptions) for statistical processing.

4.2 Benefits, Limitations and Challenges

The key benefit of the VRE is that by sharing a well-engineered computational environment for NPF event classification and statistical analysis, the research community produces FAIR derivative data without giving it any thought. In contrast to the current practice in this research community where derivative (in particular secondary) data are of high syntactic and semantic heterogeneity and impossible to integrate easily, in the VRE derivative data are automatically identified and catalogued, and thus meet key data findability and accessibility principles. Furthermore, by using languages for knowledge representation and a plurality of descriptive attributes according to domain-relevant FAIR vocabularies, derivative data also meet key data interoperability and reusability principles. Since these features are built into the infrastructure, they appear invisible to the individual researcher who can thus focus on data analysis without being exposed to the complexity of data FAIRification.

A second benefit is that derivative data are FAIR at birth rather than FAIRified retrospectively, e.g. by data curators of a research infrastructure data centre or a data publisher. FAIRification is a complex process that requires considerable domain expertise and often relies on tacit information known only to the researcher. FAIRifying early rather than later makes good sense and ensuring data are FAIR at birth is arguably the most attractive option.

A third benefit is that the VRE eliminates the need to download and upload data to and from a local computing environment (e.g. a workstation). The specialised Python functions ensure that primary data are fetched via the SmartSMEAR API and read into native Python data structures (e.g. data frames) to enable arbitrary data processing in Jupyter. Similarly, derivative data are automatically catalogued and can be retrieved into native Python data structures from the catalogue.

A fourth benefit is that individual researchers and research groups in the community can potentially collaborate on program code development and easily share a common code base, rather than implementing scripts individually. This increases efficiency and likely software quality. Furthermore, it is trivial to add a new member of a research group (e.g. a new PhD student) to such an environment. The new member can readily benefit from work done by her colleagues, potentially even from the larger research community.

While the approach has a number of important benefits that contribute significantly to FAIR research data as well as reproducibility in science, there exist limitations. First, the development of such kind of VREs is very resource-intensive. While efficiency gains may be possible, e.g. by factoring out and reusing components that are commonly required by such VREs, research data analysis is highly contextual and difficult to generalise (and thus scale) without losing efficacy. While e-Infrastructure service providers could develop services for commonly required functionality, the development of specialised (Python) scripts for data analysis in Jupyter must rely on contributions from the research community.

The development of vocabularies that meet domain-relevant community standards and follow FAIR principles is equally resourced intensive and typically relies on strong ICT specialists and research community co-development. Researchers are mostly unaware of the benefits of such vocabularies for data (machine-to-machine) interoperability and even if the benefits are acknowledged the significant resources required to develop such vocabularies compete with research activities, which (arguably rightly so) are always prioritised over good research data management.

A relatively minor technical limitation is the poor performance of retrieving data from the catalogue. While the catalogue may be an approach to deposit data, it is not ideal for fast retrieval of data needed in the analysis. To address this performance issue, the VRE system architecture should employ more efficient intermediate data storage systems.

The challenges are perhaps more important than the current limitations. The pressing challenges of the presented VRE-based approach to NPF analysis are predominantly social. A particularly pressing one is how to motivate individual researchers to use the VRE instead of their local computational environments. Probably the most important barrier to adoption by the research community is the maturity of data analysis program code. The most advanced researchers in this community have developed mature scripts that precisely serve their needs. The key objection from such researchers is thus the maturity of the code served in the VRE. Addressing this objection is non-trivial because code maturity naturally relies virtually entirely on contributions from the research community.

Furthermore, the automated cataloguing of research data on e-Infrastructure, often perceived as potentially beyond the control of the individual researcher who created the data, is an additional barrier to adoption. Trust in e-Infrastructures that the data are safe and embargoed until at least publication is not a given but must be earned. Unfortunately, trust is gained largely through experience with working with e-Infrastructures and the kind of VREs described here - an experience which, unfortunately, most researchers are unlikely to gain easily.

5 Mosquito Diseases Study

This science demonstrator illustrates how a LifeWatch researcher can easily upload and integrate an R-based algorithm in D4Science, making it available to other researches, in particular members of the VRE in which the algorithm was published. Once published, researchers can discover the algorithm and use it with their own data. It is also possible to adapt the algorithm and to share improved versions. When processing data-intensive analysis algorithms, the computation can be outsourced on federated resources, such as those provided by the EGI e-Infrastructures.

The scientific vision of this science demonstrator is to enable more efficient management of mosquito-borne diseases and nuisance mosquitoes. Mosquito-borne infections are among the most important new and emerging diseases globally and in Europe, and in order to predict diseases transmission areas, statistical correlation approaches are used.

LifeWatch RI provides advanced ICT, such as BioVel, supporting biodiversity research. However, it currently only provides standard algorithms for data processing. There is a need to support individual researchers' requests, e.g. import a new set of hydrological data layers into the analysis, add new algorithms that handle presence/absence into analysis etc., and a need for access to Cloud resources, e.g. to execute a large number of analytical cycles for many species under different climate scenarios.

These objectives should be achieved following the technical vision of supporting researchers in combining biological and hydrological data in a collaborative and evolving Virtual Research Environment (VRE) allowing intensive statistical computations: researchers should be able to easily share and use algorithms that they can adapt and use with their own data.

5.1 Architecture

The proposed service architecture is shown in Fig. 8. It combines different infrastructures: at a lower layer is the LifeWatch RI, containing the Swedish LifeWatch Portal that provides high-quality biological data for mosquito species, and the community data repositories that preserve environmental information and a series of ecological modelling algorithms. Datasets to be exploited include species data (95,730 abundance measurements from Sweden, Denmark, and Germany for 40 disease-carrying species in 2016), and hydrological data (generated by a regional hydrological model using 15 land-use types and 8 soil types).

At the middle layer is the EGI e-infrastructure, which provides cloud computation and storage resources supporting data-intensive workflow executions.

At the top layer is the D4Science VRE and the Biodiversity Virtual e-Laboratory (BioVel) portal, that provide high-level user interfaces. BioVel[6] is a software environment that assists scientists in collecting, organising, and sharing data processing and analysis

---

[6] https://www.biovel.eu/.

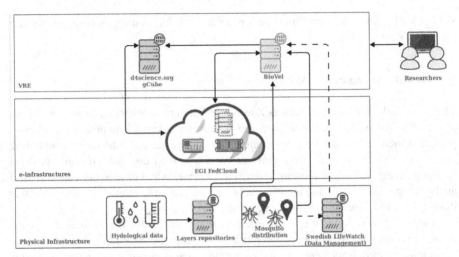

Fig. 8. Architecture includes three layers: 1) Physical Infrastructure, 2) e-Infrastructure, and 3) VRE.

tasks in biodiversity and ecological research. The service components of the platform include a Biodiversity Catalogue (a library with well-annotated data and analysis services), the data processing environments (such as RStudio for creating R programs), a workbench (for assembling data access and analysis pipelines), the myExperiment workflow library (that stores existing workflows), and the BioVel Portal (that allows researchers and collaborators to execute and share workflows).

The existing BioVel platform can generate environmental values from species occurrences, however, it only provides standard analysis algorithms. Integrating the D4Science and gCube -based VRE can enrich the functionality of the LifeWatch ICT to allow dynamic modelling.

5.2   User Interface

The D4Science/gCube-based VRE for mosquito disease study has been set up with the support from T7.1. The interfaces for the mosquito disease study are shown in Fig. 9, Fig. 10. It provides a programming environment (shown in Fig. 10), and it allows biodiversity researchers to develop and compile own/customised analysis algorithms using R, CLI, etc. A researcher can decide to share his/her data, algorithms, or workflows by publishing them in the group area (shown in Fig. 9) that enables social communications via messages, comments, etc.

5.3   Advantages

Using the VRE, there is no more need for manual sharing of data and algorithms. Information is always synchronised, and data and algorithms are joint in a single place. Users can enjoy an easy and user-friendly access interface. The D4Science/gCube-based VRE has an interface to EGI Cloud/HTC resources. If needed, it can outsource the

Fig. 9. VRE area for sharing data, algorithms, and workflows.

Fig. 10. VRE area for developing an analysis algorithm.

computation on the large-scale e-Infrastructure that can handle computation in parallel and store and share large volumes of data.

The integration service can bring added value to the Lifewatch community. It makes it possible for individual researchers to repeat and reuse algorithms at will, run trend analysis, and add new parameters and custom data. The VRE provides provenance registration that improves reproducibility. The VRE also allows retention of computation results in the user's workspace. This makes it possible to edit and adapt algorithms.

The integration service also brings added value to ENVRIplus community. Enabling individual researchers to share data and/or algorithms is common to many ENVRIplus RIs where currently data is processed using standard models. Researchers want to use different analysis models and they need a VRE to work together.

This pilot investigation tested and validated WP7 technology. The demo illustrates the integration solutions of linking gCube VRE to LifeWatch RI and to the EGI e-Infrastructure. There are also some lessons learned from the pilot activities: The D4Science/gCube VRE is easy for simple algorithms

. It needs integration efforts for complicated algorithms that request domain researchers to have technical skills to work with different techniques.

6 Conclusion

This chapter presented several diverse science demonstrators that were implemented by building on state-of-the-art D4Science e-Infrastructure to realise specific VREs. The demonstrators show that D4Science is capable of supporting the implementation of complex data processing and analysis pipelines and, more importantly, does so efficiently by ensuring the reuse of services and support extensions to VRE functionality with user-defined functions (scripts). The strong encapsulation of user-defined functions in D4Science (in contrast to, e.g. Jupyter notebooks) can at first be seen as an unwanted overhead but comes with advantages. First, the functions are automatically exposed as Web Processing Services and can be called also from third-party systems (e.g. Taverna workflows). Second, being a collaborative environment, D4Science ensures that collaborators do not inadvertently modify processing and thus potentially introduce errors. Moving individual researchers and entire research communities from their local computing environments into VREs is surely a monumental task in its own right. However, there are a lot of arguments for it, one being that infrastructures and communities of practice can ensure that research data are born FAIR instead of being FAIRyfied in a subsequent stage.

Acknowledgements. This work was supported by the European Union's Horizon 2020 research and innovation programme via the ENVRIplus project under grant agreement No. 654182.

References

1. Asmi, A., Kutsch, W.L.: ENVRI PLUS: European initiative towards technical and research cultural solutions for across-disciplines accessible Research Infrastructure products. AGU Fall Meeting Abstracts, IN31B-1764 (2015)
2. Assante, M., et al.: The gCube system: delivering virtual research environments as-a-service. Future Gener. Comput. Syst. **95**, 445–453 (2019). https://doi.org/10.1016/j.future.2018.10.035
3. Assante, M., et al.: Enacting open science by d4science. Future Gener. Comput. Syst. **101**, 555–563 (2019). https://doi.org/10.1016/j.future.2019.05.063
4. Aubinet, M., Vesala, T., Papale, D.: Eddy Covariance: A Practical Guide to Measurement and Data Analysis. Springer, heidelberg (2012). https://doi.org/10.1007/978-94-007-2351-1
5. Barker, M., et al.: The global impact of science gateways, virtual research environments and virtual laboratories. Future Gener. Comput. Syst. **95**, 240–248 (2019). https://doi.org/10.1016/j.future.2018.12.026
6. Bartling, S., Friesike, S. (eds.): Opening Science. Springer, Cham (2014). https://doi.org/10.1007/978-3-319-00026-8
7. Dada, L., et al.: Refined classification and characterization of atmospheric new-particle formation events using air ions. Atmos. Chem. Phys. **18**(24), 17883–17893 (2018). https://doi.org/10.5194/acp-18-17883-2018

8. Fratini, G., Mauder, M.: Towards a consistent eddy-covariance processing: an intercomparison of EddyPro and TK3. Atmos. Meas. Tech. **7**(7), 2273–2281 (2014). https://doi.org/10.5194/amt-7-2273-2014
9. Hellström, M., et al.: Near real time data processing In: ICOS RI. In Proceedings of 2nd International Workshop on Interoperable Infrastructures for Interdisciplinary Big Data Sciences (IT4RIs 16), Porto, Portugal, vol. 30, November 2016
10. Jones, S., Abramatic, J.-F. (eds.): European Open Science Cloud (EOSC) Strategic Implementation Plan. European Commission (2019). https://doi.org/10.2777/202370
11. Sabbatini, S.: Eddy covariance raw data processing for CO2 and energy fluxes calculation at ICOS ecosystem stations. Int. Agrophys. (2018). https://doi.org/10.1515/intag-2017-0043
12. Wilczak, J.M., Oncley, S.P., Stage, S.A.: Sonic anemometer tilt correction algorithms. Bound.-Layer Meteorol. **99**(1), 127–150 (2001). https://doi.org/10.1023/A:1018966204465

Case Study: LifeWatch Italy Phytoplankton VRE

Elena Stanca[1]([✉]) [iD], Nicola Fiore[2] [iD], Ilaria Rosati[3] [iD], Lucia Vaira[2] [iD],
Francesco Cozzoli[3] [iD], and Alberto Basset[1,2,3] [iD]

[1] Ecology-Unit, DiSteBA, University of Salento, Lecce, Italy
{elena.stanca,alberto.basset}@unisalento.it
[2] LifeWatch ERIC, Lecce, Italy
{nicola.fiore,lucia.vaira}@lifewatch.eu
[3] Research Institute on Terrestrial Ecosystems, National Research Council, Rome, Italy
ilaria.rosati@cnr.it, francesco.cozzoli@unisalento.it

Abstract. LifeWatch Italy, the Italian node of LifeWatch ERIC, has promoted and stimulated the debate on the use of semantics in biodiversity data management. Actually, biodiversity and ecosystems data are very heterogeneous and need to be better managed to improve the actual scientific knowledge extracted, as well as to address the urgent societal challenges concerning environmental issues. LifeWatch Italy has realized the Phytoplankton Virtual Research Environment (hereafter Phytoplankton VRE), a collaborative working environment supporting researchers to address basic and applied studies on phytoplankton ecology. The Phytoplankton VRE provides the IT infrastructure to enable researchers to obtain, share and analyse phytoplankton data at a level of resolution from individual cells to whole assemblages. A semantic approach has been used to address data harmonisation, integration and discovery: an interdisciplinary team has developed a thesaurus on phytoplankton functional traits and linked its concepts to other existing conceptual schemas related to the specific domain.

Keywords: Phytoplankton · Virtual Research Environment · Data management

1 Introduction

Phytoplankton plays an important role in aquatic ecosystems because it accounts for most global primary production and affects biogeochemical processes, trophic dynamics and biodiversity architecture. In order to understand ecosystem function and to improve predictions of aquatic ecosystem responses to environmental and climate change, it is strictly important that plankton physiologists and ecologists understand the phytoplankton structure.

In this chapter, we present the Phytoplankton Virtual Research Environment (Phytoplankton VRE), a collaborative working environment aimed at supporting researchers in addressing basic and applied studies on phytoplankton ecology. In particular, it allows

Z. Zhao and M. Hellström (Eds.): Towards Interoperable Research
Infrastructures for Environmental and Earth Sciences, LNCS 12003, pp. 324–341, 2020.
https://doi.org/10.1007/978-3-030-52829-4_18

researchers to analyse and share phytoplankton data at different resolutions: from individual cells to whole assemblages, data which are generally unharmonised and unavailable as online services. Moreover, it allows researchers to assess phytoplankton cell size, i.e. the biovolume, and other morphological traits.

The remainder of this chapter is organised as follows: Sect. 2 describes the architectural overview of the VRE, mainly based on a set of virtual machines that are accessible through a remote desktop connection. Section 3 is for the Phytoplankton case study and, in particular, it presents the important role played by Phytoplankton in aquatic environments, its main characteristics in terms of size, shape and other morphological traits, and the problem caused by the presence of several diverse methods to compute biovolume, surface area and other indices that do not allow researchers to compare data. The Phytoplankton VRE is then presented with all its tools and services: the atlas of taxonomy (Atlas of Phytoplankton), the atlas of morphological traits (Atlas of Shapes), the Phytoplankton Traits Thesaurus, and the Taverna workflow management system, that represents an orchestrator able to run all services composed in the workflow aimed at computing hidden dimension, biovolume, surface area, multi-metric indices of the ecological status, etc. The data lifecycle is also presented in order to illustrate all the stages involved in the management of data for their use and re-use (data acquisition, data curation, data access and data processing). Section 4 concludes the chapter.

2   The LifeWatch Italy Approach to VRE

The Italian community that works on biodiversity and ecosystem research topics is composed for the most part by a multitude of little research groups. Investigators, usually, work with the limited resources of their laboratory: they can count on a laptop, equipped with computational and storage capabilities. In conceiving the architecture of the Life-Watch Italy VRE, we took into account that there was a resistance to changing the proper way investigators are expected to work. In order to reduce the resistance to change, we tried to maintain how investigators already work while supplying them more innovative services and unique storage and computational powers. The result is an architecture that supplies to researchers a set of Virtual Machines that are accessible through a Remote Desktop Connection very similar to the environment they normally use, but able to ensure very good performance in terms of computational capabilities and storage space, with no need to install additional tools on their workspace. Figure 1 shows an overview of the architecture. The requested Virtual Machine can be equipped in different ways, in line with the researchers' goals. We give the possibility to set up the environment with an open source or Microsoft Operating System (i.e. Ubuntu or Windows Server), with all the tools needed for the data collection, curation and analysis. There is a Broker machine that is dedicated to managing all researchers' connections and assigning dynamically computational and storage resources to users.

Each user has a dedicated account; a common Authentication and Authorization Infrastructure (AAI) based on Windows Active Directory is used to control the access and to assign different authorizations and rights on the machines. This allows users to have a personal desktop/area on each machine where they can organize their work in term of personal folders where they could store documents and files coming from the analysis

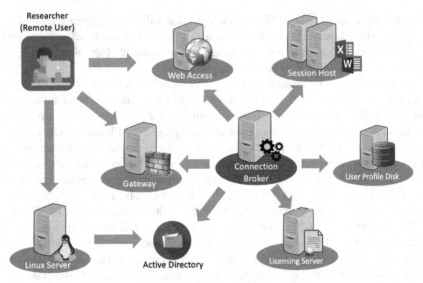

Fig. 1. Overall architecture.

done in the environment or shared folders if they need to share the work with other members in the LifeWatch network. The computational capacity of the environment is supplied by the distributed national data centres, but this is transparent to users that can just start their analysis without being worried about connection problems or being connected during analysis time. All analysis processes are run in the background and when they come back in the environment, they will find the results stored in the chosen folders.

An example of the Virtual Machine and User Desktop is introduced and described referring to the Study Case Study in the following sections.

3   The Phytoplankton Case Study

3.1   Overview

Phytoplankton is the primary autotrophic component in aquatic ecosystems, responsible for almost half of global net primary production [1]. On the one hand, it plays an important role in carbon sequestration and on the other hand, oxygen production. For this reason, this photosynthetic organism plays a key role in aquatic environments, forming the base of the food web and having a substantial function in nutrient dynamics and in the carbon biogeochemical cycle [2–4]. Therefore, considering the total phytoplankton structure, the community of phytoplankton has profound effects on higher trophic levels and key biogeochemical processes [5]. For these reasons, understanding both the role of phytoplankton features, traits and the abiotic and biotic drivers that determine phytoplankton distribution and its succession patterns is fundamental. The most important features of the phytoplankton community, organism size and elemental composition, will influence processes at the level of individuals, populations, communities and ecosystems [6, 7].

Size, shape, morphology and specific traits of these organisms provide relevant proxies for the ability to survive and coexist in response to abiotic and biotic drivers [8]. Regarding size, phytoplankton is an extremely diverse group of organisms that range over nine orders of magnitude in cell size volume and shapes [9, 10]. They show a huge scale of size from 1–2 μm in equivalent spherical diameter for the picoplankton, 2–20 μm for the nanoplankton, 20–200 μm for the microplankton, and up to 200–2000 μm for macroplankton [11, 12]. Every single phytoplankton organism is characterized also by a specific geometric shape. Currently, a well-defined number of shapes include simple and combined shapes (see Fig. 2) [13–15] that represent another morphological trait, very useful to describe and characterize phytoplankton community [16].

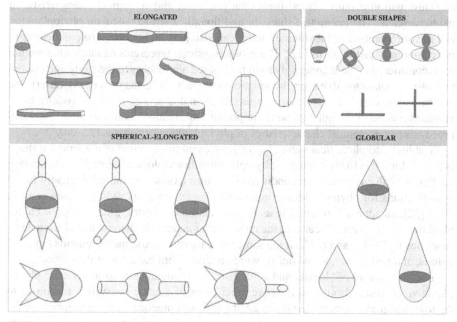

Fig. 2. The Atlas of shapes. (The Atlas of shapes - Phytoplankton Bio-Imaging by the Ecology Unit of the University of Salento http://phytobioimaging.unisalento.it/Products/AtlasOfShapes. aspx?ID_Tipo=0)

Cell size is often referred to as a master trait because body size influences the physiology, ecology, and evolution of species [17]. Phytoplankton cell size varies over three orders of magnitude [10] and is mechanistically linked to all the physiological and ecological traits: maximum growth rate, nutrient acquisition, minimum and maximum cell quota, light absorption and susceptibility to high light stress, sinking rate and susceptibility to grazing and viral attack [10]. The size, structure and elemental composition of the phytoplankton community has a cascading influence on the proportion of organic material transferred to the microbial loop, higher trophic levels or exported into the deep sea. Phytoplankton size spectra and size classes have been shown to have high information content to detect environmental condition change in transitional and coastal waters [18–20].

It is therefore essential to investigate the development of phytoplankton populations in order to understand the biological functioning of aquatic systems and detect changes in them [21].

The physiological responses induced by different cell sizes and surface areas could provide valuable information about phytoplankton distributions as well as how distributions might be altered by environmental change. Cell size and shape may be the primary drivers of variations in physiological responses and therefore provide community assemblages with the flexibility to respond from macro spatial scale to local environmental conditions [8].

Regarding the other most important morphological trait, the shape, it provides important information about essential functional processes and ecological characteristic of phytoplankton. The geometric shape is traditionally used to calculate phytoplankton cell measurements (e.g. biovolume), but it can also play an important role in determining community distributions [22]. The geometric shape represents an interesting feature to be considered in the increasingly used trait-based approaches to the study and prediction of phytoplankton dynamics in aquatic ecosystems. The shape is easily observable and measurable and its application in the functional approach does not require taxonomic affiliation, although it provides important information about essential functional processes and ecological characteristics of phytoplankton organisms [22–24].

The high morphological variability of phytoplankton in terms of a geometric shape is not random. It is likely related to phytoplankton morphological adaptations to achieve the best fit with environmental conditions [22]. Based on shape, morphologically-based classifications for phytoplankton communities have been proposed by Stanca et al. in 2013 [22], an approach referred to as Phytoplankton Geometric Shapes (PGS). Phytoplankton species were allocated to the most similar geometric shape selected from those described by [9, 13] and [14]. At the same time, morphometric measurements (surface, volume and surface to volume ratios) were obtained from basic linear dimensions.

For plankton physiologists and ecologists, it is fundamental to understand phytoplankton structure in order to improve predictions of aquatic ecosystem responses to environmental and climate change. In addition, from a practical point of view and in the context of conservation, protection and management of aquatic resources, the assessment of phytoplankton community structure is essential to understand ecosystem function [25].

Due to their short life cycle, planktonic algae respond quickly to environmental changes. Therefore, phytoplankton is considered a useful biological quality element for water quality monitoring assessment, according to the European Water Framework Directive (WFD). Phytoplankton parameters to be used for this assessment are biomass, community composition and abundance, as well as frequency and intensity of blooms. This quality element responds mainly to pressures generated by nutrient and organic enrichment and alteration of the water body's hydrological and morphological characteristics, to environmental forcing and to human-generated pressures [26].

Even though demographic traits (e.g. presence/absence, abundance and biomass) have been traditionally included in directives and monitoring programmes, morphological traits are attracting growing interest to be implemented as a descriptor of the ecological status of aquatic ecosystems [27–29]. Direct counts and measurements of algal size, in terms of biovolume, are potentially a more accurate measure of phytoplankton biomass and abundance [30]. Assessing phytoplankton cell size, i.e. the biovolume, has therefore been approached with different procedures and methodologies, each of which has aspects that need consideration and improvement.

The variety of applied methods, from sampling to counting, as well as the mode of calculation, unfortunately leads to general poor comparability of the data, which currently represents a huge problem. Indeed, to be shared and comparable, data have to answer to several data quality criteria. For this reason, they have to be sufficiently precise, accurate, representative and complete. Standardising protocols for validating and reporting data improves the comparability of data and the confidence with which one data set can be compared to another, either overtime or between research groups [31].

High-quality data is a key element for research and impacts the replicability of results. Quality checks should be performed during collection, data entry and analysis [32]. There have been many individual phytoplankton datasets collected across the world, but most of them are unavailable to the research community. The cornerstone action is to bring together data and information in a way that enables researchers to produce knowledge that yields novel insights or explanations, establishes correlations and identifies patterns [33]. Given the scale and urgency of the societal challenges related to the environment and given that data are being generated at an ever-increasing rate, better-coordinated efforts are required to enable structuring, aggregating, linking and processing of such data in a meaningful way [34]. Since quality assurance of data is an important component of the monitoring programme, the use of a standardised nomenclature list and a standardised computational model are decisive in improving the quality of the phytoplankton data and the comparability of results, at different spatial and temporal scales.

3.1.1 Comparability of Data: Taxonomy

New phytoplankton organisms are continuously being described, and changes in the naming and categorisation of organisms is common. Changes should be based on internationally accepted rules, which have been established in nomenclatural codes. It is essential to keep standardised lists, which are updated in a systematic way. Due to the inherent complexity of taxonomic, nomenclatural and systematic concepts, the quality and resolution of data are necessarily required [34]. For this reason and to optimize the management and integration of primary biodiversity data, it is necessary to develop a consistent vocabulary, semantic rules and ontologies; contribute to the harmonisation of terminology and practices; provide a synthetic guide for taxonomists and non-taxonomists involved in biomonitoring and biodiversity studies [34]. Moreover, when data from different sources, geographical areas and points in time are integrated into taxonomic inventories and databases or time series, they need a very careful critical revision, with the aim of internal consistency and quality evaluation [35] ensuring data interoperability and automated processing.

In order to do that, the "LifeWatch Taxonomic Backbone" service, a central part of the European LifeWatch Infrastructure set-up by the Flanders Marine Institute, can be exploited since it aims to (virtually) bring together different component databases and data systems, all of them related to taxonomy, biogeography, ecology, genetics and literature. By doing so, it standardises species data and integrates biodiversity data from different repositories.

3.1.2   Comparability of Data: Morphological Traits

Terminological ambiguity slows down scientific progress, leads to redundant research efforts, and ultimately impedes advances towards a unified foundation for ecological science [36, 37]. An important step of improvement of the phytoplankton analysis is the development of standard calculation procedures. There exists no unique procedure applied worldwide for all steps of phytoplankton morphological traits computation, no common set of protocols from linear dimensions measurement to biovolume calculation, which would allow inter-comparisons of data. Many countries or institutes have used their own methods for decades and may be reluctant to make changes [35]. Different measures and methodologies are in use to quantify cell size and they require unequivocal definition to ensure standardisation and comparability of measurements [37].

With the aim to solve the ambiguity issues of natural language by formalizing the construction of the terms themselves, their definitions and their inter-relationships, and in order to provide a standard set of structures that enable computers to more precisely assist data users in locating (data discovery) and processing the data of interest, we developed a specific thesaurus: the PhytoTraits thesaurus [37]. This thesaurus contains 120 terms hierarchically organized and focusing on morphofunctional traits, such as linear dimensions and shapes, which are univocally defined.

This controlled vocabulary provides a standard terminology for traits, that is essential for data integration and increasingly required in ecology. PhytoTraits is freely available[1] and can be used for different purposes.

3.1.3   Comparability of Data: Computation Processes

Biovolume estimates and conversion factors required by indirect methods increase opportunities for error because the error associated with multiple independent factors can be propagated at each stage of calculation [38, 39].

The biovolume of phytoplankton must be assessed accurately in order to identify the ecological status of water bodies in line with the WFD requirements. There are several ways to calculate cell volumes. The 'gold standard' is to determine the geometrical shape that approximates the shape of the cell and then make measurements of the dimensions to enter into the formula for that particular geometrical shape [40–43]. Some of the challenges in this approach are that different investigators may choose a different geometric shape than the recommended shape [9, 14] for the same species, especially for cells with a complex shape. In addition, the 'hidden dimension' (i.e. the depth dimension) is difficult to measure since cells are viewed in two dimensions under the microscope [44].

---

[1] PhytoTraits: http://thesauri.lifewatchitaly.eu/PhytoTraits/index.php.

Evaluating the most exact cell biovolume should help to avoid errors such as an overestimation or underestimation of phytoplankton biomass/biovolume. A properly estimated biovolume based on verified and agreed geometric shapes should lead to an accurate and comparable assessment of a phytoplankton-based ecological status [45].

To facilitate and accelerate the estimations of phytoplankton biovolume, which has become very important in the WFD-required ecological status assessment of water bodies, we revised and rearranged basic geometric shapes. Moreover, we verified and improved the precision as well as the accuracy of different formulas. Since only up to two dimensions can be visualised under the microscope, at least one dimension has to be derived from one of the others or a fixed value has to be determined from a number of specimens in a special effort. We calculated and provided conversion factors, hidden dimension factors, which are species-specific, in order to obtain dimensions that are difficult to measure, but needed for biovolume calculation. In this way, we provided a more accurate biovolume calculation, at a specific taxonomic level. We provided a set of 51 geometric models, including formulas for biovolume assessment and cell linear dimensions evaluation. There are two typology groups: Simple shape and Complex shape, with 23 and 28 shapes respectively. The models are provided in a specific Atlas, but also in a specific workflow developed for biovolume computation.

Having the opportunity to be more accurate and doing massive computation analysis allows for the reduction of mistakes and errors due to manual procedures and operator, permits the saving of time and should contribute to having fast answers in evaluating the ecological status of water bodies and providing more accurate results in line with the WFD requirements.

3.2 The Phytoplankton Virtual Research Environment

The e-Biodiversity Research Institute of LifeWatch Italy (hereafter LW ITA) has realized the Phytoplankton Virtual Research Environment (hereafter Phytoplankton VRE), a collaborative working environment supporting researchers to address basic and applied studies on phytoplankton ecology. The Phytoplankton VRE provides the IT infrastructure to enable researchers to obtain, share and analyse phytoplankton data at a level of resolution from individual cells to whole assemblages. The Phytoplankton VRE allows researchers to:

1. Obtain and share harmonised data on taxonomy and morphological traits by using the Atlas of Phytoplankton, the Atlas of Shapes and the Phytoplankton Traits Thesaurus.
2. Discover, access, integrate and export both own and other datasets (including additional metadata) held by LifeWatch Data Portal or distributed data centres.
3. Share and create workflows by means of orchestrators like Taverna Workbench[2] by using algorithms and web services.
4. Work together in a real-time environment that fosters the sharing of knowledge overcoming the limitations of traditional working practices e.g. the transfer of large datasets among users or the need for significant computational power for the analysis.

---

[2] Taverna Workbench: www.taverna.org.uk.

3.2.1 Harmonised Data on Taxonomy and Morphological Traits

The Phytoplankton VRE provides a number of features for harmonising phytoplankton taxonomic data and morphological trait data:

The Atlas of Phytoplankton: this provides a reference point for marine, transitional and freshwater scientists and students involved in phytoplankton identification and classification. It includes illustrative cards with information about i) taxonomy, with pictures and schematic drawings, information on similar species and/or synonyms, references; ii) ecological characteristics and geographical distribution of species; and iii) morphological features, such as shape association, linear dimensions association and formulae for cell volume and surface computation.

The Atlas of Shapes: this represents a reference point for marine, transitional and freshwater scientists and students involved in phytoplankton morphological traits association and measurement and provides a schematic protocol for calculating biovolume of phytoplankton species detectable with the Utermöhl method [46] in transitional ecosystems of the different world ecoregions. The Atlas includes the illustrative scheme of the shape classification subdivided in "Simple Shapes" and "Complex Shapes" (Fig. 2). Clicking on a specific shape, users are able to see: the biovolume (V) and surface area (A) computational models; and the shape views (e.g., lateral, frontal, etc.) with the corresponding linear dimensions (e.g. length indicated by alphabetical code "a", "l", etc.; width indicated by alphabetical code "b", "d", etc.). For each specific shape group there is the frontal view for the shape and the biovolume and area computational models. Clicking again on a specific shape, user is redirected to all taxonomic cards characterised by the selected shape that are on the Atlas of Phytoplankton. Both atlases are integrated and can be easily browsed, switching from taxonomic identification to morphological characterisation of phytoplankton.

The Phytoplankton Traits Thesaurus (PhytoTraits): this thesaurus reflects the agreement of a scientific expert community regarding the definition of semantic properties of approximately 120 traits [37]. Following Semantic Web standard technologies, the thesaurus has been implemented in Simple Knowledge Organisation System (SKOS), a common data model based on the Resource Description Framework (RDF). The PhytoTraits is freely available online[3], it can be queried through a SPARQL endpoint[4] and is also accessible via API[5] for integration with other systems. If adopted as a standard, and hence rigorously applied and enriched by the scientific community, PhytoTraits has the potential to significantly reduce the barriers to data discovery, integration, and exchange since it provides harmonised concepts with associated unique and resolvable URIs.

3.2.2 Data Access, Discovery, Integration and Download

A user who is registered at the LifeWatch Data Portal can access their own section titled "My Datasets" that lists all types of datasets in which he/she is involved (e.g. enabled,

---

[3] PythoTraits: http://thesauri.lifewatchitaly.eu/PhytoTraits/index.php.

[4] SPARQL endpoint: http://thesauri.lifewatchitaly.eu/PhytoTraits/sparql.php.

[5] PythoTraits API: http://thesauri.lifewatch.eu/PhytoTraits/services.php.

pending, refused, disabled or owned by users). For a specific dataset, the user can perform two main actions: download the dataset (RDF or CSV format) and/or visualize it in a separate window in JSON format.

Data Search Interface: this interface allows researchers to search species according to several dimensions of analysis. The searching criteria are: the geographic area (by drawing a polygon or a circle on the map); the biogeographic regions; the country; the ecosystem type; the habitat type; the organism group; and the scientific name. The resulting datasets are characterised by a title and an author. They are enriched with metadata and the user has to request authorization in order to access them. An advanced data search can be performed after that the administrator gives access to the dataset.

3.2.3   Sharing and Creating Workflows

The Phytoplankton VRE allows researchers to use Taverna, a workbench for the design and execution of scientific workflows. This tool enables the interoperation among databases and tools by providing a toolkit for composing, executing and managing workflow experiments. Taverna Workbench Biodiversity is an edition of Taverna that includes support for building and executing scientific workflows targeting biodiversity services. Taverna workflows show intermediate results of the execution, are easy to use for inexperienced users, and very flexible for the skilled ones. Taverna Workbench Biodiversity allows the use of a set of local and remote services to analyse and manage data, create nested workflows and use automatic iteration.

In order to facilitate the computation of phytoplankton traits and to investigate their distribution patterns, we developed a workflow, which allows automating a set of operations that were originally written in the R language[6]. Two R scripts have been developed and are incorporated in the PhytoTraitsComputationAndDistribution workflow:

- the *Phytoplankton Traits Computation*, which computes morphological and demographic traits, such as hidden dimension, biovolume, surface area, surface-volume ratio, cell carbon content, density, carbon content and total biovolume;
- the *Phytoplankton Size Distributions*, which performs Modality (Hartigans' dip test), Normality or LogNormality (Anderson-Darling test, Cramer-von Mises) tests of phytoplankton biovolume (expressed as μm_3) or cell carbon content (expressed as $pgC*cell_{-1}$) distributions, at different levels of data aggregation (i.e. spatial, temporal, taxonomic).

The workflow is represented in Fig. 3. By default, "input ports" are shown on the top, and "output ports" are shown on the bottom. Boxes represent processing nodes, and the solid directed arrows between them are data connections. User has to specify for each input port:

1. *CompTraits*: the input port for entering traits to be computed. Users shall select the "add value" button and enter one or more of these options in the box to compute: Biovolume; Surface Area (typing SA); Surface/Volume ratio (typing SV); Cells/Liter

---

[6] R website: https://www.r-project.org.

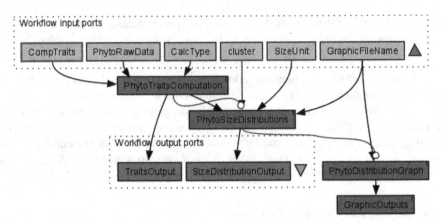

Fig. 3. PhytoTraitsComputationAndDistribution workflow.

(typing CL); Biovolume/Liter (typing BVL); Carbon content (typing CC); Carbon content/Liter (typing CCL).

2. *PhytoRawData*: the input port for entering raw data in CSV format. The workflow runs only with data resources structured according to the LifeWatch Italy Data Schema.

3. *CalcType*: the input port for entering the computation type. Users can choose between two computation modalities:

   a. "Simplified" mode approximates the taxa specific-biovolume calculation/computation based on two linear dimensions only, length and width. The mandatory fields for this calculation/computation type are scientific name, measurement remarks (e.g. vision of the organism, dimension more or less than 20 μm), length and width;

   b. "Advanced" mode allows a more accurate estimate of taxon-specific biovolume, but it requires more information.

For each shape, at least two measured basic linear dimensions need to be provided by the user. The mandatory fields for this calculation/computation type are scientific name, measurement remarks and linear dimensions. The latter must be measured according to the Phytobioimaging Atlas of Shapes.

4. *Cluster*: the input port for entering the level of aggregation for size distributions. The aggregation could be done at spatial (e.g. eventid, paraeventid, locality, country and Eunis habitat type-name), and/or temporal level (e.g. day, month and year), and/or using taxonomic categories (e.g. Phylum, Order and scientific name). Size distributions will be aggregated according to a unique combination of the provided criteria (e.g. using three countries, four eventids for country and three dates for eventid as aggregation criteria, users will aggregate data in a single bin for each combination of country, eventid, and collection date, resulting in $3 \times 4 \times 3 = 36$ bins).

5. *SizeUnit*: the input port for entering the morphological trait that will be used to perform Modality, Normality or LogNormality tests of distributions.
6. *GraphicFileName*: the input port for entering the name that will be used to create a PDF distribution file that will be visible on a web page once the workflow is completed.

Once the user has inserted the input values, the input dialogue window will close and users will be directed to the "results" screen, where it is possible to monitor the workflow execution progress in real-time. The first iteration of "PhytoTraitsComputation" will produce as output the dataset "TraitsOutput" in CSV format that contains all input data and computed traits, while the second iteration "PhytoSizeDistribution" will produce another CSV file "SizeDistributionOutput" reporting a summary of the distribution tests and "PhytoDistributionGraph". At the end of the workflow process, users will automatically obtain also the distribution calculation graph.

3.3 Data Lifecycle

The data lifecycle illustrates the stages involved in the management of data for their use and re-use. There exists a wide range of data lifecycle models, each with a different focus or perspective. Starting from the DataONE Data Life Cycle framework [47] we customised the cycle according to our needs as represented in Fig. 4.

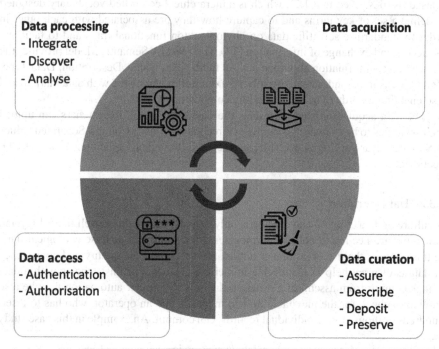

Fig. 4. Data lifecycle.

3.3.1 Data Acquisition

Data from the observations and experiments of individual investigators usually occur in heterogeneous formats and terminology (the same variable is reported with different terms or abbreviations) and are stored in flat files or spreadsheets with minimal formal structure and few or even without metadata information.

Data compilation is, therefore, an important and necessary step of the procedure for morphological and demographic traits calculation. We compiled a data template in which ancillary information (e.g. sampling site, longitude and latitude) and phytoplankton features (e.g. linear dimensions and surface area) are well-structured in a Microsoft Excel file. This procedure is important in order to import the file in the Phytoplankton Bio-Imaging System, to allow the traits calculation and to make the data interoperable.

The Excel file is structured in different fields related to a semantic model proposed by LW ITA (LifeWatch Data Management System[7]), which clearly defines semantics that can be understood by researchers or interpreted by machines making it possible to determine appropriate uses of the data encoded therein.

According to this model, the ancillary data inherits concepts from the Darwin Core standard[8] and from the EnvThes vocabulary[9]. The Darwin Core standard includes a glossary of terms, which aims to create a common language for sharing biodiversity primary data and related information, while EnvThes consists of lists of standardised terms for the description of data and information within geological, ecological and hydrological sciences. Data regarding phytoplankton traits are related to the Phytoplankton Traits Thesaurus (described in 3.1.2), which is a hierarchical controlled vocabulary designed to define a set of key terms and to capture how they are associated with each other in order to standardise scientific data on phytoplankton functional traits and to facilitate the access and exchange of information [37]. The LW ITA Semantic Model provides the relevant meta-information about the dataset fields (e.g., Name, Description, Data Type, Unit of Measure, Standard etc.) by solving ambiguities associated with data markup and also enabling records to be interpreted by computers.

The data acquisition step represents the data entry stage. Researchers can upload their own files to be shared, that can be in three main forms: a Comma-Separated values (CSV) file, a Darwin Core (DwC) file, an Access to Biological Collections Data (ABCD) document.

3.3.2 Data Curation

A culture of data curation and sharing is only recently establishing itself in ecology and new tools are needed to collect, harmonise, store, share and analyse ecological data. In this context, the use of computer automation to control the quality and consistency of data is of great help in identifying numerical or lexical inconsistencies within data strings coming from assembling different datasets. Computer automation may be also applied to check for the inevitable human mistakes that an operator, who has to insert hundreds or thousands of individual records, can commit. An example in this case study

---

[7] LifeWatch Data Management System: http://www.servicecentrelifewatch.eu/home.

[8] Darwin Core standard: http://rs.tdwg.org/dwc.

[9] EnvThes vocabulary: http://vocabs.ceh.ac.uk/evn/tbl/envthes.evn.

is a short code used to check for disproportionate numerical values that may potentially arise from typos or inconsistency of units of measurements.

The collection, correction and harmonisation of data is only a first step towards understanding the investigated ecosystem or biological processes investigated. The data curation stage is to guarantee accuracy and to assess the quality and includes all steps required to clean and validate data uploaded by researchers that have to comply with FAIR (*Findability, Accessibility, Interoperability, and Reusability*) principle [48] for sharing purposes. This stage includes 4 sub-stages:

- Data assurance: the quality of data is assured by means of checks, inspections and validation procedures that allow to detect format errors, nomenclatural errors, numeric warnings, taxonomic warnings, and semantic warnings. Data assurance is performed by means of different automatic or semi-automatic tools available on the LifeWatch Data Portal.
- Data description: data are accurately described by using an appropriate metadata standard to ensure understanding and long-term control. The LW Data Model gives the relevant meta information about the dataset fields (e.g., Name, Description, Data Type, Unit of measure, etc.).
- Data deposit: data are published and hence made available to other researchers by providing adequate provenance (allowing to achieve authoring).
- Data preservation: data are stored and preserved to be available in real-time for usage. This step allows also to manage and administer all curation lifecycle actions.

3.3.3 Data Access

Access control rules and authentication procedures are applied to ensure that only allowed users can access and use data. The technology used is Microsoft Active Directory.

- Authentication: to access LW IT VRE, users need to have a LW VRE account. This account also gives access to the service catalogue and data resources.
- Authorisation: first access to any Virtual Lab needs to be authorised by the LW administrator. Users will be "pending" until the administrator authorises their access. After that, users' access will be enabled.

3.3.4 Data Processing

Ecologists collectively produce (and have historically produced) large volumes of data through diverse individual projects. Furthermore, the recent developments in Information and Communication Technologies give to ecologists the possibility to access two new types of data: new information created from new technology applications (e.g. remote sensing observations and advanced microscopy) and existing information that was previously unavailable (e.g. existent data that were not publicly available or simply not previously uploaded to an online source). It is difficult and often even impossible to characterise the functioning of a complex system by means of direct measurements. The size of the system and the complexity of the involved interactions often make necessary the use of descriptors able to summarise the collected information. In the case of large datasets, this need has to be extended also to the tools necessary to analyse data and produce summary indicators. Biomass, composition, abundance and size spectra of the phytoplankton community, as well as frequency and intensity of phytoplankton blooms, have been considered as fundamental summary descriptors to be included in the assessment of the aquatic ecological status. The data processing stage includes 3 sub-stages:

- *Data integration*: data coming from heterogeneous data sources are combined in order to form homogeneous sets of data that can be easily and readily analysed.
- *Data discovery*: data are provided to interested users for knowledge discovery purposes which are enriched with relevant and structured information (metadata).
- *Data analysis*: data are explored and analysed by researchers according to the needs to create derived results useful for research, teaching and learning purposes.

Within the phytoplankton case study, we provided computational tools to calculate in a fast and automated way and at any chosen level of spatial and temporal aggregation:

- The biovolume and biomass of any recorded individual cell starting from its linear dimensions measured at the microscope; considering the high variety of geometrical shapes that phytoplankton cells have, this tool is associated with a species-specific inventory of shapes and mathematical formulation needed for the calculation of the biovolume.
- Different biodiversity indices, including taxonomic indices of richness, diversity, evenness and dominance and indices based on the size spectra of the phytoplanktonic community.
- Inferences on the distribution of body mass across phytoplankton individuals (normality, log-normality, bimodality tests).

4   Conclusion

In this chapter we demonstrated how the LifeWatch Italy experience can be exploited by a group of researchers to address basic and applied studies on phytoplankton ecology. The Phyto VRE is able to reduce the chance of error and to optimise the whole process, the analysis and the processing and computational time. One of the challenges in computing the biovolume of a phytoplankton organism is represented by the fact that different investigators may choose a different geometric shape with respect to the recommended one for the same species, especially in the case of cells having a complex shape. The proposed Atlas of Shapes allows to have a reference point for marine, transitional and freshwater scientists and students interested in phytoplankton biodiversity and ecology. Moreover, the added value of the proposed approach is represented by the fact that it can be reproduced and exploited also for other studies (e.g. for alien species).

   In conceiving the architecture of the proposed VRE, we considered the typical "resistance to change" of most researchers and we tried to maintain their way of working providing them innovative services and unique storage and computational powers. The result is an architecture that supplies to researchers a set of virtual machines that are accessible through a remote desktop connection that is very similar to the environment they normally use at work, but able to ensure very good performances in terms of computational capabilities and storage space. This is the main advantage of the proposed VRE with respect for convenience aspects, but it also represents the main limit of the proposed approach. As future work, we plan to provide web-based access to the VRE and hence to design and develop user-friendly interfaces able to answer to different users' needs and expertise.

Acknowledgements. This work was supported by the European Union's Horizon 2020 research and innovation programme via the ENVRIplus project under grant agreement No 654182. The work was also funded by LifeWatch Italy, the Italian node of the escience European infrastructure for biodiversity and ecosystem research and also supported by "POR PUGLIA Progetto Strategico 2009–2012" (grant agreement CIP.PS-126).

References

1. Field, C.B., Behrenfeld, M.J., Randerson, J.T., Falkowski, P.: Primary production of the biosphere: integrating terrestrial and oceanic components. Science **281**(5374), 237–240 (1998)
2. Graham, L.E., Wilcox, L.W.: Algae. PrenticeHall, Upper Saddle River, New Jersey (2000)
3. Sarmiento, J.L., Gruber, N.: Ocean Biogeochemical Dynamics. Princeton University Press, Princeton (2006)
4. Almandoz, G.O., Hernando, M.P., Ferreyra, G.A., Schloss, I.R., Ferrario, M.E.: Seasonal phytoplankton dynamics in extreme southern South America (Beagle Channel, Argentina). J. Sea Res. **66**, 47–57 (2011)
5. Litchman, E., Klausmeier, C.A., Schofield, O.M., Falkowski, P.G.: The role of functional traits and trade-offs in structuring phytoplankton communities: scaling from cellular to ecosystem level. Ecol. Lett. **10**, 1170–1181 (2007)
6. Sterner, R.W., Elser, J.J.: Ecological Stoichiometry: The Biology of the Elements from Molecules to the Biosphere. Princeton University Press, Princeton (2002)
7. Hessen, D.O., Elser, J.J.: Elements of ecology and evolution. Oikos **109**, 3–5 (2005)
8. Bestová, H., Munoz, F., Svoboda, P., Škaloud, P., Violle, C.: Ecological and biogeographical drivers of freshwater green algae biodiversity: from local communities to large - scale species pools of desmids. Oecologia **186**, 1017–1030 (2018)
9. Hillebrand, H., Durselen, C.D., Kirschtel, D., Pollingher, U., Zohary, T.: Biovolume calculation for pelagic and benthic microalgae. J. Phycol. **35**, 403–424 (1999)
10. Finkel, Z.V., Beardall, J., Flynn, K.J., Quigg, A., Rees, T.A.V., Raven, J.A.: Phytoplankton in a changing world: cell size and elemental stoichiometry. J. Plankton Res. **32**, 119–137 (2010)
11. Sieburth, J.M., Smetacek, V., Lenz, J.: Pelagic ecosystem structure: heterotrophic compartments of the plankton and their relationship to plankton size fractions. Limnol. Oceanogr. **23**, 1256–1263 (1978)
12. Beardall, J., et al.: Allometry and stoichiometry of unicellular, colonial and multicellular phytoplankton. New Phytol. **181**(2), 295–309 (2009)
13. Vadrucci, M.R., Cabrini, M., Basset, A.: Biovolume determination of phytoplankton guilds in transitional water ecosystems of Mediterranean Ecoregion. Transit. Water. Bull. **2**, 83–102 (2007)

14. Sun, J., Liu, D.Y.: Geometric models for calculating cell biovolume and surface area for phytoplankton. J. Plankton Res. **25**, 1331–1346 (2003)
15. Atlas of Shapes - Phytoplankton Bio-Imaging by the Ecology Unit of the University of Salento. http://phytobioimaging.unisalento.it/en-us/products/AtlasOfShapes.aspx. Accessed 17 Dec 2019
16. Salmaso, N., Naselli-Flores, L., Padisak, J.: Functional classifications and their application in phytoplankton ecology. Freshw. Biol. **60**, 603–619 (2015)
17. Finkel, Z.V., et al.: A universal driver of macroevolutionary change in the size of Marine phytoplankton over the Cenozoic. PNAS **104**, 20416–20420 (2007)
18. Sabetta, L., Basset, A., Spezie, G.: Marine phytoplankton size–frequency distributions: spatial patterns and decoding mechanisms. Estuar. Coast. Shelf S. **80**, 181–192 (2008)
19. Lugoli, F., et al.: Application of a new multi-metric phytoplankton index to the assessment of ecological status in marine and transitional waters. Ecol. Indic. **23**, 338–355 (2012)
20. Vadrucci, M.R., et al.: Ability of phytoplankton trait sensitivity to highlight anthropogenic pressures in Mediterranean lagoons: a size spectra sensitivity index (ISS-phyto). Ecol. Indic. **34**, 113–125 (2013)
21. Hötzel, G., Croome, R.: A phytoplankton methods manual for Australian Freshwaters. LWR-RDC Occasional Paper 22/99. Land and Water Resources Research Development Corporation, Canberra, Australia (1999)
22. Stanca, E., Cellamare, M., Basset, A.: Geometric shape as a trait to study phytoplankton distributions in aquatic ecosystems. Hydrobiologia **701**, 99–116 (2013)
23. Naselli-Flores, L., Padisák, J., Albay, M.: Shape and size in phytoplankton ecology: do they matter? Hydrobiologia **578**, 157–161 (2007)
24. Salmaso, N., Padisák, J.: Morpho-functional groups and phytoplankton development in two deep lakes (Lake Garda, Italy and Lake Stechlin, Germany). Hydrobiologia **578**, 97–112 (2007)
25. Choudhury, A.K., Bhadury, P.: Phytoplankton study from the Sundarbans ecoregion with an emphasis on cell biovolume estimates–a review. Indian J. Mar. Sci. **43**(10), 1905–1913 (2014)
26. Varkitzi, I., et al.: Pelagic habitats in the Mediterranean Sea: a review of Good Environmental Status (GES) determination for plankton components and identification of gaps and priority needs to improve coherence for the MSFD implementation. Ecol. Indic. **95**, 203–218 (2018)
27. Olenina, I., et al.: Biovolumes and size-classes of phytoplankton in the Baltic Sea HELCOM Balt. Sea Environ. Proc. n. 106 (2006)
28. OSPAR: OSPAR Integrated Report on the Eutrophication Status of the OSPAR Maritime Area Based Upon the First Application of the Comprehensive Procedure. Eutrophication Series. OSPAR Commission (2003)
29. HELCOM: Development of tools for assessment of eutrophication in the Baltic Sea. Baltic Sea Environmental Proceedings No. 104. Helsinki Commission (2006)
30. Carvalho, L., et al.: Strength and uncertainty of phytoplankton metrics for assessing eutrophication impacts in lakes. Hydrobiologia **704**, 127–140 (2013)
31. King County: Marine Phytoplankton Monitoring Program Sampling and Analysis Plan. Prepared by A. Kolb, G. Hannach, L. Swanson, Water and Land Resources Division. Seattle, Washington (2016)
32. Sarmiento Soler, A., Ort, M., Steckel, J.: An Introduction to Data Management Reader_GFBio_BefMate_20160222, BEFmate, GFBio Project (2016)
33. Koureas, D., et al.: Unifying European biodiversity informatics. Res. Ideas Outcomes **2**, e7787 (2016)
34. Sigovini, M., Keppel, E., Tagliapietra, D.: Open Nomenclature in the biodiversity era. Methods Ecol. Evol. **7**, 1217–1225 (2016)
35. Zingone, A., et al.: Increasing the quality, comparability and accessibility of phytoplankton species composition time-series data. Estuar. CoastShelf S **162**, 151–160 (2015)

36. Madin, J., Bowers, S., Schildhauer, M., Krivov, S., Pennington, D., Villa, F.: An ontology for describing and synthesizing ecological observation data. Ecol. Inform. **2**, 279–296 (2007)
37. Rosati, I., et al.: A thesaurus for phytoplankton trait-based approaches: Development and applicability. Ecol. Inform. **42**, 129–138 (2017)
38. Baguley, J.G., Hyde, L.J., Montagna, P.A.: A semi-automated digital microphotographic approach to measure meiofaunal biomass. Limnol. Oceanogr. Methods **2**, 181–190 (2004)
39. Di Mauro, R., Cepeda, G., Capitanio, F., Viñas, M.D.: Using ZooImage automated system for the estimation of biovolume of copepods from the Northern Argentine Sea. J.- Sea Res. **66**, 69–75 (2011)
40. Mullin, M.M., Sloan, P.R., Eppley, R.W.: Relationship between carbon content, cell volume, and area in phytoplankton. Limnol. Oceanogr. **11**, 307–311 (1966)
41. Strathmann, R.R.: Estimating the organic carbon content of phytoplankton from cell volume or plasma volume. Limnol. Oceanogr. **12**, 411–418 (1967)
42. Taguchi, S.: Relationships between photosynthesis and cell size of marine diatoms. J. Phycol. **12**, 185–189 (1976)
43. Wheeler, P.A.: Cell geometry revisited: realistic shapes and accurate determination of cell volume and surface area from microscopic measurements. J. Phycol. **35**, 209–210 (1999)
44. Harrison, P.J., et al.: Cell volumes of marine phytoplankton from globally distributed coastal data sets. Estuar. Coast. Shelf S. **162**, 130–142 (2015)
45. Napiorkowska-Krzebietke, A., Kobos, J.: Assessment of the cell biovolume of phytoplankton widespread in coastal and inland water bodies. Water Res. **104**, 532–546 (2016)
46. Edler, L., Elbrächter, M.: The Utermöhl method for quantitative phytoplankton analysis. In: Karlson, B., et al. (eds.) Microscopic and Molecular Methods for Quantitative Phytoplankton Analysis. Intergovernmental Oceanographic Commission Manuals and Guides 55, pp. 13–20. UNESCO, Paris (2010)
47. Michener, W.K., et al.: Participatory design of DataONE - Enabling cyberinfrastructure for the biological and environmental sciences. Ecol. Inform. **11**, 5–15 (2012)
48. Wilkinson, M.D., Dumontier, M., Aalbersberg, I.J., et al.: The FAIR Guiding Principles for scientific data management and stewardship. Sci. Data **3**, 160018 (2016). https://doi.org/10.1038/sdata.2016.18

Sustainability and Future Challenges

Towards Cooperative Sustainability

Wouter Los[(⊠)] [iD]

Faculty of Science, University of Amsterdam, 1098XH Amsterdam, The Netherlands
w.los@uva.nl

Abstract. The inescapable question in the ENVRIplus project is how to sustain all achievements after the end of a collaborative project. Considering that each individual research infrastructure as cooperating in the ENVRIplus project has its own legal entity with dedicated governance and management, the challenge is to agree on modes of cooperation to keep tools and services of common interest up to date and operational. This chapter starts with the views of stakeholders, more specifically the views of scientific bodies, policy bodies, and of the infrastructure managers and operators who have to keep a lot of balls in the air. The sustainability plan has to consider the influences from external developments such as the European Strategy Forum on Research Infrastructures (ESFRI), and the emergence of a European Open Science Cloud (EOSC). The chapter discusses ENVRI strategic views, position, and future challenges.

Keywords: Sustainability · Governance · Planning

1 Challenges

As demonstrated in this book, many experts from many different research infrastructures worked together to develop novel tools supporting users of environmental research infrastructures in data production or to access, retrieve, and analyse existing data. All these data represent the dynamics of our environment with respect to the planetary crust, marine and freshwater bodies, the atmosphere, the impact of solar fluctuations, and the buffering role of the living environment/biodiversity. Apart from enhancing the services of each individual research infrastructure as they pertain to these environmental components, it is increasingly important to benefit from their combined services. The ambition of the cooperating research infrastructures is to provide scientists with the tools to work across traditional disciplinary boundaries and to discover, extract and analyse/model selected data as dispersed across many different sources and in many different formats. The cooperating research infrastructures tackled these ambitions by developing operational tools and services. Moreover, this provided the basis for bringing together the facilities, resources and services in support of the scientific community for innovative research and improved informed environmental policies.

Z. Zhao and M. Hellström (Eds.): Towards Interoperable Research
Infrastructures for Environmental and Earth Sciences, LNCS 12003, pp. 345–359, 2020.
https://doi.org/10.1007/978-3-030-52829-4_19

1.1 Expectations of Scientific Bodies

Developing the new services was strongly promoted by European funding under contracts for two successive projects ENVRI (2011–2014) [5] and ENVRIplus (2015–2019) [6], and was leveraged by national and other domestic funding. Such catalysing incentives proved to be crucial to foster cooperation between previously separated research infrastructure communities. Since the beginning of the 21st century, the European Strategy Forum on Research Infrastructures (ESFRI) brought together national policy and scientific stakeholders to consider and to promote the establishment of new research infrastructures in scientific areas requiring innovations for breakthroughs[1]. ESFRI showed itself to be very effective in promoting intergovernmental agreements for supporting the construction and operation of new research infrastructures at European and sometimes at the global scale in all scientific disciplines. Despite the challenging process to come to agreements within the scientific community and between supporting governments, a growing number of European research infrastructures managed to reach an agreement with and amongst funding countries, and started construction and successively entered their operational phases.

The ENVRI and ENVRIplus projects were conditional for sharing experiences, for sorting out solutions for common technical problems, and for providing guidance to support users requesting data and services from more than a single data source. In addition, the cooperation in these projects tackled common issues with respect to for example access services to multiple research infrastructures, ethical issues, training and exchange of staff, and common strategy development at the management level. These could not be developed in isolation, because attention had to be paid to the relationship with parallel developments in GEOSS[2] and Copernicus, the European Union's environmental observation programme offering information services based on satellite observations and in situ data. Joint workshops assisted in defining each one's work area with at one side the research infrastructures as data generators and at other sides GEOSS and Copernicus as portals for access to the data as relevant for their successive missions. Still, a seamless connection in between them is not obvious as the requirements by the scientific users of research infrastructures, and the policy-oriented requirements of GEOSS and Copernicus users are not similar.

The *League of European Research Infrastructures* (LERU) published in 2017 a report on four golden principles for enhancing the quality, access and impact of research infrastructures [1]. First, a smart funding strategy is needed for research infrastructures to remain competitive, and to be leading and agile regarding further technological developments. Second, mechanisms should be developed to encourage cooperation, especially cross-border, thereby helping to avoid unnecessary duplication. Third, it is required to create a more robust and better-connected European network of research infrastructures. Fourth, the academic community should play a leading role in planning and operating the research infrastructures. Indeed, these four principles touch on important and also sensitive issues. Scientists expect that the research infrastructures are leading in new technical developments and related user services. However, such a risk-bearing

---

[1] https://www.esfri.eu.

[2] http://www.geoportal.org.

approach is very different from the more conservative risk-avoiding attitude of funding bodies (often ministries). This intrinsic tension complicates the sustainability of novel tools and services as described in this book, especially when this would require funding from different resources (countries, institutions and individual infrastructures). Most environmental research infrastructures are inherently distributed, dictated by the need to collect data and observations locally from around the planet. This may be an advantage for financing research infrastructures as it allows for domestic funding solutions rather than (only) funding a not-national European or international organisation. The disadvantage is however that it often prohibits the second principle above on avoiding unnecessary duplication through cross-border cooperation.

1.2 Expectations of Scientific Bodies

Several policy bodies addressed the challenges of sustaining research infrastructures, and of international cooperation. A strong example is the OECD Global Science Forum (GSF) with its mandate to address scientific issues that require global solutions[3]. Much attention has been and still is on research infrastructures, especially internationally funded infrastructures, and the international access to their facilities. The GSF promotes the principle that the world's best researchers should have access to the best infrastructure facilities, and that these facilities benefit from their innovative engagement. Continuous activity of the GSF is to identify policies and procedures that can strengthen the sustainability and the effectiveness of the functioning of research infrastructures during their entire lifecycle (including their dismantling or potential reuse). Priority targets are to lower initial investment and operating costs by improved standard planning and business models; to accelerate construction and implementation phases by better-adapted planning procedures; and to realise effective planning, budgeting and implementation of human-resources and controlled optimization of the running costs for RIs, considering that operational performance will lead to sustainable, attractive and productive research environments at both single-site and distributed RIs. However, this on-going activity did not give much attention to the implications of cooperation to promote the interoperability of research infrastructures.

In Europe, the initiative to implement a European Open Science Cloud (EOSC) puts emphasis on generic services for open access to data and related data interoperability[4]. This initiative is focusing on research outputs in general, and not specifically targeting research infrastructures. Nevertheless, the collaborative data services from research infrastructures are offering the strongest use cases in the EOSC development. The European Commission published in 2018 a "Call for Action Report" asking attention for critical issues to ensure the sustainability of research infrastructures [2]. The report, drafted as a Staff Working Document of the European Commission, recognises that sufficient funding is crucial for sustaining research infrastructures and their cooperation, but concludes that sustainability of RIs goes well beyond funding only, touching upon several dimensions such as scientific excellence, socio-economic impact and innovation. This is quite true for the environmental domain since societal and scientific challenges

---

[3] http://www.oecd.org/sti/inno/global-science-forum.htm.

[4] https://www.eosc-portal.eu.

require full attention by the cooperating research infrastructures for tackling environmental complexity and its socio-economic impact of global change. Regrettably, the EC Staff Working Document only addresses the sustainability of individual research infrastructures and not their cooperative efforts which are beneficial from the viewpoints of interoperability, user support, and efficiency. Clearly, ENVRI enters unknown territory in the process to sustain its cooperative efforts, tools and services. ENVRI does not enter this unknown territory alone; however, since cluster-projects in other scientific domains meet similar challenges, for example, the CORBEL cooperative project of research infrastructures in the life sciences, which offers more integrated access to data resources required for biomedical research[5].

1.3 Keeping a Lot of Balls in the Air

In the context explained above, sustaining the growing interoperability of research infrastructures is not an easy dedication for the management of collaborating individual infrastructures. One should consider that the emphasis in the last decade was on establishing new research infrastructures in Europe, and the understandable implication is that the infrastructure's management is primarily concerned with its own internal business. The EC financial support for fostering cooperation in between environmental research infrastructures was attractive to explore and develop common practices, tools, and services, but sometimes this was considered a distraction from the internal infrastructure's challenges. It takes time to appreciate efficiency benefits from cooperation or joint external marketing, as well to manage expectations in regard to keeping control of internal power versus advantages of delegating tasks to cooperative efforts. Currently, the infrastructure's view on organising data is changing. From putting data processing within a 'walled garden' (where many tools are brought together within a single curated platform), the new approach is a data 'marketplace' approach, where many tools are made available as micro-services within a wider market allowing users to find, compare, learn from others, and negotiate their preferred solutions or service level agreements. Apart from dealing with user demands, the management also has to deal with diverging interests of its national funders. Some of these funding bodies prefer to focus their policies on one or a few research infrastructures, while others don't want to bother with individual initiatives and are encouraging merging efforts to avoid fragmentation. The next paragraph reports about the process that the European environmental research infrastructures followed to sustain their cooperative efforts.

2 The Making of a Sustainability Plan

Environmental research infrastructures in Europe gained in the last decade more insight on how to work together and to benefit from cooperative efforts. A key principle was to foster engagement of all European-level research infrastructures, independent of any European or other funding. The management of such infrastructures engaged in regular

---

[5] https://www.corbel-project.eu/home.html.

meetings of a common advisory panel, the Board of Environmental Research Infrastructures (BEERi)[6] [9]. This Board entered a process to agree on sustaining the results of the ENVRIplus project after the end of this project. One approach was to identify crucial common services, products and other results to be sustained after the end of the project, and by which organisations. This "bottom-up process" focused on the results relevant for all research infrastructures. A number of ENVRIplus partners are prepared to sustain these results with continued service provision. However, not all required tasks could be taken up by individual organisations. Covering these tasks brings into consideration a kind of collaborative organisational structure. This is a "top-down" process, focusing on the future ENVRI: the perspective for the infrastructures at the level of joint cooperation after the end of the funded ENVRIplus project.

The bottom-up process required the analysis of each ENVRIplus result in order to assess its relevance for all cooperating research infrastructures, and to negotiate with developers who have ownership and responsibility for sustaining them. In contrast, the top-down approach implied that infrastructure managers had to consider modes of future cooperation.

2.1   The Bottom-Up Process to Identify Tools and Services of Common Interest

The partners in the ENVRIplus project are best positioned to identify the common services, products and other results to be sustained after the end of the project. The focus here is primarily on work package (WP) results that are relevant for all research infrastructures. Results that are only relevant for one or a few Research Infrastructures will have to be dealt with by these infrastructures themselves. As a first step, the WP leaders were asked to consider which developed products/services should be sustained, and why. The results were categorised as follows:

- **Intellectual:** e.g. standards, concepts and reference docs.
- **Networks:** e.g. active communities, expert groups, BEERI and peer review mechanisms.
- **(Online) Services/Distribution channels:** e.g. interoperable data, web services, training and helpdesks.
- **Physical infrastructure and software:** e.g. computing environments, e-infrastructure and common virtual labs.
- **Branding:** Selected products/services requiring ENVRI branding as well as joint publicity and communication.

As a next step, the WP leaders were asked to clarify the ownership of results, applicable licenses, value propositions, required maintenance costs, and related annual costs. The identification of ownership of project results is important since the owners are in principle expected to take up the responsibility to sustain these results after the end of the project. Sustaining results does not only imply the maintenance and updates of services and products but also their accessibility and user support including training. The project's Grant Agreement states that (joint) owners may grant non-exclusive licenses to third

---

[6] https://www.envriplus.eu/beeri/.

parties to exploit jointly-owned results. Each project partner had to seriously consider what would the wisest and most attractive way to proceed.

The process to collect the required information followed a number of steps as visualised in Fig. 1.

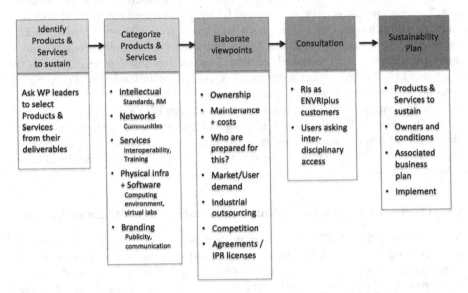

Fig. 1. Process steps to identify project results to be sustained.

Finally, this process ended up in an extensive overview of key results to be sustained at the ENVRI cluster level, and with the following details:

Who are the owners of products/services to sustain?

- Who are the owners (organisations) for each identified product or service?
- Are these owners ready to take organisational responsibility for sustaining the products or services?
- Alternatively, are agreements in place to transfer ownership (or license it) to a different organisation?

Value proposition

- Why is the product or service important/valuable for the research infrastructures?
- Which are the key user groups?
- How to advertise and provide support for deploying the product or service.

Required maintenance and other activities in addition to annual costs

- Which kinds of activities would be required to maintain, update, advertise, and provide user support and to provide training for the product or service?

- What would approximately be the total annual costs for these activities?

This process was important to get all involved partners thinking about the future of their developed products. It was necessary to create awareness on any required agreements with the concerned partners to sustain their results for future use in the ENVRI community. But also on the question, if such agreements should be established with each separate research infrastructure, or with a joint ENVRI structure acting on behalf of the research infrastructures. The latter simplifies the picture and would allow for a common license model for ENVRIplus results. With this perspective, a joint indexed catalogue of all services and tools may become a reality. The catalogue should also provide contact details for getting the specific information and on associated training.

2.2   The Top-Down Process to Conclude on a Joint ENVRI Service

Different from the bottom-up analysis of the ENVRIplus results to be sustained, the top-down view considers how the research infrastructures want to proceed as a cooperative community. As for the question on the purpose and structure of a sustained ENVRI, the Implementation Roadmap for the European Open Science Cloud (EOSC) offers interesting considerations for establishing a strong ENVRI community [3]. This Roadmap identifies six main action lines to ensure the successful implementation of EOSC:

- **Architecture:** Create a pan-European federation of existing and future data infrastructures and resources. Bringing European research data infrastructures together will be a great improvement in their current state of fragmentation.
- **Data:** Foster the development of professional practices in research data management based on Findable, Accessible, Interoperable, and Reusable (FAIR) principles.
- **Services:** Make EOSC services available through a single access channel to all European researchers regardless of their discipline or location.
- **Access and Interfaces:** Simplify the use of data across different disciplines.
- **Rules of participation:** Set out the rights, obligations and accountability of EOSC stakeholders.
- **Governance:** Ensure EU leadership in data-driven science and adapt to new governance challenges.

The 2017 ESFRI report on the long-term sustainability of research infrastructures also presents arguments to foster a continued ENVRI community. Three recommendations are specifically relevant for the ENVRI cluster cooperation:

- Harmonise and integrate the operation of research infrastructures and e-infrastructures.
- Demonstrate the economic and wider benefit to society of research infrastructures.
- Coordinate at National and European levels.

A top-down view may focus on promoting the impact of the ENVRI cooperation for multidisciplinary research, on addressing global change challenges, and on influencing policy and political decisions. It implies that prior to investigating any preferred organizational ENVRI structure, a common view should exist on the future purpose of the ENVRI community. Agreed priorities based on shared visions, values, goals and roles, will assist in sorting out preferred options for the future structure of ENVRI. Another top-down view is related to the efficiency benefits that the cooperating research infrastructures are expecting from a continued ENVRI cooperation. Which operational activities could be better organised within a common ENVRI community, rather than being addressed in each infrastructure separately?

2.2.1 ENVRI Collaborative Strategy

ENVRI, as a cooperation of European environmental research infrastructures, started in the first half of the 2010 decade with a common strategic view serving as guidance for organising joint project activities. The strategy was based on the vision to provide scientific support for a holistic understanding of our planet and its behaviour, processes, feedbacks, and fluxes. The challenge is then to contribute to developing an environmental system model, a framework of all interactions within the Earth System, from solid earth to near space. Many of the urgent challenges we are facing (such as climate change, energy use, water availability, food security, land degradation, hazards and risks, life in megacities, and human health) are closely related to complex interactions in the environment. Whilst each individual research infrastructure is concerned with its own domain of interest, it was thought imperative to find robust yet lightweight means to integrate various operations across research infrastructures to serve an increasingly multidisciplinary scientific community and to help to address the urgent societal challenges. To this end, three resources 'capitals' were identified as strategic targets within a conceptual model.

- **Technological Capital:** Capacity to measure, observe, compute, and store data, with technologies, software, and analytical and modelling capabilities.
- **Cultural Capital:** Open access to data, services in between RIs, requiring rules, licenses, citation agreements, IPR agreements, machine-machine interactions, workflows, metadata, data annotations, etc.
- **Human Capital:** The specialists to make it all work, with also generalists overseeing more than only their own discipline.

The ENVRIplus project focused on implementing these 'capitals', and this book reports the results in relation to common data issues. The next step is to benefit from the developed 'capitals' to address the above grand challenges. Discussions between the infrastructure managers revealed two main messages for an updated strategy. First, reconfirmation that ENVRI – as a large-scale cluster of European environmental research infrastructures -contributes to the societal challenges by providing high-quality multidisciplinary research data, services and expertise in a systemic way to mitigate societal risks. Second, that ENVRI aims to become a globally recognised cluster with strong international links and an attractive service portfolio for researchers, the private sector and policy-makers. This position has its implications for sustaining the common tools, services and other products as described in this book. As such, it assists in concluding about any future ENVRI organization. The next paragraph reports about the process followed to conclude on this.

2.2.2   The Process Towards a Joint Position on the ENVRI Organization

Section 1.3 explained why it is for the management of research infrastructures, not an easy question to conclude about the role and organization of ENVRI in the next decade. While many research infrastructures are primarily concerned with their own construction and/or are consolidating their operations, it is understandable that they are less focused on common ENVRI interests. Nevertheless, they all appreciate the advantage of positioning some common activities at the ENVRI level. In addition, they feel that an ENVRI entity should take care of overseeing the agreements to sustain the results from the finished ENVRIplus project (the bottom-up process in Sect. 2.1).

In order to facilitate the discussion about alternative views, a consultation and interactive process was organised to promote consensus building on a common perspective. This process addressed the future of ENVRI as a cooperation of all recognised European research infrastructures. Subsets of research infrastructures may agree on other cooperation modes, but ENVRI is meant for the full infrastructure community. As such "mixed" models are always possible. The discussions focused on preferred approaches from 2025 on. Two consecutive anonymous surveys collected the preferences and opinions of each research infrastructure as associated with the ENVRIplus project. In between the two surveys, a moderated workshop discussed the pros and cons of alternative options to organise the future ENVRI, while taking into account the growing consensus of a common ENVRI strategy.

2.2.3   Considered Organisational Options

There are different options to consider for a future ENVRI structure: finishing the ENVRI cooperation, proceeding with a network, or federated cooperation, forming a modest or more extensive common legal entity, or even merging into a joint ENVRI research infrastructure. The collaborating research infrastructures were asked to express their preferences and views in two subsequent surveys, with in between a workshop to discuss the options. The first survey dealt with all options summarised below, and the second one focused on the most favoured ones.

A)   Finishing the ENVRI cooperation

This option would imply that each research infrastructure goes separately, as shown in Fig. 2. Or, that some may conclude to proceed in a cooperative way, while others are leaving the ENVRI community. Of course, there may be irregular joint meetings, but bringing an end to the ENVRI cooperation will most likely reduce the voice and impact of the environmental cluster.

B)   ENVRI network, without required commitments

An ENVRI network could be considered when enough research infrastructures would prefer to meet regularly and consider joint activities, either at the management level or at operational levels, as shown in Fig. 3. A simple network can be organised on the basis of a Memorandum of Understanding (MoU) with some agreements on how to run the network. The MoU may have provisions on for example a rotating chair and secretariat, on the subjects to be discussed or initiated, and on how each infrastructure may offer

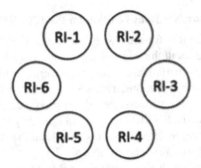

Fig. 2. No cooperation; each research infrastructure goes its own way.

in-kind contributions for the benefit for all. The latter may include the preparation of joint project proposals.

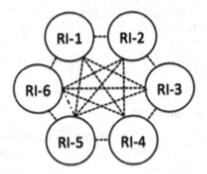

Fig. 3. Network cooperation without commitments.

C) *Federated structure with a joint consortium agreement*

   Similar to collaborative projects, the research infrastructures agree to work together within a Consortium Agreement on specific topics. Figure 4 shows the basic idea. It will state in which areas will be cooperated, how this will be governed and managed, and how any financial contributions from partners for joint activities will be arranged. Depending on the agreement, one or more individual infrastructures may be assigned with the task of chair, secretary and treasurer.

D) *Establish an ENVRI legal body with limited roles*

   This would be a legal body with limited tasks and powers, mainly for organising meetings and for organizing publicity, as shown in Fig. 5. It is a cheap option, and has the advantage of a united "place", both physically and on the Internet with a clear signal of the presence of ENVRI. Such a 'small' legal body should at least facilitate a secretariat and repository of joint undertakings. It may be considered to have in the legal body an independent director in charge, overseen by a board drawn from the cooperating research infrastructures.

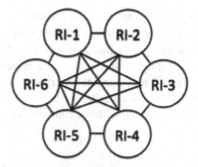

Fig. 4. No cooperation; each research infrastructure goes its own way.

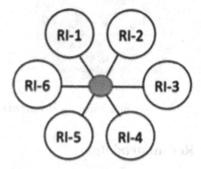

Fig. 5. Common legal body (in red) with limited roles.

E) *An ENVRI legal body taking up common tasks for the research infrastructures*

The common tasks are the ones that can more efficiently and/or less costly be operated from the legal entity. As a consequence, the legal body will employ its own staff, and costs are to be covered by fees paid by the involved RIs. The Board of the legal entity (with representatives of the cooperating RIs) decides on the tasks that can better be operated by the legal entity. Figure 6 shows the basic idea.

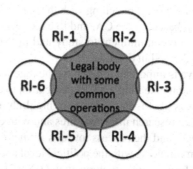

Fig. 6. A legal body (in red) taking up common tasks.

356   W. Los

F) *Merge the individual infrastructures in a joint research infrastructure*

This would a united legal ENVRI acting as an umbrella organisation of federated research infrastructures as working units, as shown in Fig. 7. Each individual infrastructure will continue its normal operations, while an overarching board takes care of the common interests. This board may appoint supervisory committees from the relevant communities for each RI. This is a similar set-up as for CERN and EMBL where different research facilities are operated within a common legal body.

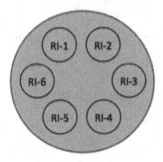

Fig. 7. A joint research infrastructure with merged facilities.

3   Conclusions and Recommendations

3.1   Next Steps in the ENVRI Community

The surveys and discussions about collaboration in a cluster of environmental research infrastructures revealed that nobody is in favour of finishing the ENVRI cooperation. In the long term, this would imply an agreement to establish an ENVRI cooperative body. The far-reaching option of merging all infrastructures within a single overarching research infrastructure was rejected. As for the intermediate considered options, the favoured approach is to plan for a legal body taking up a few common tasks, but to proceed through a step-by-step process: first, to collect signed Letters of Intent by each infrastructure, confirming the willingness to continue cooperation and the agreement to draft a Consortium Agreement (CA) matching with the above option c; next to agree on entering a joint consortium based on the CA, and to experience how this would evolve. This has to underpin a future commitment to study and prepare a possible legal body with allocated tasks to be defined in detail.

Clearly, this perspective is important to promote the togetherness within the ENVRI community and to demonstrate the joint ENVRI dedication to external stakeholders. To this end an ENVRI business plan is in preparation, highlighting its focus, the benefits for collaborating in between research infrastructures and with external stakeholders, the commitments to agree upon, and the process towards further consolidation. In parallel, it is considered how to communicate with the political level about adapted and appropriate funding arrangements for the infrastructures in the ENVRI community. Considering the steps taken by the ENVRI community to reduce fragmentation in its infrastructure landscape, it makes sense to address the current fragmented arrangements for funding the European-scale research infrastructures. These are mainly funded through financial support by different combinations of individual countries. This picture is the result of elaborate negotiations with potentially interested countries, but with the complication that not any country will financially contribute to all environmental research infrastructures. This implies that huge transaction costs result from all these negotiations, and to be repeated every five years for another cycle of financial security. Another implication is that this fragmented funding structure makes it for the ENVRI community difficult to pursue a joint scientific and technical policy to address the grand societal and scientific questions.

3.2 The ENVRI-FAIR Project

The focus of the developing ENVRI business plan is primarily on securing the sustainability of jointly-developed tools and services, through securing service-level-agreements with the infrastructures or other organisations maintaining these tools and services, and by taking up some common tasks in the envisaged common ENVRI organisation. As for the tools and services mentioned in this book, the cooperating infrastructures are in a good position with the project ENVRI-FAIR, running for four years from the beginning of 2019[7]. This project is working on the uptake of the developed tools and services by individual research infrastructures and will facilitate the planned process towards a legal ENVRI entity. The business of the ENVRI-FAIR project is very much concerned with the parallel European developments for establishing a European Open Science Cloud (EOSC) [7]. The overarching goal of ENVRI-FAIR is to establish sustainable, transparent and auditable data services, for each step of the data lifecycle, compliant with the FAIR principles in the ENVRI community and connecting them to EOSC. Common policies, open standards, interoperability solutions, operational services, and stewardship of data on the basis of FAIR (Findable, Accessible, Interoperable, Reusable) principles require a common approach. The final goal is to prepare the open-access platform for interdisciplinary environmental research data in the European Research Area utilising EOSC. More specifically, the high-impact ambition of ENVRI-FAIR is to establish the technical preconditions for the successful implementation of a virtual, federated machine-to-machine interface to access environmental data and services provided by the contributing research infrastructures. This ENVRI-hub is planned as a federated system of data policies and management, access platforms and virtual research environments. The system will be completely open-source, modular and scalable and build on the experience available in the consortium and already operational systems.

3.3 Future Challenges

The ENVRIplus project focused on bringing into place the capacities required for assisting researchers and other stakeholders in dealing with environmental challenges and providing key products for societal needs. Much progress has been made to improve cooperation in technology, and in a joint culture and in human capital. The next question is how to best benefit from the achievements as presented in this book. Clearly, the focus is than on supporting users in addressing the grand environmental challenges from solid earth to

[7] http://envri.eu/envri-fair/.

near space. Many of the present urgent challenges are closely related to complex interactions in the environment. Whilst each research infrastructure is primarily concerned with its own domain of interest, another challenge is to integrate various operations across infrastructures, supporting the growing multidisciplinary scientific community. Indeed, the Earth system and problems related to the grand challenges are far too complex and interdependent to be studied from only a single perspective and supported by a single research infrastructure. New scientific developments require measurements covering the entire interlinked Earth system, and more integrated and interoperable infrastructure services enabling free access to and analysis of the gathered information. The latter is what ENVRIplus put in place, and was recognised by the ESFRI in its 2018 Roadmap [4], stated as follows (p. 14):

"The concept of *multi-messenger* research relies on exploiting diverse sources of information from different research methodologies to yield an integrated complementary ensemble of data that becomes the true insight on the phenomenon studied. Generalizing to all fields of research, we can recognise that a *multi-messenger* approach is already at work in domains like environmental sciences and life sciences, and that there is a high potential to address complex phenomena like grand societal and scientific challenges – e.g. climate change, population increase and differential ageing, food and energy sustainability – by using synergistically research infrastructures from all fields".

Possible approaches in this regard, and partly tackled by the new ENVRI-FAIR project, are the following:

- Challenge the scientific community to address the grand societal challenges with support by research infrastructures. An independent ENVRI scientific advisory body might be considered to define interdisciplinary research challenges, and to open up calls for inviting innovative researchers requiring advanced integrated services from the research infrastructures.
- Showcase the strengths and significance of ENVRI though user options to benefit from multiple sites and laboratory facilities, and of cross-use of experimental research platforms and vessels. An interesting consideration is to optimise the collaboration between industry, policy-makers and research infrastructures to promote a stronger impact of the research and innovation system, as also suggested in the ESFRI 2018 Roadmap.
- Be prepared to support interdisciplinary research, for example by providing data for a common minimal set of measurements/observations relevant for environmental variables regarding the Earth system, and a joint strategy to fill geographical gaps. An additional service could be to provide capacity for guiding and supporting interdisciplinary researchers requiring support from more than a single RI.

The benefit for the cooperating ENVRI research infrastructures is that they also will be challenged to provide joint services at the forefront of scientific discovery and societal impact.

Acknowledgements. This work was supported by the European Union's Horizon 2020 research and innovation programme via the ENVRIplus project under grant agreement No 654182.

References

1. Väänänen, J., van Tienderen, P.: Four Golden Principles for Enhancing the Quality. Access and Impact of Research Infrastructures. LERU (2017). https://www.leru.org/publications/four-gol den-principles-for-enhancing-the-quality-access-and-impact-of-research-infrastructures
2. The European Commission: Sustainable European Research Infrastructures – a call for action. The European Commission (2017). https://ec.europa.eu/research/infrastructures/pdf/ ri_policy_swd-infrastructures_2017.pdf
3. The European Commission: Implementation roadmap for the European Open Science Cloud (EOSC). The European Commission (2018). http://ec.europa.eu/research/openscience/pdf/ swd_2018_83_f1_staff_working_paper_en.pdf
4. European Strategy Forum on Research Infrastructures: Strategy Report on Research Infras- tructures (Roadmap 2018). ESFRI (2018). http://roadmap2018.esfri.eu/media/1066/esfri-roa dmap-2018.pdf
5. Chen, Y., Martin, P., Schentz, H., Magagna, B., Zhao, Z., Hardisty, A., Preece, A., Atkinson, M., Huber, R., and Legre. R.: A common reference model for environmental science research infrastructures. In: Proceedings of EnviroInfo 2013 (2013). http://enviroinfo.eu/sites/default/ files/pdfs/vol7995/0665.pdf
6. Zhao, Z., Martin, P., Grosso, P., Los, W., Laat, C. de, Jeffrey, K., Hardisty, A., Vermeulen, A., Castelli, D., Legre, Y., Kutsch, W.: Reference model guided system design and implementation for interoperable environmental research infrastructures. In: 2015 IEEE 11th International Conference on e-Science, pp. 551–556. IEEE, Munich (2015). https://doi.org/10.1109/eSc ience.2015.41
7. Petzold, A., Asmi, A., Vermeulen, A., Pappalardo, G., Bailo, D., Schaap, D., Glaves, H.M., Bundke, U., Zhao, Z.: ENVRI-FAIR - interoperable environmental FAIR data and services for society, innovation and research. In: 2019 15th International Conference on eScience (eScience), pp. 277–280. IEEE, San Diego (2019). https://doi.org/10.1109/escience.2019. 00038, https://zenodo.org/record/3462816

Towards Operational Research Infrastructures with FAIR Data and Services

Zhiming Zhao[1]([⊠]) [iD], Keith Jeffery[2] [iD], Markus Stocker[3,4] [iD], Malcolm Atkinson[5] [iD], and Andreas Petzold[6] [iD]

[1] Multiscale Networked Systems, University of Amsterdam,
1098XH Amsterdam, The Netherlands
z.zhao@uva.nl
[2] Keith G Jeffery Consultants, Faringdon, UK
keith.jeffery@keithgjefferyconsultants.co.uk
[3] TIB Leibniz Information Centre for Science and Technology, Hannover, Germany
markus.stocker@tib.eu
[4] MARUM Center for Marine Environmental Sciences, PANGAEA Data Publisher for
Earth & Environmental Science, Leobener Strasse 8, 28359 Bremen, Germany
[5] University of Edinburgh, Edinburgh, UK
malcolm.atkinson@ed.ac.uk
[6] Forschungszentrum Jülich GmbH, Jülich, Germany
a.petzold@fz-juelich.de

Abstract. Environmental research infrastructures aim to provide scientists with facilities, resources and services to enable scientists to effectively perform advanced research. When addressing societal challenges such as climate change and pollution, scientists usually need data, models and methods from different domains to tackle the complexity of the complete environmental system. Research infrastructures are thus required to enable all data, including services, products, and virtual research environments is FAIR for research communities: Findable, Accessible, Interoperable and Reusable. In this last chapter, we conclude and identify future challenges in research infrastructure operation, user support, interoperability, and future evolution.

Keywords: Research infrastructure · Virtual research environment · System-level science

1 Introduction

Natural and anthropogenic factors lead to environmental changes on all scales from local to global. Environmental data provides the scientific basis for analysing the physical, biological, and economic processes in the earth system, which are affecting all sectors of society as well as wildlife and biodiversity. Such data-related activities can be highlighted in several scenarios drawn from research communities, as shown in Fig. 1:

© The Author(s) 2020
Z. Zhao and M. Hellström (Eds.): Towards Interoperable Research
Infrastructures for Environmental and Earth Sciences, LNCS 12003, pp. 360–372, 2020.
https://doi.org/10.1007/978-3-030-52829-4_20

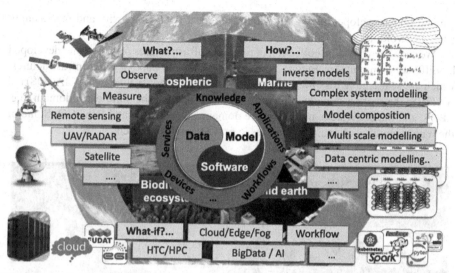

Fig. 1. Some typical research activities in the environmental RI communities. (Zhao Z. presentation in the final ENVRIplus event in Brussel, June, 2019.)

1. **Observing** the phenomena of the environmental and earth system, via distributed sensors, monitoring network or human observers [1]. Such observations are often guided by specific conceptual models of the parameters related to the earth system, or connected with the experimentations in the laboratory (e.g. rock mechanics) or in the fields (agricultural studies).
2. **Modelling** the behaviour of the environmental systems, understanding their evolution, and investigating the causality among different events by scientists [2]. These models can be developed based on physical models, e.g. the Navier–Stokes equation for modelling fluid dynamics [3], machine learning methods like neural networks [4], or combinations thereof [5].
3. **Applying** existing assets from observations, simulations, and earlier experiments to complex data-centric workflows to explore the solution space of hypotheses or discover the consequences of different conditions [6]. Scientific workflows, e.g. [7] or Jupyter notebook [8] are used to integrate different processes in the data pipeline, which may involve big data processing platforms, e.g. Spark [9], across different infrastructures, e.g. Cloud and HPC clusters [10].

As important facilities to enable scientists to perform advanced research in environmental and earth sciences, environmental research infrastructures aim to make their digital assets, including data, models and software Findable, Accessible, Interoperable and Reusable (**FAIR**). In a broad sense, all the **application workflows, the sensors** that obtain data, **the operational services** that manage those assets, **and all high-level knowledge** derived from those assets, collectively constitute valuable material for user communities to conduct scientific research with, as indicated in Fig. 1. However, the tools and infrastructures to manage, document, provide, find, access, and use all such assets are still underdeveloped owing to a combination of data complexity and increasingly large data volumes.

We have seen the large collection of research infrastructures proposed and developed by different communities. Figure 2 provides a basic landscape of those infrastructures, which are scattered across four subdomains: the atmospheric domain, e.g. IAGOS and ACTRIS, the marine domain e.g. Euro-Argo and EMSO, the solid earth domain, e.g. EPOS, and the ecosystem and biosphere domains, e.g. AnaEE and DISCCO. Some of them cross multiple domain boundaries, e.g. ICOS and Lifewatch.

Fig. 2. The landscape of ENVRI research infrastructures across domains. (Image source: https:// envri.eu/communications/)

In this last chapter, we will first give a short summary of the main topics discussed in the book, and then look at the next phase of research infrastructure evolution: new challenges that RI may face after they are developed and deployed.

2 ENVRI: Development Activities at the Cluster Level

The development of an environmental RI is often driven by the interests of a specific domain, and many RIs are funded via separate research projects. But the bottom-up development paradigm of the RIs does not come with a naturally embedded interoperability concern for guiding the evolution of those different projects. Therefore, dedicated cluster projects are funded since 2011 to specifically inspect common problems that these environmental RIs face, to recommend reusable solutions to developers of RIs, and to tackle the interoperability challenge among RIs. Figure 3 shows the three cluster projects funded for ENVRI RIs: ENVRI, ENVRIplus [11] and ENVRI-FAIR [19]. The work presented in this book is mainly based on the activities carried out in the second project.

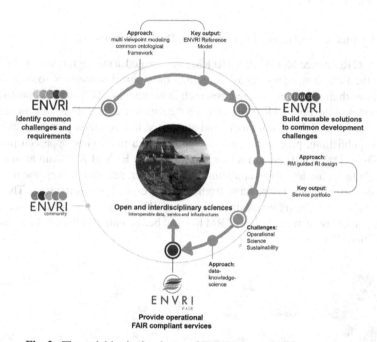

Fig. 3. The activities in the cluster of ENVRI research infrastructures.

1. In the FP7 ENVRI project (between 2011 and 2014), we analysed the initial design requirements, architecture design, and the existing assets of participating research infrastructures[1]. We abstracted a common vocabulary (ENVRI RM V1) for describing data management activities and architecture of research infrastructure.
2. Using the initial ENVRI RM, we analysed the design requirements from more than 20 research infrastructures[2] in the follow-up project H2020 ENVRIplus and applied a reference model guided approach to design and prototype six prioritised six common operations. During the practice, the ENVRI RM has also been refined as V2.

---

[1] FP7 ENVRI project contains six ESFRI projects: EMSO, Euro-Argo, ICOS, LifeWatch, EISCAT_3D and EPOS.

[2] H2020 ENVRIplus project contains 21 RIs: ACTRIS, ANAEE, EISCAT_3D, ELIXIER, EMBRC, EMSO, EPOS, ESONET, Euro-Argo, EUROFLEETS, EUROGOOS, FIXO3, IAGOS, ICOS, INTERACT, IS-ENES, JERICO, LifeWatch, LTER, SeaDataNet2 and SIOS.

3. In the H2020 ENVRI-FAIR project, we focus on the operational challenges, in particular, the FAIRness of the assets[3]. By the moment we finalise the book, the ENVRI-FAIR project just finished its initial self-assessment of FAIRness.

In this section, we will summarise our development activities during the past years via a number of highlights.

2.1 A Common Vocabulary for Describing Data Management

The ENVRI Reference Model[4] (ENVRI RM) was created at the beginning of the ENVRI project (the first cluster project) as a common ontological framework to enhance the information sharing among different research infrastructures [12] (see Chapter 4). The development of the ENVRI RM started from the data management lifecycle of research infrastructures in ENVRI community and abstracted five common phases: acquisition, curation, publishing, processing and use. Following a multi-view approach provided by the ODP (Open Distributed Processing) model, the ENVRI RM team abstracts the key vocabularies for describing communities, behaviour, data flow management, service interfaces and architectures patterns from ENVRI research infrastructures. The initial ENVRI RM has been refined with a big set of RIs in the ENVRIplus project. The ontological representation of ENVRI RM has also been created for the machine-readable specification (Fig. 4).

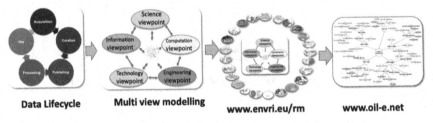

Data Lifecycle   Multi view modelling   www.envri.eu/rm   www.oil-e.net

Fig. 4. The development approach for the common ENVRI vocabulary.

2.2 Reference Model Guided Engineering

Using the common ontological framework, we analysed the requirements for research infrastructure and common challenges in the ENVRIplus project. After several iterations, we highlighted the six key common challenges: identification/citation, data processing, infrastructure optimisation, curation, cataloguing, and provenance, as discussed in Chap. 5.

The reference model enables the development team of common operations (i.e. data for science theme in ENVRIplus) effectively interact with the developers from different

---

[3] H2020 ENVRI-FAIR project contains 13 RIs: ACTRIS, ANAEE, DANUBIUS-RI, DiSSCo, EISCAT_3D, EPOS, EMSO, Euro-Argo, IAGOS, ICOS, LifeWatch, eLTER and SIOS.

[4] http://envri.eu/rm.

research infrastructures, and from the eInfrastructures to 1) analyse requirements, 2) review technologies and gaps, 3) design solutions to the common problem and 4) validate the prototypes via use cases. The details of the approach are discussed in Chapter 5.

The development team developed or recommended the key technologies for tackling the common problem identified from the research infrastructure:

1. Reference model and relevant training materials (in Chapter 4);
2. Ontological representation of the reference model (in Chapter 6);
3. Data curation services and recommendations (in Chapter 7)
4. Data cataloguing services (in Chapter 8)
5. Data identification services and citation recommendation (in Chapter 9)
6. Data processing framework and technologies (in Chapter 10)
7. Virtual infrastructure for data-centric sciences (in Chapter 11)
8. Data provenance services and recommendation (in Chapter 12)
9. Metadata and semantic linking (in Chapter 13)
10. Authentication, Authorisation, and Accounting (in Chapter 14)
11. Virtual Research Environment (in Chapter 15)

During the ENVRIplus project, the key results [11] have been documented and collected as a portfolio.

2.3 Use Case-Based Community Engagement

To best engage user communities in the development, ENVRI follows an **Agile development methodology**. Selected use cases follow a continuous procedure for accepting and reviewing proposals, prioritising specific use case projects, setting up agile use case projects, monitoring progress and exploiting results.

Based on the size and scope of individual cases, we identified three types of use cases, as shown in Fig. 5.

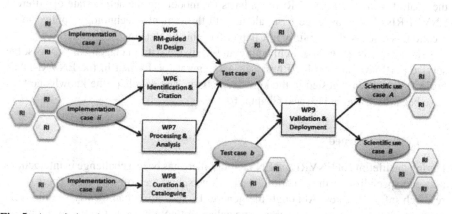

Fig. 5. Associations between science, implementation and test cases with core ENVRIplus activities. (http://www.envriplus.eu/wp-content/uploads/2015/08/D9.1-Service-deployment-in-computing-and-internal-e-Infrastructures.pdf)

1. **Science cases** often have clearly defined scientific problems, and require a big development effort for technical components, e.g. integrating data or services from different infrastructures.
2. **Test cases** focus on specific problem research infrastructures are facing, and often require the implementation of 1 or two critical components in the case. It can be implemented within typically a half year.
3. **Implementation cases** focus on specific technologies (e.g. customisation, integration or minor modification), and most of the components involved in the case are already available. The test cases can be implemented within typically 2–3 months.

In practice, the outputs of the implementation cases provide useful input for implementation cases and finally contribute to the development of science cases. Each use cases project often has members from different task teams and execute in parallel with the project task teams. In this way, the developers of the common data services participate in one or more agile case projects and closely collaborate with members from the research infrastructure communities.

Within each use case team, regular telcos are organised. By reviewing the progress, the developers can adapt the action points to meet the changing demands from the RI communities. In this book, three typical use cases have been presented in Chapter 16, 17 and 18.

2.4   A Community Knowledge Base

By the end of ENVRIplus, most of the RIs in the cluster have either finished their preparation phase or their implementation phase and are ready for final implementation or operation respectively. Collecting information about the RI's implementation status and the tools and technologies that they were using (including software, standards and vocabularies) was deemed vital for coordinating collaboration and identifying key commonalities. To this end, an ENVRI Knowledge Base is prototyped in the latter stages of ENVRIplus, for further development during the ENVRI-FAIR successor project, with the goal of using the ENVRI RM as a basis for modelling the active state of different ENVRI RIs. The knowledge base, along with the semantic technology applied in its creation, is discussed more fully in Chapter 6 of this book (Fig. 6).

The RI status, including architecture and available data management services, the service portfolio, and the FAIRness self-assessment (performed in the ENVRI-FAIR project) have been ingested in the knowledge base. The details of the knowledge base have been discussed in [13] and Chapter 6.

2.5   Lessons Learned

During the lifetime of ENVRI and ENVRIplus, there has been a challenge in interactions between specialists of the generic IT technologies, and the software developers of the research infrastructures. Although the generic IT specialists can clearly see the technical problems and gaps; since those specialists are not embedded in the development

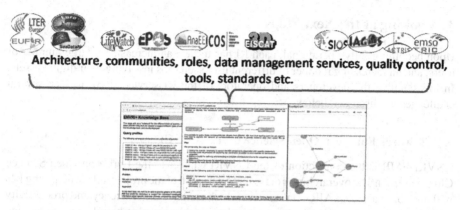

Fig. 6. Knowledge for sharing best practices.

context of each individual RI, the proposed solutions often miss matches the development priorities and the user practices of an individual RI. The interactions are thus often time-consuming.

The classical waterfall model of software engineering did not work in this context. The interactions between generic IT specialists and RI developers need to be spiral and iterative. Figure 7 shows how the other key highlights during the interactions. Besides what we have discussed in this section, two summer schools have been organised for transferring the technical knowledge to the RI developers, by the time when the book finishes. The key output has also been exploited to the third cluster project ENVRI-FAIR for the further development of the RI data management services, to make them FAIR compliant and operational.

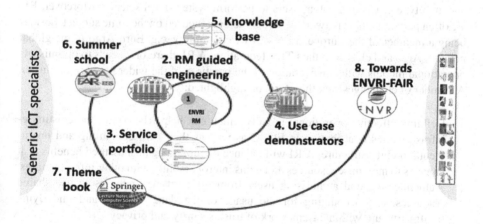

Fig. 7. Key highlights between Specialists and the RI communities.

3   Looking at the Next Steps

Upon becoming fully operational, each RI has to face an increasingly large number of users, and an increasing number of different user scenarios involving their digital assets. In this context, the developers and operators of the RI services have to face several challenges, as discussed below.

3.1   Towards European Open Science Cloud (EOSC)

ENVRI-FAIR is the connection of the ENVRI Cluster to the European Open Science Cloud (EOSC). The overarching goal is that by the end of the project, all participating RIs will have built a set of FAIR data services which enhance the efficiency and productivity of researchers, support innovation, enable data- and knowledge-based decision making and connect the ENVRI Cluster to EOSC and its core services. This goal is reached by (1) well-defined community policies and standards addressing all steps of the data lifecycle, aligned with the wider European policies and international developments; (2) each participating RI having sustainable, transparent and auditable data services for each step of data lifecycle, compliant with the FAIR principles; (3) focusing the proposed work into the implementation of prototypes for testing pre-production services at each RI, with the catalogue of prepared services defined for each RI independently depending on the maturity of the involved RIs; and (4) exposing the complete set of thematic data services and tools provided by the ENVRI cluster under the EOSC catalogue of services.

3.2   Operational Challenges

When operated as online services, RIs collaborate naturally as part of a service ecosystem, wherein each RI has to serve users from much bigger user communities than its own community, e.g. when enabling users to perform system-level science. Moreover, RIs are often part of a global network of infrastructures focused on the same subject, besides being a member of the European ENVRI RI community, e.g. Euro-Argo in the global network Argo and eLTER in the LTER federation. A RI, therefore, needs to optimise its operational models with consideration of the practices of the wider network or cluster. A number of challenges can, therefore, be highlighted:

1. Defining effective operational models which can help RIs exploit the existing e-infrastructures contributing to EOSC as well as their own computing and instrumentation infrastructure. A RI will balance disruption against assured benefits as it engages to maximise resources and gains interoperability with other infrastructures.
2. Authenticating and authorising users from different communities to use shared resources, and accounting for the usage of the data, services and underlying e-infrastructure within a framework of trust, security and privacy.
3. Allowing technical coordination across RIs through appropriate interfaces; this entails adopting interfaces for supporting shared VREs [14, 15], contributing rich (FAIR-compliant) metadata to community catalogues.
4. Ensuring the performance and quality of service and user experience required by scientists, in a manner that scales with the user base and data assets.

5. Effectively provisioning RI resources, including data and tools offered by RIs and services delivering underlying data infrastructure, to serve a broad range of demands from research developers, service managers, engineers, and researchers themselves.
6. Integrating EOSC with Fog/Edge computing scenarios and IoT (Internet of Things); some RIs have extensive sensor networks and technology which needs to be connected to the broader e-infrastructure.

3.3 Science Challenges

Many ENVRIplus stakeholders have stated that the community of environmental research infrastructures should be closely involved in EOSC developments. The ENVRIplus approach may help shape the EOSC ecosystem. More importantly, significant parts of the ENVRI community stand to benefit from EOSC. This transition will be incremental, as relevant services become available, affordable and sustainable, and when they combine well with current investments and agreed practices. ENVRIplus stakeholders are in a good position to bring in crucial views and development actions to support open science in the whole research process.

As the basis for open science, FAIR (or more appropriately FAIR+R where the additional R is reproducible) data, services and other relevant resources require not only incentives for sharing and exploiting data on the part of data producers and users, but also the development of effective technologies and standards that will enable RIs to achieve connectivity and interoperability of their data and services at any stage in the data management lifecycle.

To enable system-level and interdisciplinary science, future RIs have to face the following challenges:

1. Enabling interdisciplinary research activities to meet environmental research goals and societal challenges; not only sharing research data and software assets from different RIs, but also co-developing and using methodologies and models drawing expertise from multiple domains within and outside of environmental science [17].
2. Ensuring that the data and resources needed by scientists follow FAIR principles; this means the services, methods and metadata to make these assets FAIR.
3. Supporting user-specified and steered data processing, and automated workflows. For example, many user requests result in a workflow to download one or more selected datasets. As services local to data become well-supported then users develop and use more complex workflows involving multiple datasets, software components, computing resources and even sensors with processing partitioned and local to the data assets. The generation of workflows from user requests and their optimal deployment will grow in importance for environmental research.
4. Recording and providing provenance information for user assessment of relevance and quality of an asset, auditing, and reproduction.
5. Reusing the data and knowledge from different RIs effectively; this requires effective data and knowledge mining tools and a cohesive support knowledge infrastructure.

6. Providing support for data-intensive, compute-intensive and urgent data analysis and simulation. Frequently, complex workflows using such simulations need to inter-work between HPC (High-Performance Computing) and HTC (High Throughput Computing) platforms.
7. Providing support for working across multiple e-infrastructure environments within EOSC and beyond (e.g. DataOne in the USA [16]). RI workflows may utilise EOSC and other e-Is (including sensor networks) together and the interface should allow 'plug and play'.
8. RIs may be involved in activities with RIs on other continents and so may need to access e-Is in those other continents (and vice-versa) through appropriate gateways.

3.4 Sustainability Challenges

In the previous chapter, sustainability was specifically discussed. The operators of the RIs have to face several challenges to keep their services sustainable, including:

1. Providing sustainable business models that service data contributors, service developers, researchers, innovation makers and other payers into EOSC can use to ensure their continued participation [18].
2. Providing sustainable data management and stewardship, including the curation, long-term preservation and access of assets (information and software including associated libraries and operational environment).
3. Providing sustainable technical decisions, including standards and interfaces, so that they fit with the evolution of the digital ecosystem and operational models of RIs.
4. Providing sustainable system architecture and accompanying engineering to meet demands for scaling technical solutions for larger numbers of users.
5. Choosing effective underlying infrastructure for provisioning RIs and deploying services to achieve sustainable service quality and reliability avoiding 'lock-in' to any particular set of e-Is.
6. Educating RI researchers, managers, developers, curators and other actors on how to utilise EOSC through their RI appropriately.

4 Concluding Remarks

The ENVRIplus project ended in July 2019. Although the main content of this book is based on the output of the ENVRIplus project, the community effort put into ENVRI continues into the ENVRI-FAIR project and other collaborative and interoperability initiatives. We hope this book provides a valuable summary of the knowledge we developed in the project and enhances the transfer of knowledge to the development and user communities of the ENVRI and other scientific infrastructures.

Acknowledgements. This work was supported by the European Union's Horizon 2020 research and innovation programme via the ENVRIplus project under grant agreement No 654182.

References

1. Tanhua, T., et al.: Ocean FAIR data services. Front. Mar. Sci. **6**, 440 (2019). https://doi.org/10.3389/fmars.2019.00440
2. Brunner, D., et al.: Comparison of four inverse modelling systems applied to the estimation of HFC-125, HFC-134a, and SF6; emissions over Europe. Atmos. Chem. Phys. **17**, 10651–10674 (2017). https://doi.org/10.5194/acp-17-10651-2017
3. Woodring, J., Petersen, M., Schmeiber, A., Patchett, J., Ahrens, J., Hagen, H.: In situ eddy analysis in a high-resolution ocean climate model. IEEE Trans. Visual. Comput. Graphics. **22**, 857–866 (2016). https://doi.org/10.1109/TVCG.2015.2467411
4. Kurth, T., et al.: Exascale deep learning for climate analytics. In: SC18: International Conference for High-Performance Computing, Networking, Storage and Analysis, pp. 649–660. IEEE, Dallas (2018). https://doi.org/10.1109/SC.2018.00054
5. Kutz, J.N.: Deep learning in fluid dynamics. J. Fluid Mech. **814**, 1–4 (2017). https://doi.org/10.1017/jfm.2016.803
6. Hey, T., Tansley, S., Tolle, K. (eds.): The Fourth Paradigm: Data-Intensive Scientific Discovery. Microsoft Research, Albuquerque (2009)
7. Atkinson, M., Gesing, S., Montagnat, J., Taylor, I.: Scientific workflows: past, present and future. Future Gener. Comput. Syst. **75**, 216–227 (2017). https://doi.org/10.1016/j.future.2017.05.041
8. Prathanrat, P., Polprasert, C.: Performance prediction of Jupyter notebook in JupyterHub using machine learning. In: 2018 International Conference on Intelligent Informatics and Biomedical Sciences (ICIIBMS), pp. 157–162. IEEE, Bangkok (2018). https://doi.org/10.1109/ICIIBMS.2018.8550030
9. Stocia, I.: Conquering big data with spark. In: 2015 IEEE International Conference on Big Data (Big Data). p. 3. IEEE, Santa Clara (2015). https://doi.org/10.1109/BigData.2015.7363734
10. Evans, K., et al.: Dynamically reconfigurable workflows for time-critical applications. In: Proceedings of the 10th Workshop on Workflows in Support of Large-Scale Science - WORKS 2015, pp. 1–10. ACM Press, Austin (2015). https://doi.org/10.1145/2822332.2822339
11. Ari, A., et al.: Final ENVRIplus project report, (2019). Zenodo https://zenodo.org/record/3517905
12. Martin, P., et al.: Open information linking for environmental research infrastructures. In: 2015 IEEE 11th International Conference on e-Science, pp. 513–520. IEEE, Munich (2015). https://doi.org/10.1109/eScience.2015.66
13. Zhao, Z., et al.: Knowledge-as-a-service: a community knowledge base for research infrastructures in environmental and earth sciences. In: 2019 IEEE World Congress on Services (SERVICES), pp. 127–132. IEEE, Milan (2019). https://doi.org/10.1109/SERVICES.2019.00041
14. Martin, P., Remy, L., Theodoridou, M., Jeffery, K., Sbarra, M., Zhao, Z.: Mapping heterogeneous research infrastructure metadata into a unified catalogue for use in a generic virtual research environment. Future Gener. Comput. Syst. **101**, 1–13 (2019). https://doi.org/10.1016/j.future.2019.05.076
15. Hu, Y., et al.: Deadline-aware deployment for time critical applications in clouds. In: Rivera, F.F., Pena, T.F., Cabaleiro, J.C. (eds.) Euro-Par 2017. LNCS, vol. 10417, pp. 345–357. Springer, Cham (2017). https://doi.org/10.1007/978-3-319-64203-1_25
16. Sandusky, R.J.: Computational provenance: DataONE and implications for cultural heritage institutions. In: 2016 IEEE International Conference on Big Data (Big Data), pp. 3266–3271. IEEE, Washington DC (2016). https://doi.org/10.1109/BigData.2016.7840984

17. Casale, G., et al.: Current and future challenges of software engineering for services and applications. CloudForward (2016). http://dx.doi.org/10.1016/j.procs.2016.08.278

18. Petzold, A., Asmi, A.: ENVRI-FAIR EOSC Position Paper (2020). Zenodo http://doi.org/10.5281/zenodo.3666806

19. Petzold, A., et al.: ENVRI-FAIR - interoperable environmental FAIR data and services for society, innovation and research. In: 2019 15th International Conference on eScience (eScience), pp. 277–280. IEEE, San Diego (2019). https://doi.org/10.1109/escience.2019.00038, https://zenodo.org/record/3462816

Author Index

Printed in the United States
By Bookmasters